Sustainable Energy

This comprehensive guide to sustainable energy builds robust connections between abstract theory and practical applications, providing students with a global perspective on this most timely subject. Includes a succinct refresher on essential thermodynamics, fluid mechanics and heat transfer, giving students a solid foundation on which to build. Introduces technologies for hydropower, biomass, geothermal, ocean, solar and wind energy, and fuel cells, with material on nuclear energy, fossil fuel generation and future energy directions, providing a consistent framework for analyzing past, present and future energy systems. Provides coding examples, and real-world case studies, giving students experience in applying theory to practice. Supported by topics for classroom debate, video solutions, and links to online resources, to interactively engage students and inspire further exploration. With a consistent structure and pedagogy, over 160 end-of-chapter problems, and solutions for instructors, this is the ideal introduction for senior undergraduate and graduate students, and a resource for energy professionals.

SERDAR ÇELIK is a professor in the Department of Mechanical and Mechatronics Engineering, and Chair of the Climate and Sustainability Advisory Board, at Southern Illinois University Edwardsville (SIUE). His research fields include renewable energy, building energy efficiency, green roofs, and HVAC technologies. In 2018, he was selected as the Paul Simon Outstanding Teacher-Scholar at SIUE, and he is the Founding Chair of Ilgaz Energy Symposia and SIUE Energy Symposia.

Sustainable Energy

Engineering Fundamentals and Applications

Serdar Çelik

Southern Illinois University Edwardsville

Shaftesbury Road, Cambridge CB2 8EA, United Kingdom

One Liberty Plaza, 20th Floor, New York, NY 10006, USA

477 Williamstown Road, Port Melbourne, VIC 3207, Australia

314–321, 3rd Floor, Plot 3, Splendor Forum, Jasola District Centre, New Delhi – 110025, India

103 Penang Road, #05-06/07, Visioncrest Commercial, Singapore 238467

Cambridge University Press is part of Cambridge University Press & Assessment, a department of the University of Cambridge.

We share the University's mission to contribute to society through the pursuit of education, learning, and research at the highest international levels of excellence.

www.cambridge.org
Information on this title: www.cambridge.org/highereducation/isbn/9781316517383

DOI: 10.1017/9781009043113

First published 2023

Printed in the United Kingdom by TJ Books Limited, Padstow Cornwall

A catalogue record for this publication is available from the British Library.

Library of Congress Cataloging-in-Publication Data
Names: Celik, Serdar, 1979- author.
Title: Sustainable energy : engineering fundamentals and applications / Serdar Celik, Southern Illinois University, Edwardsville.
Description: First edition. | Cambridge, United Kingdom ; New York, NY, USA : Cambridge University Press, 2023. | Includes bibliographical references and index.
Identifiers: LCCN 2022023449 (print) | LCCN 2022023450 (ebook) | ISBN 9781316517383 (hardback) | ISBN 9781009043113 (epub)
Subjects: LCSH: Renewable energy sources.
Classification: LCC TJ808 .C44 2023 (print) | LCC TJ808 (ebook) | DDC 621.042–dc23/eng/20220902
LC record available at https://lccn.loc.gov/2022023449
LC ebook record available at https://lccn.loc.gov/2022023450

ISBN 978-1-316-51738-3 Hardback

Additional resources for this publication at www.cambridge.org/celik

Brief Contents

Contents

Preface

World energy consumption has been rising significantly, not only due to our expanding global population, but also due to evolving living standards and energy consumption habits. Urbanization and industrialization in developing countries have also been contributing to this increase. These changes indicate that there is a need for understanding and evaluating potential future problems and identifying feasible solutions. This is where *sustainability* becomes inevitable. The concept, as described by the United Nations, refers to (a) the use of the biosphere by present generations while maintaining its potential yield for future generations; and/or (b) the non-declining trends of economic growth and development that may be diminished by natural resource consumption and environmental degradation.

The notion of sustainability is procured from the interdependence between humans and the environment. Tools such as ecological footprinting, the Human Development Index, or green national net product are examples of ways of measuring and evaluating the impact of humans on the environment.

Sustainable energy is one of the components under the wide umbrella of sustainability. While it can be perceived to be associated with science and engineering only, it is influenced by entities beyond that. Governments, policymakers, industry, stakeholders, NGOs, academia, and local communities play a significant role in prioritization of research and development (R&D) studies, investments, small- and large-scale applications, and assessments that determine the progress of sustainable energy projects.

This textbook focuses on the general topic of sustainable energy, including production and generation of energy at regional and global scales, sustainability, fossil and non-fossil energy resources, renewable energy technologies, energy storage systems, energy efficiency, and future prospects. The book is aimed at senior and/or beginning graduate-level students as a textbook, or at energy professionals as a comprehensive resource. As the title implies, the book is designed to have a balance between theory and applications. The engineering students and energy professionals using this book will have a resource that portrays the big picture of energy systems from their origins to real-world applications. Chapters have been organized to cover an introduction, a global overview, and a spectrum of existing technologies pertaining to the topic, followed by theory, world applications, a case study, and economic analysis. Theory sections utilize a quantitative approach while ensuring qualitative remarks where appropriate.

It is my expectation that the readers of this book will:

- have a clear understanding of energy sources and generation, energy demand, and energy issues with a global perspective;
- gain knowledge on renewable and non-renewable energy systems, energy storage technologies, energy efficiency, and principles associated with these topics;
- develop an appreciation for the fundamental analysis methodology of energy systems;

- expand on problem-solving skills and acquire the ability to solve related engineering problems by (1) identifying the problem, (2) formulating the solution, and (3) interpreting the results with an understanding of the impact of engineering solutions in a global, economic, environmental, and societal context.

Energy with Pedagogy

Learning taxonomies typically offer a hierarchical flow of stages, with each stage being dependent on the previous one. This textbook was designed to follow a different approach for learners. This approach was influenced by *Fink's taxonomy of significant learning*, which unlike other taxonomies is not hierarchical but is instead *interactive*. Fink's taxonomy recognizes six aspects of learning: foundational knowledge, application, integration, human dimension, caring, and learning how to learn. These are all critical for engineers, architects, and energy professionals of the twenty-first century, who are expected to design systems and find solutions to energy and sustainability challenges. All of these learning stages are interactive and each one can stimulate the other stages, which is well fitting and beneficial for energy topics. Moreover, inclusion of *human dimensions* and *caring* stages enhances the value of the learning taxonomy by considering topics such as climate, ecosystems, and sustainability. To address all the learning aspects suggested by Fink, a number of features have been added to this textbook with the hope of enhancing learning while making it interactive and enjoyable. These features are (1) learning objectives, (2) global perspectives, (3) sidebar images, (4) in-chapter examples, (5) Python programming language, (6) applications, (7) case studies, (8) further learning resources, (9) classroom debates, (10) end-of-chapter exercises, and (11) video recordings of selected problem solutions.

Learning Objectives Each chapter starts with a list of learning objectives. This helps instructors to prepare for a lecture effectively and enables students to be aware of the expected outcomes.

Global Perspective Through illustrations, in-chapter examples, and end-of-chapter examples, over 60 countries, ranging from small island countries to major energy producing and consuming nations, have been highlighted in the book. As such, the textbook has been designed to enhance the understanding of globality and to encourage students to approach problems with a holistic perspective.

Sidebar Images Illustrations of scientists, technologies, or applications are provided in the margins in harmony with the topics discussed. These images are intended to enhance the reader's interest in the topic while making it enjoyable.

In-Chapter Examples Step-by-step problem solutions are provided through in-chapter examples. Besides covering appropriate use of inputs, assumptions, and equations, use of units is also emphasized. Some examples include discussion and interpretation of results.

Python Programming can be an instructive tool in obtaining and interpreting quantitative results for energy applications, utility sector, and trends. Python is an effective programming language which is freely available and easy to learn. Coding examples in the chapters include both theoretical solutions and plotting graphs based on data retrieved from reliable sources.

Applications All energy source chapters include two global applications pertinent to the technologies discussed. These applications are expected to give students an understanding of projects of diverse scale with various design conditions in different parts of the world.

Case Studies Each energy source chapter also includes a case study. The case studies consist of *Introduction*, *Objective*, *Method*, *Results*, *Discussion*, and *Recommendations* sections, which are expected to help students actively engage with both the basic principles and the applied examples, while reducing the potential for any bias.

Further Learning Resources Online resources for further learning or engaging with interactive tools are provided in the sidebars where appropriate within the chapters. These resources help students learn the discussed material in more depth. Some of these resources provide tools to interactively engage with the topics.

Classroom Debates This feature originated from the *Energy Debate Series* that I have been organizing in my energy courses, which proved to be effective and enjoyable for students. Classroom debates pointed out in this textbook are expected to (1) keep students interested in the topics covered, (2) encourage them to do research and further reading, and (3) enhance their soft skills, which include teamwork, written and oral communication, and interpersonal skills.

End-of-Chapter Exercises Problems at the ends of the chapters are mostly quantitative and can be solved by traditional numerical methods discussed in each chapter. Some exercises, however, require additional research to make connections between different systems, to retrieve data for further analysis, or to come up with interpretations. Such a variety of exercises aligns with the learning aspects of Fink's taxonomy.

Video Recordings of Problem Solutions A unique feature of this textbook is the videos of solutions for selected problems on the lightboard. The solutions are carried out following the steps for solving engineering problems. These videos are expected to promote in-depth comprehension in a quasi-classroom environment.

Ancillaries

Additional resources for instructors to support this text include:

Solutions Manual, which includes the solutions for all the end-of-chapter exercises in the book.

Lecture Slides with important figures, tables, and equations from each chapter for use in lectures.

Acknowledgements

I am thankful to all the skilled people who have helped make this book possible. First of all, I am indebted to my students who have taken courses on energy and thermal/fluid systems at Southern Illinois University Edwardsville. Their interactions and feedback during these courses over the years have provided invaluable insights that helped me structure the book. I am also grateful to Baris Sanli, who has been of tremendous help with the Python component of the book. His vision and suggestions have been a great contribution. In addition, I would like to thank the following reviewers who provided indispensable and inspiring comments on the chapters of this book:

Fatih Birol, *International Energy Agency*, France
Francesco La Camera, *International Renewable Energy Agency*, UAE
Jennifer Bousselot, *Colorado State University*, USA
Ibrahim Dincer, *Ontario Tech University*, Canada
Orhan Ekren, *Ege University Solar Energy Institute*, Turkey
Riccardo Guidetti, *University of Milan*, Italy
Susan Morgan, *Southern Illinois University Edwardsville*, USA
William Retzlaff, *Southern Illinois University Edwardsville*, USA
Ugur Soytas, *Technical University of Denmark*, Denmark

The editorial team of Helen Shannon, Charles Howell, Jane Adams, and Lisa Pinto has been one of the keys to the success of this textbook. Their understanding of the scope and vision of this book and recommendations on the format resulted in a fruitful and fulfilling collaboration.

Finally, thanks to my wife, Dr. Anastasia Celik, for her understanding and continuous support during every phase of this long project.

CHAPTER 1
A Broad Spectrum Introduction

LEARNING OBJECTIVES

After reading this chapter, you should have acquired knowledge on:

- The relation between energy and power
- Energy units and dimensional analysis
- Global energy production and consumption
- Electricity generation, transmission, and distribution
- Environmental concerns and potential remedies
- The connection between energy, environment, and engineering
- Life cycle energy analysis
- Using social media for energy topics

1.1 The Topic of Interest: Energy

In every aspect of life, during all ages, energy has been one of the major essentials for human beings. When we eat, the carbohydrates, proteins, and fats in food furnish the calories needed by our bodies. Livestock utilize energy from the food that they eat in terms of producing labor and products such as meat, milk, or eggs. There has always been a need for energy to carry objects from one point to another. Invention of the wheel made this easier. But the task still required energy, and it still does.

Today we use energy in all aspects of our lives: transportation, cooking, heating and cooling, manufacturing, and lighting. We need it for healthcare, entertainment, recreation, agriculture, and communication. Electrical home appliances, which have contributed to our improved standard of living, require energy. We cannot get online on social media on our phone, nor can we prepare a presentation for a project if the battery of our phone or laptop is dead and there is no energy to charge it.

The study of energy topics involves both theory and practice. These topics include energy generation, transmission, distribution, storage, efficiency, consumption, economics, and feasibility. Environmental aspects of energy are also of great significance. An energy project should account for the expected environmental impacts and potential remedies for the anticipated consequences. Engineering can be defined as "the application of scientific, economic, social, and practical knowledge and *ethics* in order to invent, design, build, operate, maintain, and improve structures, machines, systems, and materials." The ethical component here for those working in the field of energy systems involves not only being ethical to employers, coworkers, clients, and community, but also being fair and ethical to the environment. As the native American proverb says, "*We do not inherit the earth from our ancestors, we borrow it from our children.*"

Appropriate use of terminology, especially for engineers and technicians, is important whether they have an office job or they are in the field. One popular confusion is the nuance between energy and power. Energy is a property that is required to do work on an object, and/or to add or remove heat from it. It can be transferred from one object to another. It can also be converted from one form to another form. Power on the other hand is the rate at which energy is transferred. Energy is a property that relates to a time component. Power, however, is an instantaneous quantity. An example to better illustrate the difference between energy and power is the relation between distance and velocity. Distance is measured in meters and velocity is defined in units of meters per second. Their relation is that velocity is distance divided by time. A similar relation is seen between energy and power, where power is energy divided by time. Another difference between the two terms is that energy can be stored, power cannot be stored.

1.2 Units and Dimensional Analysis

In all categories of science (**Figure 1.1**), analysis of problems, cases, domains, or systems comes with some form of quantitative inputs and outputs. Use of units and dimensional analysis not only signifies professionalism, but also provides a double-check mechanism to make sure the right variables were used in the analysis.

The importance of conducting dimensional analysis and using units becomes more obvious in applied sciences, such as engineering. Every engineering student should make it a habit to always provide the unit of the quantity demonstrated. Another important point is consistency with the unit system being used. There are two major unit systems: the *International Unit System* (SI – abbreviation from the French term

FIGURE 1.1 Branches of science. Source: Efbrazil, Wikimedia Commons author, https://en.wikipedia.org/wiki/File:The_ Scientific_Universe.png.

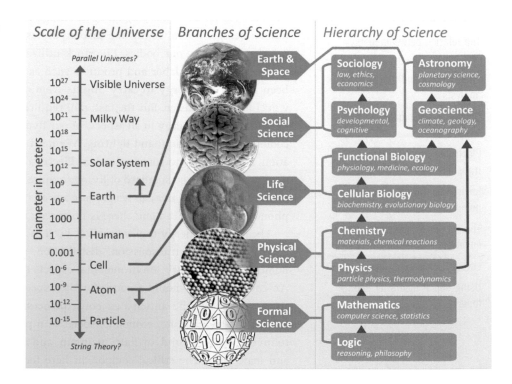

Table 1.1

Basic dimensions and units		
Dimension	**SI unit**	**Imperial unit**
Length (*L*)	meter (m)	foot (ft)
Mass (*M*)	kilogram (kg)	pound mass (lb$_m$)
Time (*t*)	second (s)	second (s)
Temperature (*T*)	degree Celsius (°C)	degree Fahrenheit (°F)
Thermodynamic temperature	kelvin (K)	degree Rankine (°R)
Electric current (*I*)	ampere (A)	ampere (A)
Luminous intensity	candela (cd)	candela (cd)

FURTHER LEARNING

"A Turning Point for Humanity: Redefining the World's Measurement System"
www.nist.gov/si-redefinition/turning-point-humanity-redefining-worlds-measurement-system

FURTHER LEARNING

Energy Units and Calculators Explained
www.eia.gov/energyexplained/units-and-calculators/energy-conversion-calculators.php

Système International d'unités) and the *Imperial Unit System* (**Table 1.1**). There is also another group of units called the US Customary Units, derived from the imperial units and redefined. In the United States, measuring dimensions in terms of imperial units is customary and they are used in many daily life and commercial applications.

In addition to making sure to include units in an analysis and being consistent with the unit system selected, it is also important that the symbols for units are written properly. If a value is in kilowatts, the symbol should be kW, not KW or Kw.

Some common thermal energy units in both SI and imperial unit systems are listed in **Table 1.2**. Prefixes and symbols used in SI are given in **Table 1.3**.

Table 1.2

Common thermal energy units in the SI and imperial unit systems		
Quantity	SI unit	Imperial unit
Heat	J, kJ, cal, kcal	Btu
Specific heat	J/kg °C, J/kg K	Btu/lb °F, Btu/lb °R
Enthalpy	kJ/kg	Btu/lb
Cooling energy	J, kJ	Btu
Cooling capacity	W, kW	Btu/h, ton

Table 1.3

Prefixes used in the SI unit system			
Prefix	Symbol	Scientific notation	Example
Nano-	n	10^{-9}	Visible light has a wavelength range of ~380–740 nanometers.
Micro-	μ	10^{-6}	Energy of a proton at the Large Hadron Collider at CERN is about 1.1 microjoules.
Milli-	m	10^{-3}	Average resting membrane potential of a neuron in our body is 70 millivolts.
Kilo-	k	10^{3}	A 24,000 Btu/h air-conditioner has a cooling capacity of 7 kilowatts.
Mega-	M	10^{6}	A megawatt-hour is equivalent to the amount of electricity consumed by about 1380 people in the Netherlands during one hour.[1]
Giga-	G	10^{9}	One ton of green wood chips yields approximately 14 gigajoules of biomass energy.
Tera-	T	10^{12}	Hoover Dam generates 4 Terawatt-hours of energy annually.
Peta-	P	10^{15}	One petajoule is equivalent to the amount of electrical energy consumed by 868,000 domestic refrigerators per year.[2]
Exa-	E	10^{18}	Annual global energy consumption is estimated to be 580 exajoules which is about 550 quads.[3]

[1] Electrical energy use for the Netherlands is 6346 kWh per person per year (www.cia.gov/the-world-factbook/countries/netherlands/#energy).
[2] A typical 2.5 star refrigerator uses 320 kWh of electricity annually.
[3] 2022 data (www.theworldcounts.com/challenges/climate-change/energy/global-energy-consumption).

Table 1.4

Units used in the oil and natural gas industry	
Btu	British thermal unit
toe	1 tonne of oil equivalent
Mtoe	1 million tonnes of oil equivalent
Bcm (natural gas)	1 billion cubic meters
MMCF (natural gas)	1 million cubic feet
MCF (natural gas)	1 thousand cubic feet
MMBtu (natural gas)	1 million Btu
quad	1 quadrillion Btu

Table 1.5

Energy unit conversions
1 kJ = 0.948 Btu
1 GJ = 947,817 Btu
1 GJ = 277.8 kWh
1 kWh = 3412 Btu
1 TWh = 0.086 Mtoe

1 W = 3.412 Btu/h
1 hp = 2545 Btu/h
1 hp = 746 W

1 therm = 100,000 Btu
1 toe (15 °C) = 41.85 GJ
1 toe (15 °C) = 39.66 MMBtu
1 quad = 1015 Btu
1 quad = 1.055 EJ

1 ton of refrigeration = 12,000 Btu/h
1 ton of refrigeration = 3.517 kW

There are certain units commonly used in the oil and natural gas industry. Being familiar with these units is important to be able to analyze and interpret books, articles, and technical reports published in these fields. **Table 1.4** lists some common units associated with oil and natural gas applications, and **Table 1.5** lists unit conversions.

1.3 Energy Production and Consumption

The required capacity for energy production is defined by energy consumption. The growth in energy consumption is based on the growth in population and wealth, as well as changes in the energy consumption habits of societies. For example, there are parts of the world where using an air-conditioner at home was not standard practice two decades ago, but now they have adopted the use of AC at home. If we assume that population growth was 20% in the past two decades for such a country, the growth in energy consumption will be much higher as now the per capita consumption has also increased. Industrialization of a country is another example. The more a society becomes industrialized, the more energy consumption will be observed. This goes hand in hand with wealth. **Figure 1.2** depicts the history of global population growth from AD 0 projected to 2050 based on data from the United Nations, and **Figure 1.3** illustrates the total world population from 1950, projected to 2100 with the deterministic high and low variants of ± 0.5 child [1]. The growth in energy

FIGURE 1.2 Long-term global population growth from AD 0, projected to 2050. Source: United Nations DESA, Population Division, World Population Prospects 2019.

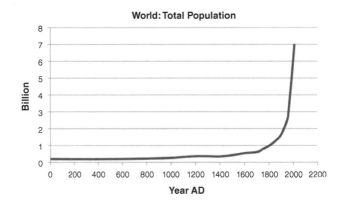

FIGURE 1.3 Global population growth from 1950, projected to 2100. The figures illustrate the probabilistic median, 80% and 95% prediction intervals of the probabilistic population projections, and the (deterministic) high and low variant (± 0.5 child). Source: United Nations DESA, Population Division, World Population Prospects 2019.

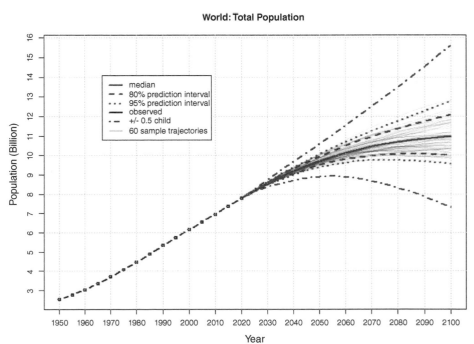

FIGURE 1.4 Growth of global energy mixed with technological milestones. Source: *Global Energy Assessment: Toward a Sustainable Future*, 2012, Cambridge University Press.

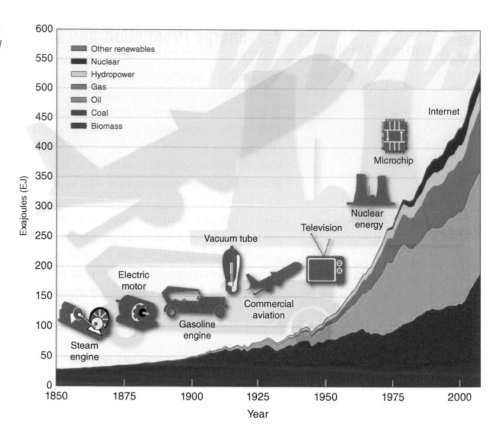

FIGURE 1.5 Per capita electricity consumption in the world in 2020. Source: Our World in Data based on BP Statistical Review of World Energy and Ember Global Electricity Review (2021).

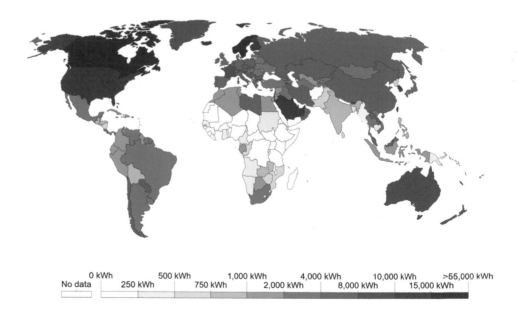

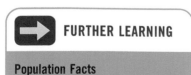

FURTHER LEARNING

Population Facts
https://populationmatters.org/the-facts

consumption corresponding to the increase in population and wealth can be seen in **Figure 1.4**. As mentioned before, changes in energy consumption habits also play a role in the increasing trend in this graph [2]. Global per capita electricity consumption is presented in **Figure 1.5** on a choropleth map. Energy consumption by fuel types in the building sector in OECD and non-OECD countries by 2050 is shown in **Figure 1.6** [3].

FIGURE 1.6 Building sector energy consumption by fuel types in OECD and non-OECD countries through 2050. Source: U.S. Energy Information Administration, International Energy Outlook 2021.

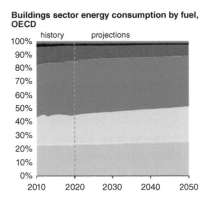

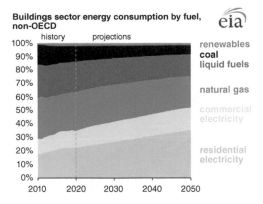

1.4 Electrical Energy: Generation, Transmission, and Distribution

Electricity is about the charge of atoms, which are the building blocks of everything. Atoms are made up of protons, neutrons, and electrons. Electrons move between atoms because of electric and magnetic forces. This motion of electrons yields electricity. Electrical energy is the form of energy resulting from the flow of electric charge or electricity. It can be in the form of kinetic or potential energy. It is often considered to be potential energy where the potential is defined by the stored energy due to the charged atoms or electric field. Electrical energy goes through three main stages from its production at a power plant to being ready to be utilized by the end-user. These stages are **generation**, **transmission**, and **distribution**, as illustrated in **Figure 1.7**.

FIGURE 1.7 Electrical grid diagram. Source: Federal Energy Regulatory Commission.

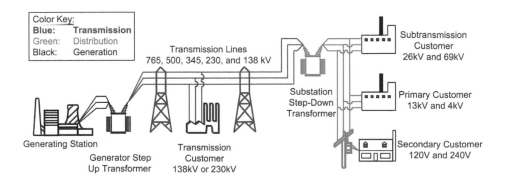

FIGURE 1.8 Electric generator. Source: U.S. Energy Information Administration.

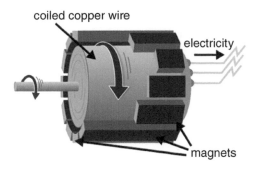

FIGURE 1.9 Michael Faraday (1791–1867) was an English physicist and chemist whose studies contributed to the understanding of electromagnetism and electrochemistry. His interest in science surfaced when he was working as an apprentice at a bookseller where he read many books. He is a great example of a self-educated scientist of his time. While he is known mostly for his studies in electromagnetism and his discovery of electromagnetic induction, he has many other scientific achievements such as discovering benzene, producing optical glass, demonstrating that diamagnetism is a property of matter, and inventing the Faraday cage (or Faraday shield) which is used in blocking electromagnetic fields. Source: Pictore/DigitalVision Vectors via Getty Images.

FIGURE 1.10 Step-down transformer on utility poles. Source: Robert Brook/Science Photo Library via Getty Images.

Generation

Electricity is generated in power plants where the mechanical energy of a turbine is harvested into electrical energy using generators (**Figure 1.8**). The turbine blades are forced to rotate by means of high-pressure steam (coal, natural gas, nuclear plants), water (hydropower), or air (wind turbines). These generators work on the principles of electromagnetic induction discovered in 1831 by *Michael Faraday* (**Figure 1.9**), who found that electric fields can be formed when an electrical conductor is moved in a magnetic field. This movement yields a voltage gradient between the two ends of a wire, resulting in an electric current.

Solar panels also generate electricity; however, they do not require a turbine and a generator. Their working principle is based on the photovoltaic effect, where semiconductors interact with photons coming from the Sun and generate electric current.

Transmission

Once generated, electricity passes through a transmission station where the voltage is stepped up with aid of transformers. This results in a decrease in electric current. The increase in voltage is essential as electrical energy will be lost when electricity is traveling long distances through conducting wires. Stepping up the voltage via transformers enables the flow of electricity for further destinations. The energy lost during the travel of electricity is proportional to the square of the electric current. Hence the decrease in current becomes crucial as, for instance, halving the current results in a decrease in energy loss by a factor of four. Therefore, high-voltage transmission lines are an important part of the grid as they transport electricity for long distances with relatively small amounts of energy loss.

Distribution

High-voltage electricity that reaches residential, industrial, or commercial areas where the end-users need electrical energy has to go through a voltage reduction. This is achieved using step-down transformers this time (**Figure 1.10**). These transformers operate in an opposite manner to the step-up systems as the high voltage from the transmission lines needs to be reduced to appropriate levels for different applications for which consumers require electrical energy. The reduction in voltage is performed at distribution substations housing step-down transformers. The distribution grid connects these substations to the end-users, hence fulfilling the task of transferring energy all the way from power generation plants to the consumers.

1.5 Environmental Concerns and Remedies

Environmental Concerns

The environment is everything that surrounds us. It involves the lithosphere (land), hydrosphere (water), biosphere (living things), and atmosphere (air). Different types of natural events or human-made processes can cause environmental changes. As the ecosystems are connected to the environment, environmental changes can result in ecological changes. These changes are mostly uninvited, and they can lead to environmental concerns such as climate change, air pollution, and water shortage and quality (**Figure 1.11**).

Among these concerns, global climate change is one of the top issues that need to be addressed. The climate has been changing constantly over geological time, with

FIGURE 1.11 Major global environmental concerns.

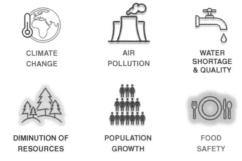

CLIMATE CHANGE

AIR POLLUTION

WATER SHORTAGE & QUALITY

DIMINUTION OF RESOURCES

POPULATION GROWTH

FOOD SAFETY

noteworthy alterations of global average temperatures. However, the most fluctuations have been caused by humans in the past century, mainly by releasing greenhouse gases through burning of fossil fuels. The increases in the atmospheric concentration of carbon dioxide (CO_2) and other greenhouse gases lead us to the global warming problem. Changes in the global average temperature and atmospheric CO_2 concentration from 1880 to 2019 are illustrated in **Figure 1.12** [4]. As depicted in the graph, changes in both quantities are important, with the CO_2 level requiring more attention as it exceeded the 300 ppm threshold in the early 1900s. This level had not been reached for millennia in the fluctuating pattern of CO_2 concentration history.

Remedies

Fixing a problem requires a thorough diagnosis, clear understanding of the sources of the problem, and a committed plan of action to solve the problem. Looking at the big picture, one can see that greenhouse gases accelerate climate change, and CO_2 is the amplest greenhouse gas. Hence, reduction in CO_2 emissions should be a major task in setting a plan to address the global climate change issue. It is important to note that everyone on the globe is in the same boat, and this action plan to reduce CO_2 emissions requires a global commitment, rather than individual nations trying to do their best while others are ignoring the problem, or not paying sufficient attention.

The *Brundtland Report* was published in 1987 by the World Commission on Environment and Development. This comprehensive report covered both environmental and economic aspects for sustainable development. Ten years later, the *Kyoto Protocol* as part of the United Nations Framework Convention on Climate Change (UNFCCC) was agreed upon; state parties accepting that global warming is caused by humans and needs action. It was adopted in 1997 and became effective in 2005. Another

FIGURE 1.12 Temperature differences compared to the twentieth-century average (blue and red bars) between 1880 and 2019 based on data from NOAA NCEI; and atmospheric CO_2 concentrations (gray line): 1880–1958 data from IAC, and 1959–2019 data from NOAA ESRL. Source: Original graph by Dr. Howard Diamond, adapted by National Oceanic and Atmospheric Administration.

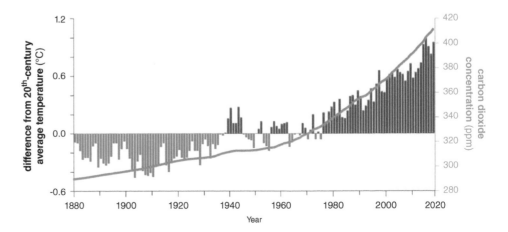

FIGURE 1.13 Reforestation not only increases the amount of CO_2 absorbed, but also helps mitigate climate change and reverse biodiversity loss. Source: Maica/E+ via Getty Images.

agreement within the UNFCCC is the *Paris Agreement*, which was signed by almost all nations of the world in 2016. The agreement has a long-term goal of keeping the increase in the global average temperature to well below 2 °C above pre-industrial levels and pursuing efforts to limit the increase to 1.5 °C above pre-industrial levels, acknowledging that this would notably reduce the risks and impacts of climate change. The pre-industrial levels are determined using the 1850–1900 period as a baseline as this is the earliest period with near-global monitoring.

These agreements and reports include actions to address the climate change issue. Some of the actions to combat CO_2 emissions with the intention of (i) stopping the increase and (ii) reducing the total concentration of the gas in the atmosphere are:

- **Reducing energy consumption per capita:** this can be achieved by conserving energy, changing old habits where energy is consumed more than is required, and implementing more energy-efficient products into people's lives.
- **Substitution of coal with other fuels:** switching to natural gas or nuclear fuels in power plants will result in lesser amounts of CO_2 emissions for the same amount of power output.
- **Implementation of renewable energy technologies:** wider use of renewable energy technologies such as solar, wind, geothermal, and hydro will result in less consumption of fossil fuels in generating power.
- **Improving carbon sequestration techniques:** this will enable long-term removal, capture, or sequestration of CO_2 from the atmosphere, contributing to mitigation or deferral of global warming.
- **Reforestation:** natural or intentional expansion of forests and woodlands will help increase the amount of CO_2 absorbed by trees and plant canopies.

1.6 Connecting the 3E: Energy, Environment, and Engineering

As the global population and energy consumption per capita are increasing, the responsibility of engineers in protecting and enhancing the environment is also increasing. This is where the relation between the 3E (energy, environment, and engineering) becomes obvious.

According to the World Federation of Engineering Organizations (WFEO), engineers play a crucial role in improving living standards throughout the world. Hence, engineers are the key elements in leading and coordinating projects to achieve the UN Sustainable Development Goals for a more sustainable globe [5].

Engineers' responsibility is not only in protecting the environment but should also be in enhancing it. This should happen in collaboration with other sections of professional society including sociologists, economists, and political leaders. Technological advancement is only one leg of a holistic approach to improve environmental quality. A thorough systematic method involving reform in society's energy consumption culture will help this. This reform in consumer culture should not be confused with sacrifice of quality of life, but instead should be a re-evaluation of what is essential and what is not for consumers.

Engineers should have the mindset to be able to think outside the box to advance environmentally friendly technologies to control energy consumption without

FIGURE 1.14 Two young engineers are checking a wind farm field in a rural landscape in Turkey. The 3E are in one photo: energy, environment, and engineering. Source: serts/E+ via Getty Images.

sacrificing consumers' living standards. As *Albert Einstein* said, *"We cannot solve the problems by using the same kind of thinking we used when we formed them."* This thinking can be applied to both sides of energy-related systems: energy-consuming systems and energy-generating systems. Enhancement of the energy efficiency of systems will lead to lower amounts of energy consumption for the same amount of work output, while advancement of efficiency in energy-generating systems (whether a hydroelectric power turbine, a photovoltaic panel, or a wind turbine) will yield higher energy outputs for the same amount of input available. When these improvements are combined with a society of consumers with increased awareness of appropriate use of energy, the outcome will be promising for the environment and the ecosystem.

In summary, the 3E that we have discussed involve a producer (engineer), the product (energy system), and the surroundings (environment). The producer should design the product in such a way that the negative impacts of the product on the surroundings are minimal as the surroundings are the essential component for a better life for the producer and the consumers.

1.7 Life Cycle Energy Analysis

In the previous section, the importance of protecting the environment and the responsibilities of engineers was emphasized through connecting the 3E. It is important for engineers and scientists to consider a system that is being designed and built, and its impacts on the environment. There is a tool that becomes very handy in quantifying this impact: Life Cycle Analysis (LCA) is a systematic analysis of the potential environmental impacts of products, systems, or services during their entire life. This analysis sheds a light on whether or not an application will help the environment, in both a quantitative and a qualitative manner. Life Cycle Energy Analysis (LCEA), on the other hand, is a tool for analyzing the energy used to manufacture and transport a system and the energy it consumes throughout its entire life. This analysis investigates the environmental impact of the system due to the energy input it requires and its carbon footprint.

An example set of LCEA steps that could be performed on a photovoltaic (PV) panel can be listed as:

1. Determining the energy required to extract the raw materials
2. Transportation of the raw materials to the processing facility
3. Energy use in processing the raw materials into useable materials to manufacture the PV panel
4. Transportation of the processed material to the PV manufacturing facility
5. Energy use in all stages of manufacturing the PV panel and associated components
6. Transportation and distribution of the panels to the consumers
7. Energy use by the product throughout its entire lifetime in use (cooling, lighting, maintenance, etc.)
8. Energy required for the removal/recycling/disposal of the PV panel at the end of its operational time

Once the total energy required for all the stages above and the total energy the panel can generate throughout its entire operational time are known, one will have an idea of how much the panel brings in and how much it takes away in terms of energy and carbon footprint.

FIGURE 1.15 LCEA phases for a building. Source: Y. Dong, T. Qin, S. Zhou, L. Huang, R. Bo, H. Guo, and X. Yin, "Comparative whole building life cycle assessment of energy saving and carbon reduction performance of reinforced concrete and timber stadiums: A case study in China," *Sustainability*, vol. 12, no. 4, p. 1566, 2020.

The LCEA of buildings is also a hot topic as buildings possess a significant portion of total energy consumption. **Figure 1.15** illustrates the components of the construction, operational, and end-of-life phases of a building in performing the LCEA.

1.8 Energy and Social Media

Social media have become an important part of our lives. These online tools help people discover and find out new information, interact with peers in their profession or with people from other professions, share ideas and brainstorm, and initiate collaborations or projects.

Social media platforms come in a variety of forms such as professional forums, podcasts, photo sharing, blogs, product reviews, etc. There are millions of people using social media sites today, such as Facebook, Instagram, Twitter, LinkedIn, YouTube, WhatsApp, WeChat, and Snapchat. While some of these sites are for sharing photos and videos for fun, some are extensively used for business, professional, and personal development. The number of users keeps rising and new sites continue to join the fleet. According to Statista, an estimated 3.6 billion people were using social media in 2021, and the number of users is projected to reach 4.41 billion in 2025 [6].

The popular social media sites are a rich reserve for those interested or working in the energy field. Government agencies, institutions, companies, organizations, news media, and individuals share information through different sites in the form of articles, photos, videos, blog posts, presentations, news clips, or personal posts.

The U.S. Department of Energy (DOE) defines its social media use strategy based on how most effectively to use social media platforms to engage the public in open discussion about energy issues, and how the public can benefit more from the DOE's work [7].

FIGURE 1.16 Energy social media have great potential for acquiring knowledge, networking, or expressing one's own ideas if used responsibly and with professional filtering of the information whether it is being supplied or received. Source: Peter Dazeley/ The Image Bank via Getty Images.

Energy utility companies are also highly involved in social media as their customers are using media to make decisions on what services/products to get and from whom to buy them based on the reviews and recommendations on the sites. Utility companies have been integrating their conventional marketing communications with Facebook, Instagram, Twitter, and YouTube.

Energy social media can also be used as a great source for retrieving information, networking, or expressing one's own ideas. There is so much information out there that, while this ocean of shared information lies as a rich asset, there is also a risk of being misled due to misinformation (intentional or unintentional) or speculation. The cycle of source, content, and consumer can operate very effectively if the source is reliable, the content is based on facts and crowdsourced fast-checked, and the consumer is willing to analyze and filter information before adopting it.

As in any other field, it is important to use cognitive skills in processing the information before letting it pass through the filters. When using social media, professionals need to be careful about

- what they post
- what they like
- what they share
- whom they respond to and how
- how much time they spend on social media sites

We do not want to endorse a person, an idea, or any information that we would not have approved or agreed with after performing a critical thinking process. It is also essential to focus on time management and to avoid excessive use of social media unless it is the major responsibility of an individual.

Overall, social media are a great asset for users interested in the energy field, such as students, educators, engineers, and other professionals, if used with appropriate attention. They possess high potential for learning new information, yielding new collaborations, and initiating new projects.

1.9 Summary

In this chapter, a brief introduction to energy was provided, followed by the dimensions and units of measure used in energy systems. The importance of using units in engineering analyses was emphasized. The growth in global energy consumption is a result of the growth in population and wealth, as well as changes in energy consumption habits of societies. Energy production has been increasing, aligned with the national needs of countries. The GDP and electricity consumption per capita reveal information on how developed countries are. Data shows that both population and electricity consumption growth for non-OECD countries in the past two decades were greater than those for OECD countries. Electrical energy use is one of the most helpful factors in determining human development. Electricity is generated in power plants, transmitted through high-voltage lines from the power plants to urban areas, and then distributed to consumers. While the end-users benefit from the privileges brought by energy, environmental concerns such as climate change become more apparent. Global CO_2 emissions have reached record values in the past few decades. International meetings yielding key messages to address climate change and global warming issues have been organized, starting towards the end of the twentieth century. The Kyoto Protocol and the Paris Agreement are two important touchstones focusing on remedies for environmental issues. These remedies focus mainly on reducing CO_2 emissions. The role of engineers in maintaining a sustainable environment with sufficient energy supply is one of the most important takeaways readers should have from this book. Hence, connecting the 3E (energy, environment, and engineering) is of the essence for all engineers and energy professionals. Life cycle energy analysis is a nifty tool in quantifying the impact of energy systems on the environment. This analysis yields tangible outcomes for engineers with a mindset of connecting the 3E. Use of social media was also discussed in this chapter. Using energy social media (either as a source or as a receiver) can help improvements in spreading information, discussing ideas, and building networks if used appropriately. It is the source's duty to make sure to provide the most reliable information, and it is the receiver's responsibility to ensure the information is filtered before being endorsed, shared, or adopted.

REFERENCES

[1]   United Nations, Department of Economic and Social Affairs, Population Division, (2019). "World Population Prospects 2019," Online Edition. Rev. 1.

[2]   T. B. Johansson, N. Nakićenović, A. Patwardhan, and L. Gomez-Echeverri (Eds.), *Global Energy Assessment: Toward a Sustainable Future*. Cambridge: Cambridge University Press, 2012.

[3]   U.S. Energy Information Administration, "International Energy Outlook 2021," EIA, Washington, DC, Oct. 2021. Accessed: Oct. 12, 2021 [Online]. Available: www.eia.gov/outlooks/ieo.

[4]   R. Lindsey, "News & Features," NOAA Climate.gov, Feb. 12, 2020. www.climate .gov/news-features/climate-qa/if-carbon-dioxide-hits-new-high-every-year-why-isn%E2%80%99t-every-year-hotter-last (accessed Jun. 21, 2020).

[5]   United Nations Department of Economic and Social Affairs, "World Federation of Engineering Organizations Report," Accessed: Jun. 21, 2020 [Online]. Available: https://sustainabledevelopment.un.org/content/documents/241538_World_ Federation_of_Engineering_Report.pdf.

[6]   H. Tankovska, "Number of social media users worldwide 2010–2021," Statista, Jan. 28, 2021. www.statista.com/statistics/278414/number-of-worldwide-social-network-users (accessed Feb. 27, 2021).

[7]   U.S. Department of Energy, "Social Media," www.energy.gov/about-us/web-pol icies/social-media (accessed Jun. 21, 2020).

CHAPTER 1 EXERCISES

1.1   Pick five prefixes from Table 1.3, research alternative examples and list them all in a table.

1.2   Electricity consumption values for the countries below are listed based on CIA World Factbook data. Convert these values into units of Btu and mtoe, adding two separate columns.

Country	Data year	Electricity consumption		
		kWh/yr	Btu/yr	mtoe/yr
South Africa	2020	212,000,000,000		
India	2018	1,547,000,000,000		
Turkey	2020	207,400,000,000		
Belgium	2014	81,000,000,000		
Venezuela	2014	78,000,000,000		

1.3   Using a spreadsheet, obtain the most recent GDP per capita and energy consumption per capita values for the countries listed below, and plot the data on a chart with GDP per capita values on the horizontal axis. Cite the source and mention the year of the data.

   a.   Philippines
   b.   Qatar
   c.   Japan
   d.   Haiti
   e.   Austria

1.4   Repeat **Exercise 1.3**, replacing energy consumption per capita with electricity consumption per capita. Compare the two graphs and comment on the differences.

1.5   In 2020, total transportation energy consumption in the United States dropped by 15% to 24 quads, according to the U.S. Department of Energy. This drop was mainly due to the decrease in petroleum use for travel. Calculate the equivalent amount of transportation energy use in EJ.

1.6   Explain the difference between electricity transmission and distribution. Sketch a diagram including the generation, transmission, and distribution phases. Show key components for all three phases.

1.7   Define environment, ecology, and ecosystem.

1.8   Research the strengths and weaknesses of the Paris Agreement, and write a one-page essay on it, including your suggestions on how the weaknesses can be addressed and improved.

1.9   List the life cycle energy analysis (LCEA) steps for a wind turbine and discuss the steps with the most significant environmental impacts.

1.10 Look for an energy information account on social media and find a recent posting for this account. Comment on the reliability of the post after researching the information.

CHAPTER 2
The Concept of Sustainability

2.1 Introduction

Global population was 1 billion in 1800, doubled in the next 127 years to 2 billion, then doubled to 4 billion in the following 47 years. Besides the significant increase in human population, changes in living standards, evolving consumption habits, urbanization, and growth in industrialization have made it obvious that there is a need for understanding sustainability, identifying areas on which to reflect, and defining indicators for measuring and evaluating it.

A full retrospective of what has changed regarding the economy, environment, social system, and technology is crucial to have a clear interpretation of the current set of circumstances and to be able to predict the future state of affairs.

As cited by the OECD [1] sustainability refers to

(a) the use of the biosphere by present generations while maintaining its potential yield for future generations; and/or
(b) the non-declining trends of economic growth and development that may be diminished by natural resource consumption and environmental degradation.

The notion of sustainability is derived from the interdependence between humans and the natural environment [2]. Population, economic, technological, social, and habitual changes, and development impose a risk to natural resources which is a potential threat to continued welfare and peace all across the globe.

Growing concerns pertaining to economic instabilities, recessions, human health problems, poverty, unequal education opportunities, climate change, and access to clean water and energy have raised attention on sustainable development. To address these concerns, 17 global goals were identified by the United Nations General Assembly in 2015. These goals are known as the Sustainable Development Goals (SDGs). Among the SDGs, the ones related to energy can be listed as *affordable and clean energy* (SDG7), *climate action* (SDG13), and *good health and well-being* (SDG3).

Sustainable development cannot be fulfilled without sustainable energy. According to an article of the United Nations on millennium goals [3], it is an unfortunate fact that one in five people in the world have no access to electricity. About 40% of the world population use primitive heating and cooking appliances employing wood, coal, or animal waste. This makes it difficult to eliminate poverty in these areas of the world.

On the other hand, if we look at the developed countries, the problem with energy is not about shortage or access to it, but about waste and pollution. Industrialized countries contribute to climate change more than the least developed counties (LDCs). Hence, when energy concerns are examined at a global scale, it becomes apparent that

FURTHER LEARNING

The 17 SDGs
https://sdgs.un.org/goals

FIGURE 2.1 A futuristic cityscape with renewable energy and green walls. Such cities can reflect all or most of the SDGs. Source: Andriy Onufriyenko/Moment via Getty Images.

all parts of the world have an energy problem, whether due to energy shortage or due to pollution. In addition to the geography- or region-specific energy problems, there is also climate change, which is a problem for us all. In the early phases of industrialization, development steps had minimal environmental impact on the globe. They were rather local. Today, however, the impact of industry and transportation on the ecosystem is at a global scale. In this regard, the Paris Agreement is consequential due to its integrated approach to energy and sustainable development. It has a long-term goal of keeping the increase in the global average temperature to well below 2 °C above pre-industrial levels and pursuing efforts to limit the increase to 1.5 °C above pre-industrial levels. The pre-industrial levels are determined using the 1850–1900 period as a baseline as this is the earliest period with near-global monitoring. The International Energy Agency (IEA) developed a Sustainable Development Scenario (SDS) which was a significant step towards outlining major transformation of the global energy culture [4]. This scenario is pragmatic rather than theoretical and aims to evolve the global energy sector to achieve the three energy-related SDGs.

As in every change requiring investment, this global energy transformation has a financial burden. According to the International Renewable Energy Agency (IRENA), the financial system should be aligned with the suggested sustainability and energy transition requirements. Financial limitations possess the potential of hindering the investment required to succeed in the energy transition. Increasing access to finance and keeping interest rates low would help this transition. Potential sources of finance include institutional investors and community-based finance [5].

2.2 Global Overview

The United Nations has five frontiers around the world, representing five regions. These are:

FURTHER LEARNING

UN Regions
Africa www.uneca.org
Asia and the Pacific www.unescap
.org
Europe https://unece.org
Latin America and the Caribbean www
.cepal.org/en
Western Asia www.unescwa.org

- Economic and Social Commission for Asia and the Pacific (ESCAP)
- Economic Commission for Africa (ECA)
- Economic and Social Commission for Western Asia (ESCWA)
- Economic Commission for Europe (UNECE)
- Economic Commission for Latin America and the Caribbean (ECLAC)

The common objectives of these regional commissions are to promote economic integration at the subregional and regional levels, to foster regional implementation of internationally concurred goals including SDGs through means of aiding economic, social, and environmental aspects among member states [6]. Let us take a look at each of these regional commissions' efforts to address the sustainability issues in achieving the SDGs.

Asia and the Pacific

According to the Asia and the Pacific SDG progress report [7] of the United Nations Economic and Social Commission for Asia and the Pacific (ESCAP), due to the region being very diverse, a large imbalance among countries exists even in the areas where noticeable progress has been observed. Loss of biodiversity and increasing greenhouse gas emissions are the two major common problems for the countries in the region. When observed at a subregional scale, CO_2 and greenhouse gas emissions in East and Northeast Asia, domestic material consumption in Southeast Asia, lack of proper nutrition and medium and high-tech industry in South and Southwest Asia, mental health and R&D investments in North and Central Asia, and income inequality and limited educator training in the Pacific seem to be prioritized by the governments in the corresponding regions.

Africa

Progress towards achieving the SDGs in Africa was stated to be slow and uneven in the Economic Report on Africa 2019 [8].

Despite improvements in access to energy, freshwater, and sanitation services, the infrastructure remains below the global average. The report also touches on fiscal policy and financing of the SDGs in Africa and discusses both the challenges and opportunities for the continent to achieve the goals. African countries signed up to two development agendas in 2015. One of them is the global 2030 SDGs, which has a holistic approach and aims to leave no country behind. The other is the African Union's Agenda 2063, which is a master plan to achieve the "*Africa we want*" [9].

According to the IEA's World Energy Outlook 2020, more than half a billion people will be added to the urban population in Africa by 2040 [10]. Looking at the example of China during 1990–2010, it would be reasonable to expect that steel and cement production will increase significantly. Besides this, urbanization will also come with use of a diversity of home appliances including air-conditioners. This will also noticeably add to the overall energy consumption in Africa, given the hotter climate of the continent. In light of all these facts and anticipations, Africa is expected to become highly influential for global energy trends towards 2040. As of now, the continent has an installed solar photovoltaic capacity of 5 GW, which is less than 1% of the global installed capacity. Having the highest solar resources could be a solution not only to the expected energy consumption increase in Africa, but also to the electricity access problem across the continent.

Western Asia

The UN Economic and Social Commission for Western Asia (ESCWA) comprises 18 countries in the Middle east and North Africa: Algeria, Bahrain, Egypt, Iraq, Jordan, Kuwait, Lebanon, Libya, Morocco, Mauritania, Oman, Palestine, Qatar, Saudi Arabia, Somalia, Sudan, Syria, Tunisia, United Arab Emirates, and Yemen.

The ESCWA annual report includes the sustainable development efforts in the region. The two challenges addressed are integrated management of natural resources and climate change adaptation and mitigation. According to the report, while there are efforts to promote regional and subregional cooperation for achieving the SDGs in the areas of energy, food, water, and environment, there seems to be a strong need for policy dialogue and policymaking [11].

Europe

The UN Economic Commission for Europe (UNECE) documents the general overview of the European region in terms of sustainability, and the efforts towards sustainability in their 2020 annual report [12].

The report highlights efforts to support the member states in achieving the SDGs through four major areas: sustainable use of natural resources, sustainable and smart cities for all ages, sustainable mobility and smart connectivity, and measuring and monitoring SDGs.

Regional organizations that UNECE cooperates with include the *European Commission*, *Eurostat* (the statistical office of the EU), *Eurasian Economic Commission*,

FIGURE 2.2 Regional sustainable development efforts require strategic planning. Policy dialogue and policymaking are important in actualizing the project leg of strategic planning for achieving SDGs. Source: KTSDESIGN/Science Photo Library via Getty Images.

Organization of the Black Sea Economic Cooperation (BSEC), and the *Commonwealth of Independent States* (CIS).

Latin America and the Caribbean

The 2030 Agenda and the Sustainable Development Goals report of the United Nations Economic Commission for Latin America and the Caribbean (ECLAC) discusses its priorities on achieving the SDGs considering the adoption of the 2030 agenda for sustainable development [13]. The priorities in the report are listed as:

- Strengthening the regional institutional architecture
- Enhancing analysis of the means of implementation of the 2030 agenda at the regional level
- Supporting the integration of the SDGs into national development plans and budgets
- Promoting the integration of the measurement processes necessary to build SDG indicators into national and regional strategies for the development of statistics, as well as the consolidation of national statistical systems and the governing role of national statistical offices

2.3 Indicators and Measurement of Sustainability

2.3.1 Sustainability Indicators

A global indicator framework was developed by the Inter-Agency and Expert Group and adopted at the 48th session of the United Nations Statistical Commission in 2017, as well as by the Economic and Social Council (ECOSOC) in the same year. The framework consists of 232 different indicators which address all goals. These indicators are listed in the work of the statistical commission that pertains to the 2030 agenda for sustainable development [14].

The National Risk Management Research Laboratory (NRMRL) of the United States Environmental Protection Agency (U.S. EPA) worked on a framework for sustainability indicators [2]. The report plays a guiding role, shedding light on sustainability indicators to assist the EPA in decision making. According to the framework, a sustainability indicator is a measurable characteristic of social, economic, or environmental systems that can be used to monitor the changes in these systems' behaviors for the sake of maintaining human and environmental welfare (**Figure 2.3**). The three pillars of sustainability listed are social, economic, and environmental categories (**Figure 2.4**).

FURTHER LEARNING

World Development Indicators
https://datatopics.worldbank.org/world-development-indicators/

2.3.2 Sustainability Measurement

The significance of the concept of sustainability and sustainable development has become unmistakable in recent years. One challenge with sustainability is that it is difficult to measure and assess. This brings the necessity of coming up with a method to reliably quantify sustainability.

The World Commission on Environment and Development (WCED) released the *Brundtland Report* in 1987. This report, also referred to as *Our Common Future*, takes its name from the then Prime Minister of Norway, *Gro Harlem Brundtland*.

FIGURE 2.3 Examples of global sustainability indicators. Sources: United Nations, Indicators of Sustainable Development, and World Bank, World Development Indicators, as cited in J. Fiksel, T. Eason, and H. Frederickson, "A framework for sustainability indicators at EPA," United States Environmental Protection Agency, 2012.

Poverty
- Unemployment rate
- Poverty index
- Population living below poverty line

Population Stability
- Population growth rate trend
- Population density

Human Health
- Average life expectancy
- Access to safe drinking water
- Access to basic sanitation
- Infant mortality rate

Living Conditions
- Urban population growth rate
- Floor area per capita
- Housing cost

Coastal Protection
- Population growth
- Fisheries yield
- Algae index

Agricultural Conditions
- Pesticide use rate
- Fertilizer use rate
- Arable land per capita
- Irrigation % of arable land

Ecosystem Stability
- Threatened species
- Annual rainfall

Atmospheric Impacts
- Greenhouse gas emissions
- Sulfur oxide emissions
- Nitrogen oxides emissions
- Ozone depleting emissions

Generation
- Municipal waste
- Hazardous waste
- Radioactive waste
- Land occupied by waste

Consumption
- Forest area change
- Annual energy consumption
- Mineral reserves
- Fossil fuel reserves
- Material intensity
- Groundwater reserves

Economic Growth
- GNP
- National debt/GNP
- Average income
- Capital imports
- Foreign investment

Accessibility
- Telephone lines per capita
- Information access

Sources:
UN, Indicators of Sustainable Development
World Bank, World Development Indicators

FIGURE 2.4 The three pillars of sustainability.

The WCED was sponsored by the United Nations and the objective was to study and analyze the relations between social, economic, and environmental developments and challenges. The report addressed a great many issues including sustainable development, which in its own language was defined as *"development that meets the needs of the present without compromising the ability of future generations to meet their own needs"* [15].

Since the report was published in 1987, there have been numerous institutions, organizations, governments, and agencies who have worked on coming up with an effective sustainable development measurement method. Due to the high number and interdependency of indicators, it would be reasonable to expect that the task is complex.

The validity of the selected methods is crucial as the measurements or performance values obtained from these methods may endorse investments done for the sake of addressing sustainable development and help rationally prioritize areas to focus on. If the selected method lacks scientific background or utilizes input that is not reliable, then the output of the analysis will be misleading. It may result in overestimating or underestimating the impacts of the actions taken on the environment, economy, and social systems. This would in turn reduce the potential for achieving the SDGs.

FIGURE 2.5 ESS data validation workflow. Source: European Commission, Eurostat, "Data validation".

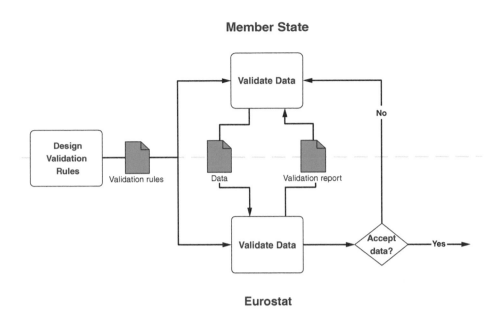

Eurostat, the statistical office of the European Union, has provided a comprehensive study on SDG indicators for all 17 goals of the United Nations. Each goal has a number of indicators and the changes in the quantities of different parameters for each indicator are available online [16].

These quantities come from archival data, statistical analyses, or theoretical models. Besides the necessity for a reliable method, the plausibility of the data used as input in the analysis is also important. This can be achieved by data validation.

Data validation means ensuring the quality and accuracy of the input values before implementing them into the model. It is a key step which needs to be utilized in all statistical analyses.

The European Statistical System (ESS) performs data validation through validation rules that are jointly designed at each statistical domain. **Figure 2.5** shows the data validation workflow of the ESS.

A joint study group formed of the UNECE, OECD, and Eurostat has made a study on measuring sustainable development. The report of this study highlights two approaches in examining sustainable development: the *policy-based approach* and the *capital approach*. The policy-based approach suggests that the relation between indicators and policy is very strong. Sustainable development indicator sets for different countries are studied within this approach where the top three most common sustainable development indicator themes are listed as management of natural resources, climate change and energy, and sustainable consumption and production. According to the capital-based approach, sustainable development is defined as "non-declining *per capita* national wealth by replacing or conserving the sources of that wealth, that is, stocks of produced, human, social and natural capital" [17].

The U.S. EPA has adopted a *Sustainability Assessment and Management* (SAM) plan which consists of a series of steps developed to achieve outcomes to inform the decision-making mechanisms [18]. The SAM process is designed to incorporate three key features to achieve a holistic approach. These features are:

1. Comprehensive and systems-based: Analysis includes an integrated evaluation of social, environmental, and economic outcomes.

FIGURE 2.6 Ecosystem vitality vs environmental health for countries. Source: Z. A. Wendling, J. W. Emerson, A. de Sherbinin, D. C. Esty, et al. (2020). 2020 Environmental Performance Index. New Haven, CT: Yale Center for Environmental Law & Policy. epi.yale.edu.

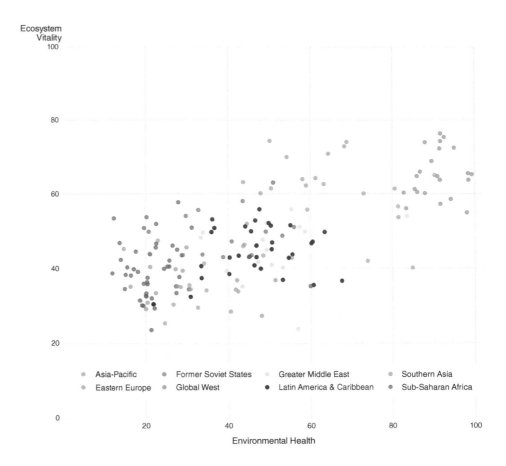

2. Intergenerational: The long-term outcomes of alternative solutions are suggested to be evaluated in addition to the outcomes with higher priority.

3. Stakeholder involvement and collaboration: Stakeholders should be involved throughout the process.

There is also a global environmental metric called the *Environmental Performance Index* (EPI), developed by the Yale University Center for Environmental Law and Policy, and the Center for International Earth Science Information Network, Columbia University, in collaboration with the World Economic Forum [19]. Evaluation of the environmental performance of countries is achieved through two policy objectives: *environmental health* and *ecosystem vitality* (**Figure 2.6**). These objectives are studied with 24 indicators and 10 issue categories.

Different indicator sets can be utilized in assessing sustainable development. Four of these sets through which sustainable development can be measured are ecological footprinting, the capital approach, green national net product (GNNP), and genuine savings [20].

Ecological (or environmental) footprinting is a method of collecting information on the required agricultural land area and water to sustain population activity. Total ecological footprint can be calculated by [21]:

$$EF_{TOT} = EF_{DIR} + EF_{INDIR} \tag{2.1}$$

The term EF_{DIR} is associated with the direct use of land for raising crops, and is given by:

$$EF_{DIR} = \left(\frac{Q}{Y_W}\right) \times EQF_{Cropland} \tag{2.2}$$

where

Q is the amount of crop harvested (in tonnes),

Y_W is the yield of production of the same crop in the world as a whole (in tonnes),

EQF (equivalence factor) is a scaling factor used to convert a specific land-use type into a universal unit of biologically productive area.

The EF_{INDIR} term in **Equation 2.1**, on the other hand, takes all indirect land uses into consideration. It is the contribution of n number of inputs (e.g., fuels, fertilizers, chemicals) required for crop production. It can be expressed as:

$$EF_{INDIR} = \sum_{i=1}^{n} EF_i \qquad (2.3)$$

In **Equation 2.3**, EF_i is the footprint resulting from each of the inputs in crop production and is given by:

$$EF_i = \sum_{j=1}^{6} \left(^R A_i\right) \times YF_j \times EQF_j = \sum_{j=1}^{6} \left(\frac{Q_i}{Y_i}\right) \times YF_j \times EQF_j \qquad (2.4)$$

where

$^R A$ is the total area required in hectares, and is the ratio of the quantity of the input (Q_i) to the yield (Y_i),

the subscript i ($=1, \ldots, n$) indicates the inputs utilized,

the subscript j ($=1, \ldots, 6$) relates to the six different land-use types of National Footprint Accounts (i.e., cropland, grazing land, fishing grounds, forest, built-up land, and carbon footprint),

YF_j is the yield factor specific to a country and j-land type,

EQF_j is the equivalence factor specific to each j-land type.

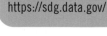

FURTHER LEARNING

Sustainable Development Indicators, United States.
https://sdg.data.gov/

FURTHER LEARNING

Sustainable Development Indicators, Europe
https://ec.europa.eu/eurostat/web/sdi/indicators

FURTHER LEARNING

Human Development Index
http://hdr.undp.org/en/content/human-development-index-hdi

Another indicator in measuring sustainable development is green national net product (GNNP) (**Figure 2.7**). Atkinson *et al.* [22] suggest the expression:

$$GNNP = GNP - (D + R \mid E) \qquad (2.5)$$

where

GNP is the gross national product,

D is the depreciation of produced assets,

R is the changes in resource stocks (depletion of non-renewable resources such as oil),

E is the environmental degradation (e.g., due to air pollution).

In 1990, the United Nations Development Programme (UNDP) launched an indicator, the *Human Development Index* (HDI) (**Figure 2.8**), which was offered as a comprehensive measure of human development in countries taking health (life expectancy), education (expected and mean years of schooling), and standard of living (gross national income per capita) into account. For defined minimum and maximum values, the dimension indices are calculated as:

$$I = \frac{(\text{Actual value} - \text{Minimum value})}{(\text{Maximum value} - \text{Minimum value})} \qquad (2.6)$$

FIGURE 2.7 Green national net product (GNNP) is an alternative indicator to gross national product (GNP) in measuring sustainable development. Depreciation of produced assets, depletion of non-renewable resources, and environmental degradation are also taken into consideration in determining GNNP. Source: Frans Lemmens/ Corbis Documentary via Getty Images.

The HDI is the geometric mean of these three calculated as follows:

$$\text{HDI} = \sqrt[3]{\left(I_{\text{Health}} \times I_{\text{Education}} \times I_{\text{Income}}\right)} \tag{2.7}$$

Maximum and minimum life expectancies are set at 85 and 20 years, respectively. Education index is the average of expected and mean years of schooling indices. For the income index, natural logarithms of the actual, minimum, and maximum values are used to prevent a surge in HDI for increasing income.

Another version of human development assessment is the *Sustainable Human Development Index* (SHDI), which also takes environmental aspects into account. Evans *et al.* [20] discuss a calculation method offered by Constantini and Manni:

$$\text{SHDI} = \frac{1}{4}\left[\left(\frac{x_1 - 0}{80 - 0}\right) + \left(\frac{1}{3}x_2 + \frac{2}{3}x_3\right) + \left(\frac{\log(x_4) - \log(100)}{\log(40,000) - \log(100)}\right) + \left(\frac{x_5 + x_6 + x_7 + x_8}{4}\right)\right] \tag{2.8}$$

where the four components inside the bracketed term on the right-hand side in order pertain to education attainment, social stability, sustainable access to resources (GNNP), and environmental quality. The variables in **Equation 2.8** are:

x_1: tertiary gross enrolment ratio, UNESCO definition,

$x_2 = (y_1 - 25)/(85 - 25)$: Health Index ($y_1$: life expectancy at birth, number of years),

$x_3 = 1 - [(y_2 - 0)/(25 - 0)]$: Employment Index ($y_2$: unemployment rate, percentage),

x_4: GNNP current purchasing power parity ($\$$PPP) per capita,

$x_5 = 1 - [(y_3 - 0)/(0.015 - 0)]$: Air Pollution Index ($y_3$: tonnes per day per worker of NO_x, SO_2, NH_3, NMVOC, CO),

$x_6 = 1 - [(y_4 - 0)/(0.35 - 0)]$: Water Pollution Index (y_4: BOD emissions in kg/day per worker),

$x_7 = 1 - [(y_5 - 0)/(1000 - 0)]$: Soil Pollution from Agriculture Index (y_5: fertilizers, herbicides, and insecticides used on arable land, kg per hectare),

$x_8 = 1 - [(y_6 - 0)/(10 - 0)]$: Energy Index ($y_6$: tonnes of oil equivalent per capita consumed per year).

FIGURE 2.8 The Human Development Index (HDI) was launched by the United Nations Development Programme (UNDP) as a comprehensive measure of human development, taking health, education, and standard of living into account. Source: hadynyah/E+ via Getty Images.

 CLASSROOM DEBATE

Which do you think is a better economic growth and development indicator: per capita GDP or the Human Development Index (HDI)?

Both HDI and SHDI calculation methods mentioned here are based on a scale from 0 (worst) to 1 (best). The models would benefit from adjusting the benchmark life expectancy and GDP per capita values in order to prevent the HDI or SHDI values exceeding unity for the highest ranked countries.

Example 2.1 Green National Net Product (GNNP)

For Algeria, depreciation of produced assets, changes in resource stocks, and environmental degradation are estimated to be 10%, 27.1%, and 4.2% of the GNP, respectively. Determine the green national net product (GNNP).

Solution

Algeria's GNP is reported as $164.8 billion. Using **Equation 2.1** on a percentage basis, we can rewrite it as:

$$\text{GNNP} = \text{GNP} - (0.1\,\text{GNP} + 0.271\,\text{GNP} + 0.042\,\text{GNP}) = 0.587\,\text{GNP}\ (58.7\%\ \text{of GNP})$$

Hence,

$$\text{GNNP} = 0.587 \times \$164.8\ \text{billion} = \$96.7\ \text{billion}$$

Note: *It is recommended that you retrieve economic data for countries from reliable sources. A good number of these sources exist. The World Bank and OECD are two of them. When researching economic data for a given country, pay attention to the difference between GNP (gross national product) and GDP (gross domestic product)!*

Example 2.2 Human Development Index (HDI)

Determine the dimension indices and the HDI value for Portugal using the data from the UNDP Human Development Reports website (http://hdr.undp.org/en/composite/HDI).

Solution

Obtaining the data from the UNDP website, we get:

Indicator	Value
Life expectancy at birth (years)	81.4
Expected years of schooling (years)	16.3
Mean years of schooling (years)	9.2
Gross national income per capita (2011 PPP $)	27,315

Now, let's calculate the dimension indices for each indicator:

$$\text{Health index} = \frac{81.4 - 20}{85 - 20} = 0.945$$

$$\text{Expected years of schooling index} = \frac{16.3 - 0}{18 - 0} = 0.905$$

$$\text{Mean years of schooling index} = \frac{9.2 - 0}{15 - 0} = 0.613$$

$$\text{Education index} = \frac{0.905 + 0.613}{2} = 0.759$$

$$\text{Income index} = \frac{\ln(27315) - \ln(100)}{\ln(75000) - \ln(100)} = 0.847$$

$$\text{HDI} = (0.945 \times 0.759 \times 0.847)^{1/3} = 0.847$$

Note: *To have a feeling of how a change in one of the indicator values affects the HDI, you can take a look at the Human Development Reports website. Compare two countries with similar indicator values, except for one value being different. For instance, in the list you will see that Poland has a higher HDI (0.865) than Portugal even though it has lower life expectancy and slightly lower GNI per capita. Its mean years of schooling (12.3 years) being higher than that of Portugal (9.2 years) yields a higher HDI. Therefore, the HDI not only gives us an idea about human development in a country, but also helps us quantify the stronger indicators and the indicators for which there is room for improvement.*

Example 2.3 Solving for HDI using Python and Performing Sensitivity Analysis

Reconsider **Example 2.2**. (a) Determine the HDI value using Python, and (b) perform a sensitivity analysis for health, education, and income indices for the ranges of input variables given below:

- Life expectancy at birth: 50 to 90 years
- Expected years of schooling: 0 to 20 years
- Gross national income per capita: $5000 to $50,000

Solution

a. Let's solve for HDI by determining health, education, and income indices first, and then implementing them into **Equation 2.7**:

```
%pylab inline
import math
import numpy

life_expectancy_at_birth=81.4
health_index = (life_expectancy_at_birth-20)/(85-20)
health_index
0.9446153846153847

expected_years_of_schooling=16.3
expected_years_of_schooling_index = (expected_years_of_schooling-
0)/(18-0)
mean_years_of_schooling=9.2
mean_years_of_schooling_index = (mean_years_of_schooling-0)/(15-0)
education_index = (expected_years_of_schooling_index + mean_years_of_
schooling_index) / 2
education_index
0.7594444444444444
```

```
gross_national_income_per_capita = 27315
income_index = (  np.log(gross_national_income_per_capita) - np.log
(100))/(np.log(75000)-np.log(100))
income_index
0.8474258395372026
```

```
HDI= ( health_index * education_index * income_index ) ** (1/3)
# ** used as exponential operator
HDI
0.8471316516932836
```

b. In certain applications, it may be essential to conduct a **sensitivity analysis** to observe how the output value for an index changes with varied inputs. Let's take a look at this for the given ranges for each input.

```
life_expectancy_at_birth=arange(50,90)
health_index = (life_expectancy_at_birth-20)/(85-20)
```

```
plot(life_expectancy_at_birth, health_index)
title("Life expectancy at birth vs Health Index")
xlabel("Life Expectancy at Birth")
ylabel("Health Index")
```

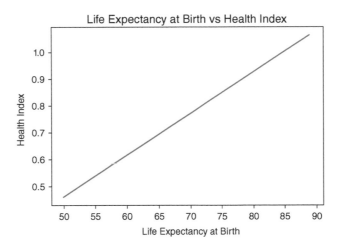

Repeating the same procedure for the education index:

```
expected_years_of_schooling=arange(5,20,0.1)
expected_years_of_schooling_index = (expected_years_of_schooling-
0)/(18-0)
mean_years_of_schooling=9.2
mean_years_of_schooling_index = (mean_years_of_schooling-0)/(15-0)
education_index=(expected_years_of_schooling_index+mean_years_of_-
schooling_index) / 2
```

```
plot(expected_years_of_schooling, education_index)
title("Expected Years of Schooling vs Education Index")
xlabel("Expected Years of Schooling")
ylabel("Education Index")
```

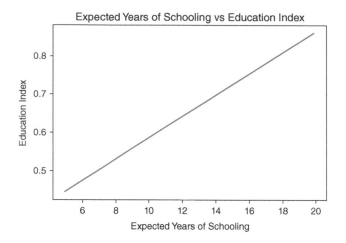

Finally, we perform the same for the income index:

```
gross_national_income_per_capita = arange(5000,50000,2500)
income_index = (  log(gross_national_income_per_capita) -
log(100))/(log(75000)-log(100))

plot(gross_national_income_per_capita, income_index)
title("Gross National Income Per Capita vs Income Index")
xlabel("Gross National Income Per Capita (2011 PPP$)")
ylabel("Income Index")
```

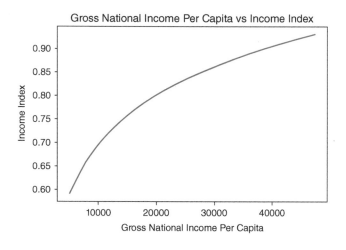

2.4 Sustainability of Energy Resources

2.4.1 Sustainability of Fossil Energy Resources

Coal, oil, and natural gas are non-renewable resources whose formation goes back to the Carboniferous period. The plants and animals that died and were buried by layers of rock in time turned into fossil fuels over millions of years. While there are different opinions on how long it will be until we run out of fossil fuels, it should be noted that

FIGURE 2.9 Years of fossil fuel reserves estimated to be left. Data source: BP Statistical Review of Energy.

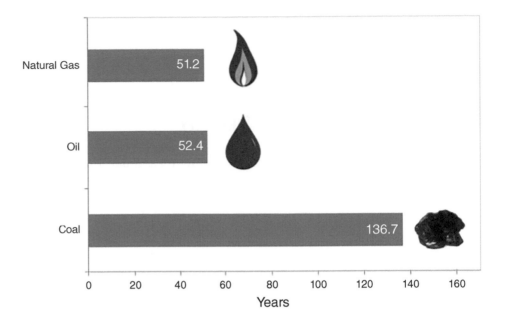

sometimes *"it is tough to make predictions, especially about the future"* as the legendary baseball player *Yogi Berra* said. Formal sciences such as mathematics, logic, and statistics help us to predict the future behavior of certain patterns based on models. There are numerous models that are able to predict or extrapolate a pattern successfully, while there also exist some models that are not as accurate due to the unpredictable behavior of some variables or addition of unforeseeable parameters into the equation over time.

According to the BP Statistical Review of World Energy 2021, the years of global coal, oil, and natural gas left can be seen in the graph in **Figure 2.9** [23].

It should be noted that the given values can change with time based on the discovery of new resources and changes in annual production and consumption.

2.4.2 Sustainability of Non-Fossil Energy Resources

The diminution of fossil fuels and economic, security, and environmental concerns spurs us on to look for alternative (non-fossil) energy resources. Non-fossil sources include nuclear (4.4%), hydroelectric (6.8%), and other renewable sources (4%) which include solar, wind, geothermal, tidal, and biomass.

Nuclear energy seems to be the most controversial resource. The OECD countries all share the same goals in achieving sustainable development; however. they sometimes may have dissimilar opinions on the role of nuclear energy towards the abovementioned goals. The advantages and disadvantages of this energy resource are discussed in more detail in Chapter 5.

Hydropower relies on the water cycle, which is driven by the Sun. Solar energy depends on solar irradiation, wind energy on atmospheric pressure differences, geothermal on the thermal energy stored in the Earth, tidal energy on the gravitational pull from both the Moon and the Sun, and biomass energy on the Sun's energy through photosynthesis. These are all sustainable resources as energy can be replenished naturally and the long-term availability is there. It is important to remember that technological developments on improving the energy efficiency of these systems play a significant role in determining the level of sustainability of these energy resources. This highlights the importance of R&D that involves both materials and systems engineering.

2.5 Summary

In this chapter, we tried to understand what sustainability and sustainable development are. The Sustainable Development Goals (SDGs) were discussed. International meetings and global commissions were covered (1) to have an idea about the objectives and structures of these institutions, and (2) to realize the efforts put in at a global scale towards maintaining a sustainable world. One important aspect of studying a topic such as sustainability, which is not tangible, is defining indicators for assessment and developing measurement tools to evaluate the defined indicators. The chapter covered global sustainability indicators, pointing out the three pillars of sustainability as social, economic, and environmental. It is important to realize that there are interactions among these indicators and pillars of sustainability. We also looked at sustainability measurement and assessment tools. Some of the approaches appeared in this chapter, including relations for *ecological footprinting* (EF), *green national net product* (GNNP), *Human Development Index* (HDI), and *Sustainable Human Development Index* (SHDI). Remembering the intangible nature of sustainability and the interactions between indicators, it is only logical to expect that there can be more than one approach in quantifying parameters. Results from reliable approaches are of great significance as they portray the strengths and weaknesses in particular sustainability indicator areas.

REFERENCES

[1]   United Nations Statistics Division, "Studies in Methods: Glossary of Environment Statistics," Series F, No. 67, United Nations, New York, 1997.

[2]   J. Fiksel, T. Eason, and H. Frederickson, "A Framework for Sustainability Indicators at EPA," United States Environmental Protection Agency (EPA), 2012. Accessed: Mar. 10, 2020 [Online]. Available: www.epa.gov/sites/produc tion/files/2014-10/documents/framework-for-sustainability-indicators-at-epa.pdf.

[3]   United Nations, "Sustainable Energy for All: An Overview," Accessed: Mar. 10, 2020 [Online]. Available: www.un.org/millenniumgoals/pdf/SEFA.pdf.

[4]   International Energy Agency, "World Energy Model," Oct. 2020. www.iea.org/reports/world-energy-model (accessed Feb. 28, 2021).

[5]   International Renewable Energy Agency, "Global Energy Transformation: A Road Map to 2050," IRENA, Abu Dhabi, 2020. Accessed: Mar. 11, 2020 [Online]. Available: www.irena.org/-/media/Files/IRENA/Agency/Publication/2019/Apr/IRENA_Global_Energy_Transformation_2019.pdf.

[6]   United Nations, "Regional Commissions," Regional Commissions New York Office. www.regionalcommissions.org (accessed Mar. 11, 2020).

[7]   Economic and Social Commission for Asia and the Pacific, "Asia and the Pacific SDG Progress Report," United Nations, Bangkok, 2020. Accessed: Mar. 12, 2020 [Online].   Available: www.unescap.org/sites/default/files/publications/ESCAP_Asia_and_the_Pacific_SDG_Progress_Report_2020.pdf.

[8]   United Nations. Economic Commission for Africa, "Economic Report on Africa 2019: Fiscal Policy for Financing Sustainable Development in Africa," UN ECA,

Addis Ababa, 2019. Accessed: Mar. 15, 2020 [Online]. Available: https://repository.uneca.org/handle/10855/41804.

[9]   African Union, "Agenda 2063: The Africa We Want," African Union Commission, Sep. 2015. Accessed: Mar. 15, 2020 [Online]. Available: https://au.int/sites/default/files/documents/36204-doc-agenda2063_popular_version_en.pdf.

[10]   International Energy Agency, "World Energy Outlook," IEA, 2020. Accessed: Mar. 17, 2020 [Online]. Available: www.iea.org/reports/world-energy-outlook-2020.

[11]   United Nations Economic and Social Commission for Western Asia, "ESCWA Annual Report 2018," United Nations, Beirut, 2019. Accessed: Mar. 17, 2020 [Online]. Available: www.unescwa.org/publications/annual-report-2018.

[12]   United Nations Economic Commission for Europe, "UNECE Annual Report 2019," United Nations, Geneva, May 2020. Accessed: Aug. 5, 2020 [Online]Available: https://unece.org/DAM/UNECE_Annual_Report_2019_Web_FINAL.pdf.

[13]   United Nations Economic Commission for Latin America and the Caribbean, "The 2030 Agenda and the Sustainable Development Goals: An Opportunity for Latin America and the Caribbean," United Nations, Santiago, Dec. 2018. Accessed: Mar. 18, 2020 [Online]. Available: https://repositorio.cepal.org/bitstream/handle/11362/40156/25/S1801140_en.pdf.

[14]   United Nations General Assembly, "Work of the Statistical Commission Pertaining to the 2030 Agenda for Sustainable Development," Jul. 2017. Accessed: Mar. 18, 2020 [Online]. Available: https://undocs.org/A/RES/71/313.

[15]   Brundtland Commission, *Our Common Future*. Oxford, UK: Oxford University Press, 1987.

[16]   European Commission, Eurostat, "Data Validation." https://ec.europa.eu/eurostat/web/main/data/data-validation (accessed Mar. 20, 2020).

[17]   UNECE/OECD/Eurostat Joint Working Group, "Measuring Sustainable Development," United Nations, New York and Geneva, 2008. Accessed: Mar. 21, 2020 [Online]. Available: www.oecd.org/greengrowth/41414440.pdf.

[18]   National Research Council, *Sustainability and the U.S. EPA*, Washington, DC: The National Academies Press, 2011. Accessed: Mar. 24, 2020 [Online]. Available: https://doi.org/10.17226/13152.

[19]   Z. A. Wendling, J. W. Emerson, A. de Sherbinin, and D. C. Esty, "2020 Environmental Performance Index," Yale Center for Environmental Law & Policy, New Haven, CT, 2020. Accessed: Jan. 23, 2022 [Online]. Available: https://epi.yale.edu/.

[20]   A. Evans, V. Strezov, and T. Evans, "Measuring tools for quantifying sustainable development," *European Journal of Sustainable Development*, vol. 4, no. 2, pp. 291–300, Jun. 2015, doi: 10.14207/ejsd.2015.v4n2p291.

[21]   S. Bastianoni, V. Niccolucci, R. M. Pulselli, and N. Marchettini, "Indicator and indicandum: 'Sustainable Way' vs 'Prevailing Conditions' in the ecological footprint," *Ecological Indicators*, vol. 16, pp. 47–50, May 2012, doi: 10.1016/j.ecolind.2011.10.001.

[22]   G. Atkinson, R. Dubourg, K. Hamilton, M. Munasinghe, D. Pierce, and C. Young, *Measuring Sustainable Development : Macroeconomics and the Environment*. Cheltenham, UK; Northampton, MA, USA: E. Elgar, 1997.

[23]   BP, "Statistical Review of World Energy," 2021. Accessed: Jan. 12, 2022 [Online]. Available: www.bp.com/statisticalreview.

CHAPTER 2 EXERCISES

2.1 Define sustainable development in your own words. Explain how this would relate to your profession in your engineering career.

2.2 On a blank global map, show the five UN Economic Commission regions. List three major SDGs for each region.

2.3 The GNP and GNNP of a country are $245.3 billion and $177.6 billion, respectively. If the country has 8.4% depreciation of produced assets, and 12.7% changes in resource stocks, determine its environmental degradation as a percentage of its GNP.

2.4 Calculate the HDI for Hong Kong, China (SAR), retrieving the indicator values from the UNDP website (http://hdr.undp.org/en/composite/HDI). Comment on the effect of its indicator values on its HDI rank.

2.5 Construct a graph for the G-20 countries with GNI per capita on the x-axis and HDI on the y-axis.

2.6 Visit the EPA website (www.epa.gov) and research sustainability indicators. Pick one indicator and write an essay on it discussing the significance of the indicator, examples of global problems, and suggested solutions. Conclude with your own words.

2.7 Pick three of the 17 SDGs and explain in your own words the potential impact of a pandemic on achieving these three goals.

2.8 Explain briefly the significance of the terms below:
  a. Brundtland Report
  b. Kyoto Protocol
  c. Millennium Goals
  d. Davos Summit
  e. Paris Agreement

2.9 Research a sustainable development project in each of the three countries listed below:
  a. Indonesia
  b. Sierra Leone
  c. Thailand

2.10 Construct (a) a bar chart showing total global coal consumption and coal production, both in million tonnes of oil equivalent (mtoe), (b) a bar chart showing total global coal production in million tonnes of oil equivalent (mtoe) and in million tonnes. Comment on the difference between mtoe and million tonnes.

CHAPTER 3

Theoretical Basics
Electrical, Chemical, Thermal, and Fluids

Energy engineering has a multidisciplinary nature including disciplines such as mechanical, electrical, chemical, civil, and environmental engineering that study the theory and applications of energy systems. The theoretical aspects of such applications stretch over these disciplines. In this chapter, the theoretical basics of five major fields are covered to help engineers and professionals have an understanding of how each discipline and field is associated with energy. The fields covered are electrodynamics, chemical energy conversion, thermodynamics, fluid mechanics, and heat transfer. It is important to keep in mind that each individual field is a wide ocean, and there are many textbooks written on each field. This chapter is intended to give the reader some brief information on each field accompanied by pertinent equations.

3.1 Electrodynamics Basics

Electrodynamics is a branch of physics that studies the interactions of electric currents amongst one another or with magnetic fields. Unlike *electrostatics* which relates to the case when electric charges are stationary or are moving very slowly such that there are no magnetic forces between them, *electrodynamics* deals with moving charges and their interactions with forces acting on them.

Ohm's Law

Ohm's law states that at a constant temperature the electric current flowing through a fixed linear resistance is directly proportional to the voltage applied across it and is inversely proportional to the resistance:

$$V = IR \tag{3.1}$$

where V is the voltage in volts (V), I is the current in amperes (A), and R is the resistance in ohms (Ω) (**Figure 3.1**).

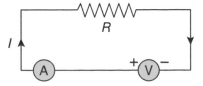

FIGURE 3.1 Electric circuit.

FIGURE 3.2 Electrical resistors on an electronic board. Source: Fred Proksch via Getty Images.

→ **FURTHER LEARNING**

Electricity 101
www.energy.gov/oe/information-center/educational-resources/electricity-101

Joule's Law of Heating

The resistance, R, of a conductor causes it to heat when a current I passes through it per unit time. The heat developed in the conductor per unit time is then expressed by *Joule's law of heating* and is given by:

$$P = VI = I^2R \tag{3.2}$$

where P has the units of watts (W) or joules per second (J/s).

Electromotive Force

Electromotive force (emf) is the electric potential produced either by a two-terminal cell (i.e., electrochemical cell) or by varying the magnetic field. The battery of an electric vehicle (EV) can provide emf by converting chemical energy into electrical energy, or a generator at a hydroelectric power plant can yield emf by converting mechanical energy into electrical energy.

Despite its name, electromotive force is not a force, it is electric potential and has the unit of volts as it is expressed as the number of joules of energy given by the source divided by each coulomb, which is the charge accumulated by a current of one ampere in one second. Electromotive force is given by:

$$\varepsilon = V + Ir = IR + Ir \tag{3.3}$$

where ε is the emf, V is the voltage of the cell, I is the current flowing through the circuit, R is the load resistance, and r is the internal resistance (i.e., of the battery).

Example 3.1 Electromotive Force and Joule's Law of Heating

Consider a battery with 12 V emf and internal resistance of 0.15 Ω.

a. Determine its voltage when it is connected to a 15 Ω load.
b. Determine its voltage when it is connected to a 0.75 Ω load.
c. Calculate the rate of heat dissipation with the 0.75 Ω load.

Solution

a. From Equation 3.3, we can deduce that the voltage can be expressed as:

$$V = \varepsilon - Ir$$

Current is:

$$I = \frac{\varepsilon}{R+r} = \frac{12\ \text{V}}{(15+0.15)\Omega} = 0.792\ \text{A}$$

Then,

$$V = \varepsilon - Ir = 12\text{ V} - (0.792\text{ A})(0.15\ \Omega) = 11.88\text{ V}$$

b. Using the same approach, with $R = 0.75\ \Omega$:

$$I = \frac{\varepsilon}{R + r} = \frac{12\text{ V}}{(0.75 + 0.15)\ \Omega} = 13.33\text{ A}$$

Hence,

$$V = \varepsilon - Ir = 12\text{ V} - (13.33\text{ A})(0.15\ \Omega) = 10.00\text{ V}$$

c. Using Joule's law of heating, we can obtain the rate of heat dissipated:

$$P = VI = I^2 R = (13.33\text{ A})^2 (0.75\ \Omega) = 133.27\text{ W}$$

Note: *As the load resistance decreases, voltage experiences a decrease as well. This is because a higher amount of current is drawn from the source when the resistance is lower.*

Heat dissipation is proportional to the square of the current. Hence, an increase in the current drawn can yield a noticeable increase in the amount of heat generated.

Faraday's Law

Electromagnetic induction is a very useful phenomenon, providing a wide variety of applications that serve people for energy generation, residential use, industry, agriculture, transportation, and many other purposes. Induction is used for power generation and transmission. A generator attached to a turbine at a thermal power plant or at a hydroelectric power station generates electricity based on induction. In an opposite manner to how generators work, an electric motor transforms electrical energy into mechanical energy, serving a wide range of purposes from small electronics to industrial motors.

Electricity generation is achieved utilizing generators wherein the spinning coil, by means of mechanical energy, induces an emf in a magnetic field. Another approach to generating electricity would be to keep the coil stationary and spin the permanent magnets around it. This is how the turbines at a hydroelectric power dam interact with the generator. The water falling spins the turbine, which in turn rotates the permanent magnets around a coil, hence inducing an electric field. An analogy to this phenomenon is an automobile placed in a wind tunnel for aerodynamic testing. The pressure distribution around the vehicle can be obtained while the automobile is stationary and air is flowing around it, rather than the car driving through the air.

Faraday's law helps quantify the emf. In 1831, he reported his findings from a set of experiments he conducted (**Figure 3.3**). In three different tests, he (a) pulled a loop of wire through a magnetic field, (b) pulled the magnet, keeping the loop of wire

FIGURE 3.3 Experiments conducted by Michael Faraday. Source: D. J. Griffiths (2018), *Introduction to Electrodynamics*, Cambridge University Press.

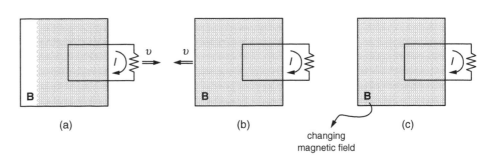

stationary, and (c) changed the strength of the magnetic field while keeping the wire and the magnet fixed. In all three experiments he observed a current flow through the wire loop [1].

From these experiments, the following relations were derived:

$$\varepsilon = -\frac{d\mathbf{\Phi}}{dt} \tag{3.4}$$

Equation 3.4 simply states that the induced electric field accounting for the emf is equal to the time rate of change of magnetic flux, $\mathbf{\Phi}$. For a coil of N identical turns of wire, each being exposed to the same magnetic field, **Equation 3.4** becomes:

$$\varepsilon = -N\frac{d\mathbf{\Phi}}{dt} \tag{3.5}$$

For the first two experiments, the emf would be equal as it is a matter of the relative motion of the wire loop and the magnet, rather than whichever is moving. And, for a magnetic field changing:

$$\nabla \times \mathbf{E} = -\frac{\partial \mathbf{B}}{\partial t} \tag{3.6}$$

where $\mathbf{E}(\mathbf{r}, t)$ is the electric field, and $\mathbf{B}(\mathbf{r}, t)$ is the magnetic field.

Maxwell's Equations

Maxwell's equations together with the Lorentz law can be considered as a summary of the entirety of classical electrodynamics. **Equations 3.7–3.10** are *Maxwell's equations*, along with **Equation 3.11** which describes the *Lorentz force law*:

$$\nabla \cdot \mathbf{E} = \frac{1}{\varepsilon_0}\rho \;(\text{Gauss's law}) \tag{3.7}$$

$$\nabla \cdot \mathbf{B} = 0 \;(\text{Gauss's law for magnetism}) \tag{3.8}$$

$$\nabla \times \mathbf{E} = -\frac{\partial \mathbf{B}}{\partial t} \;(\text{Faraday's law}) \tag{3.9}$$

$$\nabla \times \mathbf{B} = \mu_0 \mathbf{J} + \mu_0 \varepsilon_0 \frac{\partial \mathbf{E}}{\partial t} \;(\text{Ampere's law with Maxwell's correction}) \tag{3.10}$$

$$\mathbf{F} = q(\mathbf{E} + \mathbf{v} \times \mathbf{B}) \tag{3.11}$$

where ρ is the total electric charge density (total charge per unit volume), $\mathbf{J}$ is the total electric current density (total current per unit area), μ_0 is the vacuum permeability (henry per meter), and ε_0 is the vacuum permittivity (farad per meter).

3.2 Chemical Energy Conversion Basics

Chemical energy is the potential energy that is stored within the molecules of matter. To break the chemical bonds within the molecules requires energy, while energy is released when new bonds are formed. The stronger these bonds get, the more energy is required to break them, or the more energy is released if they are formed. These molecules can undergo chemical reactions where energy is released. Combustion is a

FIGURE 3.4 Schematic of a natural gas-fueled power plant with the key components including the compressor, combustion chamber, gas turbine, and generator. Source: Tennessee Valley Authority (TVA).

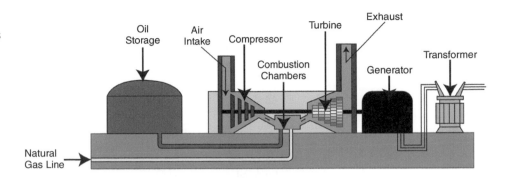

FIGURE 3.5 A natural gas-fueled power plant with cooling towers. Source: Ron and Patty Thomas via Getty Images.

very common form of chemical reaction in obtaining energy. Biomass fuels such as wood and agricultural waste have been used throughout the ages by humans for heating and cooking purposes. Combustion of fossil fuels such as coal, natural gas, or oil is used in generating electrical energy at power plants. **Figure 3.4** illustrates a natural gas-fueled power plant which produces electrical energy by means of burning natural gas in a combustion chamber.

The combustion process releases water vapor. There are techniques to recover the amount of heat the water vapor possesses by means of condensing it. Higher calorific value (or gross calorific value, aka higher heating value, HHV) of a fuel is defined as the amount of heat released by a certain amount of fuel, initially at 25 °C, once it is combusted and the products have returned to the initial temperature of 25 °C, also taking into account the latent heat of vaporization of water in the combustion products (**Table 3.1**).

The general chemical equation for stoichiometric combustion of a hydrocarbon in oxygen is given as:

$$C_xH_y + \left(x + \frac{y}{4}\right)O_2 \rightarrow xCO_2 + \frac{y}{2}H_2O \tag{3.12}$$

Stoichiometric combustion refers to a process where all the reactants are consumed. It is based on the idea of the law of conservation of mass, where the total mass of all reactants (inputs) should be equal to the total mass of all products (outputs). Let us consider combustion of natural gas (which is mostly methane, CH_4) as an example. The chemical reaction is:

$$CH_4 + 2O_2 \rightarrow CO_2 + 2H_2O$$

As combustion gives off energy, this is an *exothermic reaction*.

Table 3.1

Higher heating values of fuels* [2]	
Fuel	HHV (MJ/kg)
Wood	20.9
Coal	26.2
Natural gas	54.4
Crude oil	45.5
Diesel	44.0
Gasoline	48.4
Uranium	381,000

* Fuels consisting of a mixture of compounds can vary in HHV.

Conversion efficiency is one of the key parameters in combustion technologies employed to generate electrical energy. In producing mechanical work or electricity, efficiency can simply be determined by:

$$\eta = \frac{\text{Useful energy output}}{\text{Energy input}} \tag{3.13}$$

It is important to note that **Equation 3.13** relates only to the efficiency of conversion of chemical energy into mechanical work or electrical energy. However, a holistic engineering approach would look at the overall efficiency all the way from the extraction of the fuel to utilization of electrical energy by the end-user. In such a case, the overall efficiency can be given as:

$$\eta_{\text{overall}} = \prod_{i=1}^{n} \eta_i \tag{3.14}$$

Considering the extraction of natural gas, **Equation 3.14** can be expanded as:

$$\eta_{\text{overall}} = \eta_{\text{gas extraction}} \, \eta_{\text{gas processing}} \, \eta_{\text{gas transportation}} \, \eta_{\text{thermal power plant}} \, \eta_{\text{electricity transmission}}$$
$$\eta_{\text{electricity distribution}} \, \eta_{\text{motor}}$$

As can be seen, overall efficiency relies on many dependent and independent variables and requires efficient processes at each step from fuel extraction to electricity use by the end-user.

Electricity generation through conversion of chemical energy into mechanical work, then into electrical energy, can be achieved by execution of thermodynamic cycles. These cycles are discussed in the following section.

Example 3.2 Determining the Empirical Formula of a Chemical Compound

A precursor with a mass of 25.83 g contains carbon, hydrogen, and oxygen. After the combustion of the compound, 37.88 g of CO_2 and 15.5 g of H_2O were observed as the products of the chemical reaction. Determine the empirical formula for the compound employed.

Solution

First, we need to find the masses of each element in the compound:

Mass of C

$$37.88 \text{ g CO}_2 \times \frac{1 \text{ mol CO}_2}{44.011 \text{ g CO}_2} = 0.86 \text{ mol CO}_2, \text{ then } M_C = 0.86 \text{ mol C} \frac{12.011 \text{ g}}{1 \text{ mol C}}$$

$$= 10.33 \text{ g C}$$

Mass of H

$$15.5 \text{ g H}_2\text{O} \times \frac{1 \text{ mol H}_2\text{O}}{18.016 \text{ g H}_2\text{O}} = 0.86 \text{ mol H}_2\text{O}, \text{ then } M_H = 2 \times (0.86 \text{ mol H}) \frac{1.008 \text{ g}}{1 \text{ mol H}}$$

$$= 1.73 \text{ g H}$$

Mass of O

$$M_O = M_{\text{Total}} - (M_C + M_H) = 25.83 \text{ g} - (10.33 \text{ g C} + 1.73 \text{ g H}) = 13.77 \text{ g O}$$

An empirical formula is the chemical formula of a compound that yields the ratios of the elements in the compound, not necessarily the actual numbers of atoms. Hence, it is the lowest integer ratio of elements in the compound. The next steps are determining mole numbers of each element in the compound, dividing each mole value by the smallest number of moles, and rounding them to the nearest integer:

Mole number of C

$$10.33 \text{ g C} \times \frac{1 \text{ mol C}}{12.011 \text{ g}} = 0.86 \text{ mol C} \rightarrow \frac{0.86 \text{ mol C}}{0.86} = 1 \text{ mol C}$$

Mole number of H

$$1.73 \text{ g H} \times \frac{1 \text{ mol H}}{1.008 \text{ g}} = 1.716 \text{ mol H} \rightarrow \frac{1.716 \text{ mol H}}{0.86} = 2 \text{ mol H}$$

Mole number of O

$$13.77 \text{ g O} \times \frac{1 \text{ mol O}}{16 \text{ g}} = 0.86 \text{ mol O} \rightarrow \frac{0.86 \text{ mol O}}{0.86} = 1 \text{ mol O}$$

Hence the empirical formula of the compound is obtained as **CH₂O**.

Note: *CH_2O (formaldehyde) is an organic compound that occurs naturally in the environment and is used in many applications including building construction materials, fabrics, and paper-product coatings. Formaldehyde can also be used in manufacturing disinfectants or fungicide as it can kill bacteria and fungi.*

Besides the empirical formula, we could also have been asked to find the molecular formula of the substance had we been given the molar mass of the compound. Knowing the mass of the empirical formula, we could calculate the factor and obtain the molecular formula (e.g., if the factor was calculated as 3, then the molecular formula would be $C_3H_6O_3$). Or, to go backwards, the empirical formula of a compound can be found by dividing all the subscripts in the molecular formula by their greatest common divisor (GCD).

3.3 Thermodynamics Basics

Thermodynamics is the branch of physics that studies heat, work, and energy. The words thermo and dynamics come from the Greek words *therme* (heat) and *dynamis* (power). Thermodynamic analyses are performed and used by a wide spectrum of scientists and engineers including the fields of physics, chemistry, chemical engineering, mechanical engineering, civil engineering, environmental engineering, and aerospace engineering.

To enhance the understanding of relations that govern the transformation and utilization of energy, this section covers the first and second laws of thermodynamics, some fundamental definitions, and the principles of the Carnot, Rankine, and Brayton cycles.

First Law

Energy can neither be created, nor be destroyed. The first law has a quantitative nature. It states that we can never get more from a system than we provide into it.

$$\Delta E = \Delta U + \Delta KE + \Delta PE = Q - W \tag{3.15}$$

Second Law

- All spontaneous processes produce an increase in the entropy (S) of the universe.

$$\Delta S_{\text{univ}} = \Delta S_{\text{sys}} + \Delta S_{\text{surr}} \geq 0 \tag{3.16}$$

- Heat can never flow from a colder body to a warmer body without the aid of external work.
- No heat engine can achieve a thermal efficiency of 100%.

Enthalpy and Entropy

Enthalpy (H) is the amount of energy a thermodynamic system possesses. It is equal to the sum of its internal energy and the product of its pressure and volume:

$$H = U + pV \,(\text{kJ, Btu}) \tag{3.17a}$$

where H is enthalpy, U is internal energy, p is pressure, and V is volume. On a mass basis, specific enthalpy can be expressed as:

$$h = u + pv \,(\text{kJ/kg, Btu/lb}_{\text{m}}) \tag{3.17b}$$

For a closed system, entropy (S) can be defined as the amount of unavailable energy. It is also considered as the measure of disorder or randomness of a system.

Carnot Cycle

The Carnot cycle is an ideal cycle that operates between two reservoirs at temperatures T_H and T_L, first suggested by *Sadi Carnot* (**Figure 3.7**), who is known as the "father of thermodynamics." The cycle consists of four reversible processes, with two being adiabatic (no heat transfer) and the other two being isentropic (constant entropy, or no entropy change). **Figure 3.6** illustrates the schematic of the vapor power cycle and the T–S diagram for the Carnot cycle. Saturated liquid (water) enters the boiler (quality, $x = 0$), and saturated vapor leaves the boiler ($x = 1$), entering the turbine to experience isentropic expansion.

The Carnot cycle is the most efficient cycle that can be executed between two thermal reservoirs (source and sink) having temperatures T_H and T_L. Carnot efficiency is considered as a limit which is a useful reference for evaluating the performance of thermodynamic cycles in real-life applications:

$$\eta_{\text{Carnot}} = 1 - \frac{T_L}{T_H} \tag{3.18}$$

FIGURE 3.6 Carnot vapor power cycle.

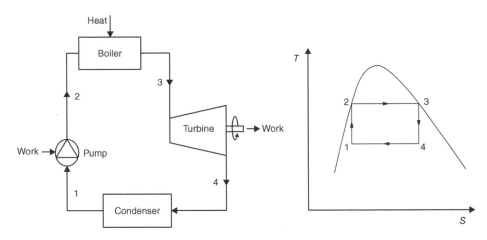

FIGURE 3.7 Sadi Carnot is known as the "father of thermodynamics" and founder of the theory of heat engines. His father chose his name after the famous Persian poet *Sadi Shirazi*. Source: Louis-Léopold Boilly/Public domain via Wikimedia Commons.

FURTHER LEARNING

Steam System Modeler Tool
www4.eere.energy.gov/
manufacturing/tech_deployment/
amo_steam_tool

FURTHER LEARNING

**Kelvin: Thermodynamic
Temperature**
www.nist.gov/si-redefinition/kelvin/
kelvin-thermodynamic-temperature

Rankine Cycle

Despite the theoretical limit the Carnot cycle reaches, there are impracticalities associated with the Carnot cycle, such as the low-quality steam flowing through the turbine blades which could pose a threat to the blades and the shaft. In a similar way, a working fluid with vapor content in it ($x > 0$) would not be good for the pump, forcing it to operate with a two-phase fluid running through it. Superheating the steam can address these concerns, as can be seen in **Figure 3.8**, which illustrates the Rankine vapor cycle with superheat, which is widely used in power generation as a combustion cycle employing fuels such as coal or fuel oil.

Thermal efficiency and relevant quantities for the Rankine cycle with a working fluid mass flow rate of $\dot{m}$ are given by:

$$\eta_{\text{th}} = \frac{\dot{W}_{\text{net}}}{\dot{Q}_{\text{in}}} = \frac{\dot{W}_{\text{turb}} - \dot{W}_{\text{pump}}}{\dot{Q}_{\text{boiler}}} \tag{3.19}$$

$$\dot{W}_{\text{turb}} = \dot{m}(h_3 - h_4) \tag{3.20}$$

$$\dot{W}_{\text{pump}} = \dot{m}(h_2 - h_1) = \dot{m}v_{\text{f}}(p_2 - p_1) \tag{3.21}$$

$$\dot{Q}_{\text{boiler}} = \dot{m}(h_3 - h_2) \tag{3.22}$$

In **Equation 3.21**, v_{f} is the specific volume of the saturated liquid at the given temperature T. Enthalpy values for any state can be obtained using the thermophysical properties tables for water in Appendix B. To determine the enthalpy of the steam leaving the turbine (h_4), the quality of the steam needs to be determined. Knowing that process 3–4 is isentropic, then $s_3 = s_4$. Hence, the quality for state 4 would be:

$$x_4 = \frac{s_4 - s_{\text{f}}}{s_{\text{g}} - s_{\text{f}}} \tag{3.23}$$

It should be noted that the Rankine cycle discussed in this section is an ideal cycle with superheat. Hence the irreversibilities are not considered. In actual vapor power systems, the irreversibilities within the turbine and the pump, and the pressure drops through the boiler and the condenser will be inevitable. There will also be heat loss from the steam to the surroundings. All these together can add up to a significant drop in the overall thermal efficiency of the cycle. The engineering challenge here is coming up with component designs with reduced irreversibilities, pressure drops, and heat losses.

FIGURE 3.8 Ideal Rankine vapor cycle with superheat.

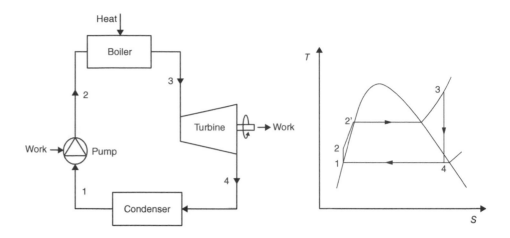

Example 3.3 Rankine Cycle with Superheat

Water is the working fluid at a power plant operating based on the Rankine cycle. Steam coming from the boiler enters the turbine at 16 MPa and 520 °C with a mass flow rate of 125 kg/s. The condensation pressure of the cycle is 10 kPa and water leaves the condenser as a saturated liquid.

a. Sketch the T–s diagram for the cycle
b. Obtain enthalpy values at all states (kJ/kg)
c. Determine the net power output of the plant (MW)
d. Calculate the rate of heat rejected from the condenser (MW)
e. Find the thermal efficiency of the plant (%)

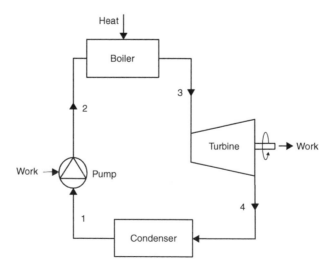

Solution

a. Assuming an ideal Rankine cycle with superheat, the T–s diagram would be:

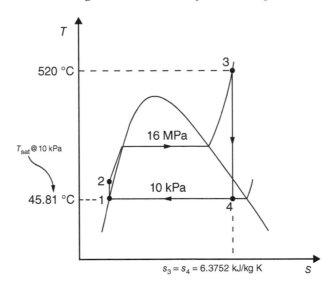

$s_3 = s_4 = 6.3752$ kJ/kg K

b. *State 1*

Saturated liquid ($x = 0$), $p_1 = 10$ kPa, then $h_1 = h_{f@10kPa} = 191.83$ kJ/kg (**Table B.2**)

State 2

From **Equation 3.21**:

$$h_2 = h_1 + \left[v_f \left(p_2 - p_1 \right) \right]$$
$$= 191.83 \, \frac{kJ}{kg} + \left[\left(1.01 \times 10^{-3}\right) \frac{m^3}{kg} \, (160 - 0.1)bar \right] \left| \frac{10^5 \, N \, m^2}{1 \, bar} \right| \left| \frac{1 \, kJ}{10^3 N \, m} \right| = 207.98 \, \frac{kJ}{kg}$$

State 3

$T_3 = 520 \,^{\circ}C$ and $p_3 = 160$ bar, then $h_3 = 3353.3$ kJ/kg (**Table B.3**)
$s_3 = 6.3752$ kJ/kg K (to be used for finding h_4 in the next step)

State 4

$s_4 = s_3 = 6.3752$ kJ/kg K, then quality of steam at the turbine exit will be:

$$x_4 = \frac{s_4 - s_f}{s_g - s_f} = \frac{6.3752 - 0.6493}{8.1502 - 0.6493} = 0.763 \,(\text{or } 76.3\%)$$

$$h_4 = h_f + x_4 \left(h_g - h_f \right) = 191.83 + 0.763(2584.7 - 191.83) = 2017.6 \, \frac{kJ}{kg}$$

c. Net power output of the plant is:

$$\dot{W}_{net} = \dot{W}_{turb} - \dot{W}_{pump} = \dot{m} \left[(h_3 - h_4) - (h_2 - h_1) \right]$$
$$= 125 \, \frac{kg}{s} \left[(3353.3 - 2017.6) - (207.98 - 191.83) \right] \frac{kJ}{kg} \left| \frac{1 \, MW}{10^3 \, kJ/s} \right|$$
$$= 164.94 \, MW$$

d. The rate of heat transfer rejection from the condenser is:

$$\dot{Q}_{condenser} = \dot{m}\,(h_4 - h_1) = 125\,\frac{kg}{s}\,(2017.6 - 191.83)\,\frac{kJ}{kg}\left|\frac{1\,MW}{10^3\,kJ/s}\right| = 228.2\,MW$$

e. Thermal efficiency of the power plant is:

$$\eta_{th} = \frac{\dot{W}_{net}}{\dot{Q}_{boiler}} = \frac{\dot{W}_{net}}{\dot{m}\,(h_3 - h_2)} = \frac{164.94\,MW}{125\,\frac{kg}{s}\,(3353.3 - 207.98)\,\frac{kJ}{kg}\left|\frac{1\,MW}{10^3\,kJ/s}\right|}$$

$$= \frac{164.94\,MW}{393.17\,MW} = 0.42$$

Hence the thermal efficiency of the given power plant is 42%.

Note: *It should be noted that turbine and pump irreversibilities were neglected in this example. If the isentropic efficiencies of both are taken into account, the thermal efficiency of the power plant will be less than 42%. There are other potential sources of losses in real-world applications, such as stray heat transfer to the surroundings. Energy analysis performed on the turbomachines in this example assumes adiabatic control volume.*

Brayton Cycle

Another combustion cycle used in power generation is the Brayton cycle, which directly converts chemical energy into mechanical energy by combusting the gas directly into the cycle, as opposed to the Rankine cycle where the combustion process first generates steam which is then sent into the turbine to do mechanical work. Hence, the Brayton cycle uses a gas turbine, unlike the Rankine cycle which employs a steam turbine. A schematic of the Brayton cycle and its *T–S* diagram can be seen in **Figure 3.9**.

The thermal efficiency of an ideal Brayton cycle assuming cold-air standard analysis is:

$$\eta_{th} = 1 - \frac{T_1}{T_2} = 1 - \left(\frac{p_1}{p_2}\right)^{\frac{k-1}{k}} \tag{3.24}$$

where k is the specific heat ratio.

CLASSROOM DEBATE

Research the Mpemba effect. Do you think this alleged phenomenon is a fact or fiction?

FIGURE 3.9 Ideal Brayton cycle.

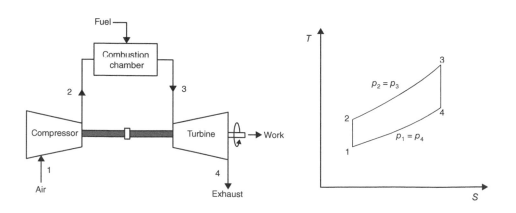

Example 3.4 Determining Thermal Efficiency of an Ideal Brayton Cycle Using Python

Assuming cold-air standard analysis, **(a)** determine the thermal efficiency of a Brayton cycle where the compressor inlet and exit pressures are 10 bar and 30 bar, respectively. **(b)** Obtain the thermal efficiency plot for compressor outlet pressure range of 10 to 50 bar. Keep inlet pressure fixed at 10 bar.

Solution

a. Thermal efficiency of the cycle can be determined using **Equation 3.24**:

```
import numpy
%pylab inline
p1=10 #inlet pressure
p2=30 #exit pressure
k=1.4 #specific heat ratio
eff_brayton = 1 - (p1/p2) ** ((k-1)/k)


eff_brayton
0.26940004435676346
```

b. To obtain the plot, first we define the range, and then type the equation and the plot command:

```
p2=arange(10,50)
eff_brayton = 1 - (p1/p2) ** ((k-1)/k)
title("Brayton cycle efficiency vs Compressor exit pressure")
xlabel("Compressor exit pressure (bar)")
ylabel("Thermal Efficiency (%)")
plot(p2,100*eff_brayton)
```

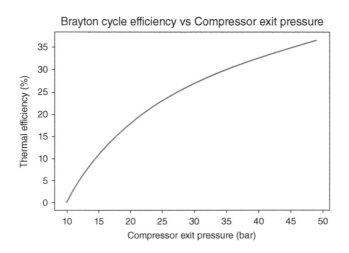

3.4 Fluid Mechanics Basics

Fluid mechanics is another key branch of physics that is used in analysis of energy systems. Just like thermodynamics, fluid mechanics also finds applications in a wide spectrum of fields such as mechanical engineering, civil engineering, environmental engineering, petroleum engineering, oceanography, and meteorology.

Some of the common equations and principles of fluid mechanics that pertain to energy analyses are the ideal gas law, Bernoulli equation, conservation of mass, Navier–Stokes equations, conservation of energy, and turbomachinery.

Ideal Gas Law

An ideal gas can be described as a fluid in which collisions between molecules are perfectly elastic (no loss in kinetic energy due to the collisions) and there are no intermolecular attractive forces. All the internal energy such a gas possesses is in the form of kinetic energy and any change in the internal energy of the gas would be reflected in a temperature gradient.

The *ideal gas equation* is:

$$pV = n\overline{R}T \tag{3.25}$$

where n is the number of moles of gas, $\overline{R}$ is the universal gas constant, and T is the absolute temperature. **Equation 3.25** can also be written on a mass basis as:

$$pV = mRT \tag{3.26}$$

where m is the mass, and R is the gas constant of the fluid of interest. Dividing both sides by mass would yield:

$$pv = RT \tag{3.27a}$$

or, with density being the reciprocal of specific volume:

$$p = \rho RT \tag{3.27b}$$

Bernoulli Equation

In energy applications, the relation between the pressure and velocity of the working fluid is essential for designing the system. This relation between pressure and velocity is described by the *Bernoulli equation*, which states that, for an incompressible (ρ = const.) and frictionless fluid, the sum of the pressures and the kinetic and potential energies per unit volume is constant along a streamline and does not vary with time. Then:

$$p + \frac{1}{2}\rho V^2 + \rho gz = \text{const.} \tag{3.28}$$

This means that for such a system the summation of all three left-hand side terms would be the same at any two points within the flow field:

$$p_1 + \frac{1}{2}\rho V_1^2 + \rho gz_1 = p_2 + \frac{1}{2}\rho V_2^2 + \rho gz_2 \tag{3.29}$$

Note that this form of the equation does not account for the losses for a flow. This is a limitation of the equation for real-life applications as it is obvious that frictional losses

FIGURE 3.10 Bernoulli equation on a museum along the Bosphorus strait in Istanbul. Source: gmaks2000 via Pixabay.

FURTHER LEARNING

Water Science School
www.usgs.gov/special-topic/water-science-school

will take place in the designed system once it is operating. To address this shortcoming, there is also an *extended* version of the Bernoulli equation which is:

$$\frac{p_1}{\gamma} + \frac{V_1^2}{2g} + z_1 + h_s = \frac{p_2}{\gamma} + \frac{V_2^2}{2g} + z_2 + h_L \tag{3.30}$$

In **Equation 3.30**, h_s is called the shaft head and refers to any work done on the fluid by means of a turbomachine (i.e., pump or turbine); h_L represents the head loss, which is the sum of major losses (due to friction) and minor losses (due to obstructions to flow such as valves, fittings, etc.). Major and minor losses can be calculated by:

$$h_{L_{major}} = f \frac{L}{D} \frac{V^2}{2g} \tag{3.31}$$

$$h_{L_{minor}} = K_L \frac{V^2}{2g} \tag{3.32}$$

In **Equation 3.31**, f is the friction factor, which can be retrieved from the Moody chart. K_L in **Equation 3.32** is the loss coefficient of the component. It can be obtained from relevant tables.

Conservation Equations

One of the main tasks in energy systems pertaining to fluid flow analysis is obtaining the velocity field that describes the flow in the domain of interest. To achieve this task, conservation equations are used. These equations are as follows.

Conservation of Mass (Continuity Equation)

As mass can neither be created, nor be destroyed, conservation of mass indicates that the mass flow throughout the flow domain will not change:

$$\frac{\partial \rho}{\partial t} + \nabla \cdot (\rho \mathbf{V}) = 0 \tag{3.33}$$

For steady flow:

$$\nabla \cdot (\rho \mathbf{V}) = 0 \tag{3.34}$$

If the flow is incompressible (ρ = const.), then:

$$\nabla \cdot \mathbf{V} = \frac{\partial u}{\partial x} + \frac{\partial v}{\partial y} + \frac{\partial w}{\partial z} = 0 \tag{3.35}$$

Conservation of Momentum (Navier–Stokes equations)

While the Navier–Stokes equations address all three conservation equations, they are mainly attributed to the momentum equation and are a particular form of the Cauchy equations. Implementation of the law of Newtonian viscous fluids given by Stokes into Newton's equation of motion yields the Navier–Stokes equations for Newtonian viscous fluids:

$$\rho \frac{DV}{Dt} = \rho \cdot g - \nabla p + \frac{\partial}{\partial x_i}\left[\mu\left(\frac{\partial v_i}{\partial x_j} + \frac{\partial v_j}{\partial x_i}\right) + \delta_{ij}\lambda \nabla \cdot V\right] \tag{3.36}$$

FIGURE 3.11 Isaac Newton (1643–1727), English physicist and mathematician. Source: Imagno/Hulton Fine Art Collection via Getty Images.

FIGURE 3.12 Augustin Louis Cauchy (1789–1857), French mathematician. Source: Bettmann/Bettmann via Getty Images.

FIGURE 3.13 Claude-Louis Navier (1785–1836), French mathematician. Source: MacTutor by JOC/EFR.

For an incompressible flow, **Equation 3.36** is appreciably simplified with the divergence term ($\nabla \cdot V$) equaling zero and the dynamic viscosity (μ) assumed constant. Then the equation reduces to:

$$\rho \frac{DV}{Dt} = \rho \cdot g - \nabla p + \mu \nabla^2 V \tag{3.37}$$

The third term on the right-hand side involves the Laplacian of the velocity, which in the Cartesian coordinate system can be expressed as:

$$\nabla^2 V = \frac{\partial^2 V}{\partial x^2} + \frac{\partial^2 V}{\partial y^2} + \frac{\partial^2 V}{\partial z^2} \tag{3.38}$$

Conservation of Energy (First Law of Thermodynamics)
The energy equation for a flow field is derived using the first law of thermodynamics, and can be given as in **Equation 3.39** for Cartesian coordinates (x, y, z) for Newtonian fluids of constant ρ and k:

FIGURE 3.14 George Gabriel Stokes (1819–1903), Irish physicist and mathematician. Source: Science & Society Picture Library/SSPL via Getty Images.

$$\rho C_p \left(\frac{\partial T}{\partial t} + v_x \frac{\partial T}{\partial x} + v_y \frac{\partial T}{\partial y} + v_z \frac{\partial T}{\partial z} \right) = k \left[\frac{\partial^2 T}{\partial x^2} + \frac{\partial^2 T}{\partial y^2} + \frac{\partial^2 T}{\partial z^2} \right] + 2\mu \left\{ \left(\frac{\partial v_x}{\partial x} \right)^2 + \left(\frac{\partial v_y}{\partial y} \right)^2 + \left(\frac{\partial v_z}{\partial z} \right)^2 \right\}$$
$$+ \mu \left\{ \left(\frac{\partial v_x}{\partial y} + \frac{\partial v_y}{\partial x} \right)^2 + \left(\frac{\partial v_x}{\partial z} + \frac{\partial v_z}{\partial x} \right)^2 + \left(\frac{\partial v_y}{\partial z} + \frac{\partial v_z}{\partial y} \right)^2 \right\}$$
$$\tag{3.39}$$

Example 3.5 Bernoulli Equation and Pump Power Input

Water from a lower reservoir at an elevation of 612 m is pumped by a centrifugal pump to a higher reservoir at an elevation of 634 m during off-peak hours to store energy by means of pumped hydroelectric energy storage (PHES). The temperature of the water is 18 °C and it flows at a rate of 750 m³/h. The steel pipe is 600 m long with a diameter of 30 cm. The entrance and exit loss coefficients for the piping are 0.5 and 1.0, respectively. Determine the pump power input if its efficiency is 80%.

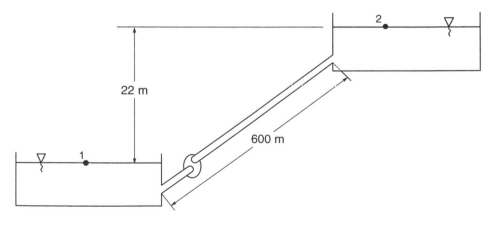

$K_{L,entrance} = 0.5$

$K_{L,exit} = 1.0$

$\varepsilon = 0.05$ mm

$\nu_{H2O@18°C} = 1.054 \times 10^{-6}$ m^2/s

$\gamma_{H2O@18°C} = 9.796$ kN/m^3

Solution

We start with the Bernoulli equation and then apply the given input to simplify it:

$$\frac{p_1}{\gamma} + \frac{V_1^2}{2g} + z_1 + h_p = \frac{p_2}{\gamma} + \frac{V_2^2}{2g} + z_2 + h_L$$

$p_1 = p_2 = p_{atm}$ and velocities at the surfaces of the reservoirs are negligible $(V_1 \approx V_2 \approx 0)$.

Then, reorganizing the general equation yields:

$$h_p = (z_2 - z_1) + h_L \text{ where the total head loss, } h_L = \frac{V^2}{2g}\left[\left(f\frac{L}{D}\right) + \sum K_L\right]$$

We need to find the velocity for the above equation, including the need to calculate Re to get the friction factor, f:

$$V = \frac{Q}{A} = \frac{750 \dfrac{\text{m}^3}{\text{h}} \left|\dfrac{1\,\text{h}}{3600\,\text{s}}\right|}{\dfrac{\pi}{4}(0.3\,\text{m})^2} = 2.98 \frac{\text{m}}{\text{s}}$$

The next step is determining the Reynolds number (Re):

$$Re = \frac{VD}{\nu} = \frac{\left(2.98\,\dfrac{\text{m}}{\text{s}}\right)(0.3\,\text{m})}{1.054 \times 10^{-6}\,\dfrac{\text{m}^2}{\text{s}}} = 8.48 \times 10^5$$

Using the calculated Reynolds number and the relative roughness (ε/D) of the piping material, the friction factor, f, is approximately 0.014 as determined from the Moody chart:

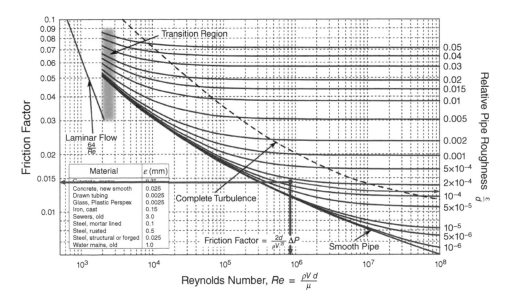

Then total head loss is:

$$h_L = \frac{V^2}{2g}\left[\left(f\frac{L}{D}\right) + \sum K_L\right] = \frac{\left(2.98\,\frac{m}{s}\right)^2}{2\left(9.81\,\frac{m}{s^2}\right)}\left[\left(0.014\frac{600\,m}{0.3\,m}\right) + (0.5 + 1.0)\right] = 13.4\,m$$

We can now plug this into the simplified Bernoulli equation to find the pump head:

$$h_p = (z_2 - z_1) + h_L = (634 - 612) + 13.4 = 35.4\,m$$

Power gained by the fluid, P_f, is:

$$P_f = \gamma\,Q\,h_p = \left(9.796\,\frac{kN}{m^3}\right)\left(750\,\frac{m^3}{h}\right)(35.4\,m)\left|\frac{10^3 N}{1\,kN}\right|\left|\frac{1\,h}{3600\,s}\right|\left|\frac{1\,kW}{10^3\,\frac{N\,m}{s}}\right| = 72.2\,kW$$

Knowing that the pump efficiency is 80%, the pump power input can be calculated:

$$\dot{W}_{pump} = \frac{P_f}{\eta} = \frac{72.2\,kW}{0.80} = 90.25\,kW$$

Note: *Pumped hydroelectric energy storage (PHES), also known as pumped-storage hydroelectricity (PSH) is a mechanical energy storage method based on charging water with potential energy. It is the most common energy storage technique and accounts for over 96% of total energy stored in grid-scale applications worldwide.*

As can be seen in the equations for the power gained by the fluid and pump power input, the power required by the pump will increase with increasing head losses and/or decreasing pump efficiency. In PHES applications, choosing the right pump will help maintain a higher round-trip efficiency for retrieving the energy back from the upper reservoir, which can be considered as a "hydro-battery."

Turbomachinery

Turbomachines are important parts of our lives in many fields. They can be used as power generation systems such as steam or gas turbines at a thermal power plant, as water turbines at a hydroelectric power plant, and as wind turbines in wind energy generation. They can also be used as energy-consuming systems such as pumps, compressors, and fans. Either way, turbomachines operate based on the interaction of a set of blades attached to a shaft with a working fluid, which would be steam for a power plant utilizing a steam generator, or water at a dam generating electricity, or air for a wind turbine. As for the energy-consuming systems, the working fluid would be a liquid for pumps, and air (or other gases) for compressors and fans. Various turbomachine types can be seen in **Figure 3.15**.

In energy systems, it is important to be able to quantify the amount of force the working fluid will exert on the blades to rotate the shaft, and hence determine the amount of power that can be generated.

To better understand this, let us take a look at **Figure 3.16** where the ideal inlet and outlet velocity triangles on both ends of an impeller blade are illustrated. The blade being rotated at a constant angular velocity will have a scalar velocity value of ωr:

$$U = \omega r \tag{3.40}$$

FIGURE 3.15 Various turbomachine types. Source: S. L. Dixon and C. A. Hall, *Fluid Mechanics and Thermodynamics of Turbomachinery.* Butterworth-Heinemann, 2010.

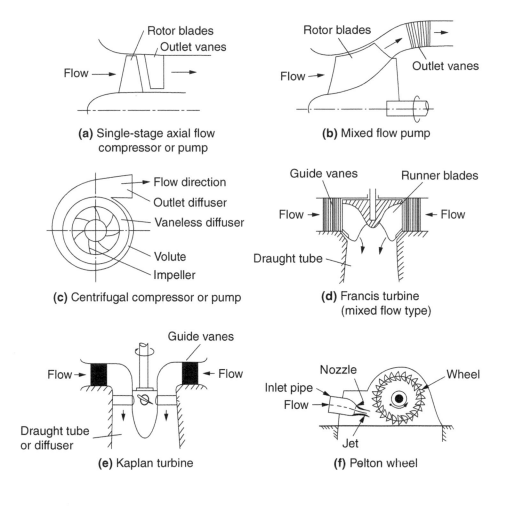

(a) Single-stage axial flow compressor or pump

(b) Mixed flow pump

(c) Centrifugal compressor or pump

(d) Francis turbine (mixed flow type)

(e) Kaplan turbine

(f) Pelton wheel

FIGURE 3.16 Ideal inlet and outlet velocity diagram. Source: Adopted with permission from A. K. Noughabi and S. Sammak, "Detailed design and aerodynamic performance analysis of a radial-inflow turbine," *Applied Sciences*, vol. **8**, no. 11, p. 2207, Nov. 2018, doi: 10.3390/app8112207.

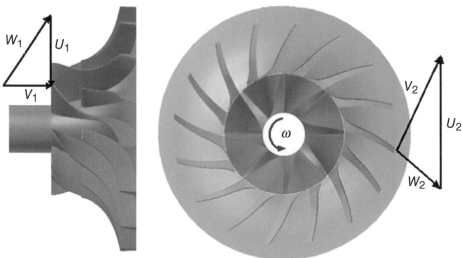

Absolute velocity **V** is equal to the vector sums of the blade velocity **U** and the relative velocity **W**.

$$\mathbf{V} = \mathbf{U} + \mathbf{W} \tag{3.41}$$

The torque can be obtained by employing the Euler turbomachine equation:

$$T_{\text{shaft}} = \dot{m}\left(r_2 V_{\theta 2} - r_1 V_{\theta 1}\right) \tag{3.42}$$

$V_{\theta i}$ is the tangential component of the absolute velocity V, and is obtained from:

$$V_{\theta i} = V_i \sin \theta \qquad (3.43)$$

where θ is the angle between the absolute and blade velocity vectors. **Equation 3.44** gives the relation for calculating the power that is transferred to a turbomachine:

$$\dot{W}_{shaft} = \mathbf{T}_{shaft} \cdot \boldsymbol{\omega} = T\omega = \dot{m}\omega(r_2 V_{\theta 2} - r_1 V_{\theta 1}) \qquad (3.44)$$

Implementing **Equation 3.40** into **3.44** gives:

$$\dot{W}_{shaft} = \dot{m} \left(U_2 V_{\theta 2} - U_1 V_{\theta 1} \right) \qquad (3.45)$$

Note that while both power-generating (i.e., turbine) and power-consuming (i.e., pump) turbomachines involve the interaction of the rotor blades and the working fluid, the sign convention is opposite. For turbines, torque exerted by the shaft on the rotor is in the direction opposite to the rotation. Energy is transferred from the fluid to the rotor. For a pump, on the other hand, the torque applied on the rotor is in the same direction as the rotation. In this case, energy is transferred from the shaft to the rotor blades, and from the blades to the fluid. The directional observation can be performed on the velocity triangle as well. If the absolute velocity vector turns in the direction of the blades, then it means that work is done *on* the fluid, and this is a pump.

Example 3.6 Sketching Velocity Triangle and Obtaining Blade Velocity

The centrifugal pump shown below has an angular velocity of 900 rpm. The absolute velocity at the inlet (V_1) makes a 90° angle with the tangential. The relative velocity at the exit (W_2) makes a 60° angle with the tangential. Volume flow rate is 90 m³/h and the blade depth at the exit (a) is 5 mm.

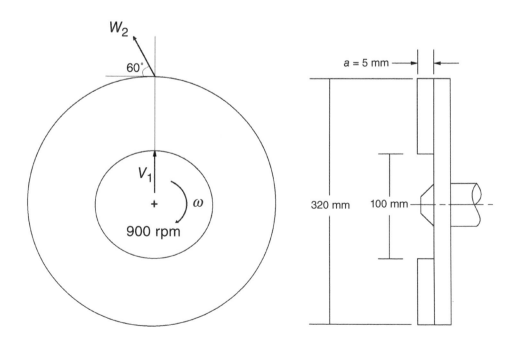

a. Sketch the exit velocity triangle
b. Find the blade velocity (U_2) at the exit (m/s)
c. Determine the shaft power (kW)

Solution

a. Velocity triangle:

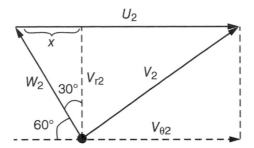

b. Blade velocity at the exit is determined by the radius and the angular velocity:

$$U_2 = \omega r_2 = \left(900 \ \frac{\text{rev}}{\text{min}}\right)(160 \times 10^{-3} \ \text{m})\left|\frac{2\pi \ \text{rad}}{\text{rev}}\right|\left|\frac{1 \ \text{min}}{60 \ \text{s}}\right| = 15.1 \ \frac{\text{m}}{\text{s}}$$

c. From **Equation 3.45**, shaft power is:

$$\dot{W}_{\text{shaft}} = \dot{m}(U_2 V_{\theta 2} - U_1 V_{\theta 1})$$

where $V_{\theta 1} = 0$ as there is no tangential component of the inlet velocity. Then the equation reduces to:

$$\dot{W}_{\text{shaft}} = \dot{m} U_2 V_{\theta 2} = \rho Q U_2 V_{\theta 2}$$

From the velocity triangle, $V_{\theta 2}$ can be expressed in terms of U_2 and V_{r2}:

$$V_{\theta 2} = U_2 - x = U_2 - (V_{r2})(\tan 30^\circ)$$

where

$$V_{r2} = \frac{Q}{A_2} = \frac{Q}{\pi D_2 a} = \frac{90 \ \frac{\text{m}^3}{\text{h}}}{\pi(0.32 \ \text{m})(0.005 \ \text{m})}\left|\frac{1 \ \text{h}}{3600 \ \text{s}}\right| = 4.97 \ \frac{\text{m}}{\text{s}}$$

Then,

$$V_{\theta 2} = U_2 - (V_{r2})(\tan 30^\circ) = \left(15.1 \ \frac{\text{m}}{\text{s}}\right) - \left(4.97 \ \frac{\text{m}}{\text{s}}\right)(0.577) = 12.2 \ \frac{\text{m}}{\text{s}}$$

Now we can solve for the shaft power:

$$\dot{W}_{\text{shaft}} = \left(999 \ \frac{\text{kg}}{\text{m}^3}\right)\left(90 \ \frac{\text{m}^3}{\text{h}}\right)\left(15.1 \ \frac{\text{m}}{\text{s}}\right)\left(12.2 \ \frac{\text{m}}{\text{s}}\right)\left|\frac{1 \ \text{h}}{3600 \ \text{s}}\right|\left|\frac{1 \ \text{kW}}{10^3 \ \text{kg m}^2 \ \text{s}^{-3}}\right| = 4.6 \ \text{kW}$$

3.5 Heat Transfer Basics

So far, we have discussed electrodynamics, chemical energy conversion, thermodynamics, and fluid mechanics principles that relate to energy systems. Heat transfer is another significant subject in conducting analyses on these systems. We could be looking at the heat exchange in the steam generator or condenser of a thermal power plant, or we could be studying the building heating and cooling loads to size an air-conditioning unit or a furnace for a given building, making sure heat transfer through the building envelope is also calculated appropriately. We might be investigating the performance of a solar-thermal system, or heat exchangers in a geothermal application. Many other applications besides the few examples listed here require a solid under-standing of heat transfer. A thermal-fluid analysis would not be complete without the heat transfer basics and principles.

There are three modes of heat transfer; *conduction*, *convection*, and *radiation*. In many energy systems, it is common to see more than a single mode of heat transfer taking place. In such cases, combined heat transfer analysis (e.g., conduction and convection) is required.

Conduction

Conduction is the mode of heat transfer by which heat flows through solid bodies. It involves the transfer of heat among adjacent molecules of a substance. It can be one-, two-, or three-dimensional. It can be independent of time (steady), it can be dependent on time (unsteady, or transitional). Fourier's law for one-dimensional heat conduction in a rectangular coordinate system is:

$$\dot{Q}_{cond} = -kA\frac{dT}{dx} \tag{3.46}$$

where k is the thermal conductivity of the material through which heat is transferring. The negative sign on the right-hand side indicates that heat is transferring in the direction of decreasing temperature. If the system has a component within it that adds or generates energy (e.g., an electrical resistance heater), then energy generation within the problem domain should also be taken into account:

$$\frac{\partial^2 T}{\partial x^2} + \frac{\dot{e}_{gen}}{k} = \frac{1}{\alpha}\frac{\partial T}{\partial t} \tag{3.47}$$

where $\dot{e}_{gen}$ is the rate of energy generation per unit volume of the solid system, and α is the thermal diffusivity of the material and is given by $k/\rho c_p$. **Equation 3.47** will reduce to the following forms for the corresponding special cases:

$$Steady\text{-}state \qquad \frac{d^2 T}{dx^2} + \frac{\dot{e}_{gen}}{k} = 0 \tag{3.48}$$

$$Steady\text{-}state, no\ heat\ generation \quad \frac{d^2 T}{dx^2} = 0 \tag{3.49}$$

$$Transient, no\ heat\ generation \quad \frac{\partial^2 T}{\partial x^2} = \frac{1}{\alpha}\frac{\partial T}{\partial t} \tag{3.50}$$

Convection

When a fluid (liquid or gas) is involved in a system, the energy transferred by means of heat due to diffusion and the bulk motion of the fluid is the convective heat transfer. It can be in the form of free (natural) or forced convection. Free convection is induced by the buoyant forces acting on the fluid, while in forced convection, external means

FIGURE 3.17 Flow through a heat exchanger pipe.

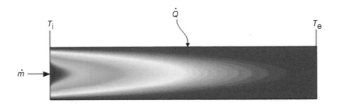

(pump, fan, blower, etc.) are involved in moving the fluid. Convective heat transfer is defined by *Newton's law of cooling*:

$$\dot{Q}_{conv} = hA\,\Delta T \tag{3.51}$$

where ΔT is the temperature gradient between the fluid and the solid that are exchanging heat; h is the convective heat transfer coefficient. Unlike thermal conductivity (k), convective heat transfer coefficient (h) is not a property of the fluid. It depends on both the physical properties of the fluid and the physical situation that involves material and flow properties, geometry, and thermal conditions.

In energy systems, convective heat transfer can be observed in heat exchangers such as the boiler or the condenser in power plants, in solar-thermal collectors, in floating photovoltaic panels, and in many other applications.

In heat exchangers, assuming there is no energy generation within the pipes, and neglecting stray heat transfer, energy given or absorbed by the fluid will equal the energy absorbed or given by the solid wall, per conservation of energy.

$$\dot{Q}_{given} = \dot{Q}_{absorbed} \tag{3.52}$$

Then, for a fluid (receiving energy from the surroundings, see **Figure 3.17**), the rate of heat absorption can be calculated by:

$$\dot{Q}_{absorbed} = \dot{m}(h_e - h_i) = \dot{m}c_p(T_e - T_i) \tag{3.53}$$

Radiation

Radiative heat transfer does not require the presence of a medium, unlike conduction and convection. In fact, it takes place more efficiently in vacuum. Thermal radiation is the energy emitted by a substance that has a temperature greater than absolute zero. It is transferred by electromagnetic waves. The energy emission rate from a body is:

$$\dot{Q}_{emitted} = \varepsilon\sigma A T^4 \tag{3.54}$$

where ε is the emissivity and σ is the Stefan–Boltzmann constant ($5.67 \times 10^{-8}\,\text{W/m}^2\,\text{K}^4$).

Radiative heat exchange between two surfaces having temperatures T_1 and T_2 is:

$$\dot{Q}_{rad} = \varepsilon\sigma A(T_1^4 - T_2^4) \tag{3.55}$$

Combined Heat Transfer

We have briefly discussed the three basic modes of heat transfer and the laws pertaining to them.

Conduction	$\rightarrow$	Fourier's law
Convection	$\rightarrow$	Newton's law of cooling
Radiation	$\rightarrow$	Stefan–Boltzmann law

FIGURE 3.18 Thermal system in a fuel cell electric vehicle (FCEV) provides cooling by means of combined heat transfer. The system maintains an appropriate operating temperature for the fuel cell stack, electric traction motor, and power electronics of the vehicle. Source: Alternative Fuels Data Center, U.S. Department of Energy.

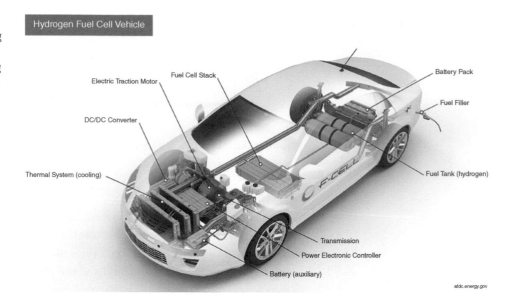

In real-life applications, however, heat transfer occurs by more than one form of mode simultaneously. For instance, one observes both convective and radiative heat exchange between the gases and the walls of a combustion gas chamber at a thermal power plant. An HVAC (heating, ventilating, and air-conditioning) engineer conducting heat transfer analysis for a building has to account for conductive, convective, and radiative heat transfer through the building envelope to be able to accurately calculate the heating or cooling load of the structure. In fuel cell electric vehicles (FCEVs), convective heat transfer occurs between the solid surfaces and the moving gases, and conductive heat transfer takes place within the solid components in the fuel cell stack. Besides the cell, the electric motor, power electronics, and other components of the vehicle also experience combined heat transfer during the cooling process to ensure operation at the desired temperature range. Therefore, depending on the application and the nature of energy flow, one needs to consider all forms of heat transfer taking place within or through the system (**Figure 3.18**).

FURTHER LEARNING

Two-Dimensional Building Heat Transfer Modeling
https://windows.lbl.gov/software/therm

Example 3.7 Convection and Radiation Heat Transfer

Water flows through the coated copper tubing of a solar collector into a hot-water tank in a residential application. The tube from the collector to the tank has a diameter of 30 mm and a length of 1.2 m. The external surface temperature of the tube is measured as 65 °C on a day when the outside air is at 18 °C. The convective heat transfer coefficient between the tube and the outside air is 7.5 W/m^2 K and the emissivity of the tube is 0.78. Determine the heat loss from the tubing between the solar collector and the hot-water tank by means of **(a)** convection, and **(b)** radiation.

Solution

a.   Convective heat loss can be calculated using **Equation 3.51**:

$$\dot{Q}_{\mathrm{conv}} = hA\Delta T = hA(T_s - T_\infty)$$

The lateral surface area of the tube exposed between the collector and the tank is:

$$A = \pi D L = \pi(0.03 \text{ m})(1.2 \text{ m}) = 0.113 \text{ m}^2$$

then:

$$\dot{Q}_{conv} = \left(7.5 \frac{\text{W}}{\text{m}^2\text{K}}\right)(0.113 \text{ m}^2)(65 - 18)^\circ\text{C} = 39.8 \text{ W}$$

b. For calculating the radiation heat loss, **Equation 3.55** is utilized:

$$\dot{Q}_{rad} = \varepsilon\sigma A(T_1{}^4 - T_2{}^4) = (0.78)\left(5.67 \times 10^{-8} \frac{\text{W}}{\text{m}^2\text{K}^4}\right)(0.113 \text{ m}^2)(338^4 - 291^4) \text{ K}^4 = 29.4 \text{ W}$$

Note: *Heat transfer by means of radiation makes up about 42.5% of the total heat loss due to both convection and radiation. The weight of radiative heat loss can be lessened by using different coatings or materials that have lower emissivity values.*

Residential solar collector application example in Izmir, Turkey. Source: Ashley Cooper/Construction Photography/Avalon via Getty Images.

3.6 Summary

A solid theoretical background and understanding of basic principles and laws is crucial in studying energy systems from design and development phases to construction, commissioning, operation, and maintenance. This chapter included a brief collection of five important fields that are fundamental to energy systems:

Electrodynamics
- *Ohm's law*
- *Joule's law of heating*
- *Electromotive force*
- *Faraday's law*
- *Maxwell's equations*

Chemical energy conversion
- *Combustion processes*
- *Conversion efficiency*

Thermodynamics

- *First law of thermodynamics*
- *Second law of thermodynamics*
- *Enthalpy and entropy*
- *Carnot cycle*
- *Rankine cycle*
- *Brayton cycle*

Fluid mechanics

- *Ideal gas law*
- *Bernoulli equation*
- *Conservation equations*
- *Turbomachinery*

Heat transfer

- *Conduction*
- *Convection*
- *Radiation*
- *Combined heat transfer*

Key theoretical relations in each field that are relevant to energy applications were discussed concisely. In approaching a given problem (which may involve multiple disciplines and fields under the umbrella of these disciplines), the important steps are:

- Understanding the problem
- Identifying the given inputs and retrieving necessary values from relevant tables, charts, or other sources that are reliable
- Selecting the right governing equations or principles
- Making reasonable assumptions (if possible) to simplify the problem
- Solving the problem to get to the output parameter(s) sought
- Verification of the results (if applicable)

REFERENCES

[1]   D. J. Griffiths, *Introduction to Electrodynamics*. Cambridge: Cambridge University Press, 2018.

[2]   R. D. Bergman, E. Oneil, I. L. Eastin, L. R. Johnson, and H.-S. Han, "Life cycle impacts of manufacturing redwood decking in Northern California," *Wood and Fiber Science*, vol. 46, no. 3, pp. 322–339, 2014.

[3]   Y. A. Çengel, M. A. Boles, and M. Kanoglu, *Thermodynamics: An Engineering Approach*. New York: McGraw-Hill Education, 2019.

[4]   B. R. Munson, D. F. Young, and T. H. Okiishi, *Fundamentals of Fluid Mechanics*. Hoboken, NJ: J. Wiley & Sons, 2006.

[5]   F. P. Incropera, D. P. DeWitt, T. L. Bergman, and A. S. Lavine, *Introduction to Heat Transfer*. Hoboken, NJ: J. Wiley & Sons, 2007.

CHAPTER 3 EXERCISES

3.1   A battery with 12 V emf has an internal resistance of 0.20 Ω. Calculate its voltage when a load with 9 Ω resistance is connected to it. Find the rate of heat dissipation with this load.

3.2   A coil with a diameter of 25 cm is wrapped with 180 loops of wire. A uniform magnetic field with a magnitude of 0.15 T perpendicular to the plane of the coil is on. Determine the average induced emf in the coil if in 0.27 s the field is reversed in direction with the same magnitude.

3.3 Current I flows through a cylindrical conductor with a radius of R. Assuming that the current flows homogeneously across the cross-section of the conductor, obtain the magnetic field (B) distribution within and outside the conductor as a function of r, which is an arbitrary distance from the axis of the cylinder, using Ampere's law.

3.4 Stoichiometric (complete) combustion of a general fuel in air has a generic oxidation relation as given below:

$$C_xH_y + \left(x+\frac{y}{4}\right)(O_2 + 3.78N_2) \rightarrow xCO_2 + \frac{y}{2}H_2O + 3.78\left(x+\frac{y}{4}\right)N_2$$

If octane (C_8H_{18}) is completely combusted in a process, determine **(a)** the fuel/air mass ratio, and **(b)** the product mole fraction for each product of the combustion.

3.5 The combustion of 2.76 g of a compound containing C, H, O, and N has 0.83 g O and results in products of 3.44 g CO_2 and 2.36 g H_2O. Determine the empirical formula of the compound.

3.6 Determine the specific enthalpy values for water for the properties of each state given below. On the same p–v diagram, mark each state and indicate its phase.
   a. $p = 18$ MPa and $T = 357.1\,°C$
   b. $v = 2.546$ m^3/kg and $u = 2779.6$ kJ/kg
   c. $T = 200\,°C$ and $v = 1.153 \times 10^{-3}$ m^3/kg
   d. $p = 45$ bar and $x = 52\%$

3.7 A vapor power plant, as shown in the figure, has water as the working fluid flowing steadily in the closed loop. The processes through the boiler and condenser are as follows:

   Process 2–3: boiling at a constant pressure of 10 MPa from saturated liquid ($x = 0$) to saturated vapor ($x = 1$)

   Process 4–1: condensation at a constant pressure of 15 kPa from $x_4 = 70\%$ to $x_1 = 32\%$

Assuming that heat transfer from the turbine and pump can be neglected (adiabatic), and kinetic and potential energy changes are negligible:
   a. Sketch the cycle on a p–v diagram.
   b. Find the actual thermal efficiency of the cycle (η_{th}).
   c. Calculate the Carnot efficiency (η_{Carnot}) of the cycle for the same boiler and condenser operating conditions.

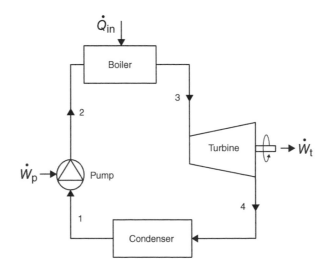

3.8 Solve the Rankine cycle problem in Example 3.3 with turbine and pump isentropic efficiencies of 88%. Compare the power output and thermal efficiency values for reversible and irreversible scenarios.

3.9 Calculate the Carnot efficiency for a power cycle operating between two thermal reservoirs at temperatures of 23 °C and 550 °C. Obtain the efficiency plot for an outside temperature range of −5 to 42 °C.

3.10 Determine the density of air flowing across a wind turbine blade if the temperature and pressure values of air are

  a. $T = 25\,°C$ and $p = 101\,kPa$

  b. $T = 18\,°C$ and $p = 100.5\,kPa$

3.11 Removing dirt and debris from solar panels increases the panel performance. A residential solar array owner is washing the panels with a garden hose nozzle. The hose carries water at a flow rate of 11 L/min with a pressure of 770 kPa leaving the tap. Diameters of the hose and the nozzle are 20 mm and 2.5 mm, respectively. Find the pressure of water at the nozzle exit if the nozzle is 3.1 m higher than the tap when the person is on top of the ladder. The hose is 30 m long, has an absolute surface roughness of 0.001 mm, and the loss coefficient (K_L) for the nozzle is 0.05.

A residential solar array owner washing the panels to enhance
photovoltaic performance. Source: Orietta Gaspari/E+ via Getty Images.

3.12 Water flows steadily between two parallel plates seated under a battery pack. The temperature variation along the direction of the flow is negligible while the temperature gradient in the vertical direction is recorded to be significant. The velocity profile of the flow between the plates is given below in terms of centerline (maximum) velocity, u_c. The pressure change along the x-axis is constant. Assuming incompressible flow with constant properties:

  a. Write the simplified conservation of energy equation for the given flow.

  b. Obtain the temperature distribution for water between the two plates.

  c. Determine the temperature of the water at the mid-plane if the upper and lower plate temperatures, T_1 and T_0, are 45 °C and 38 °C, respectively, and centerline velocity is measured as 0.3 m/s.

  d. For similar conditions as in part (**c**), determine the temperature of water 5 mm below the upper plate if the plates are 20 mm apart from each other.

$$u(y) = u_c\left[1 - \left(\frac{y}{h}\right)^2\right]$$

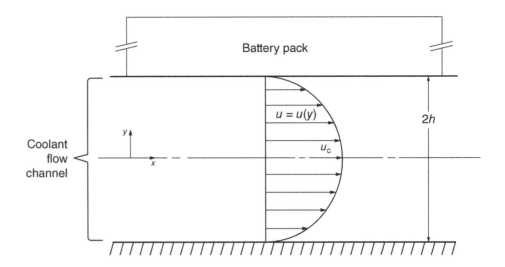

3.13 A centrifugal pump as shown in the figure operates at an angular velocity of 1200 rpm. Absolute velocity at the impeller suction (V_1) is perpendicular to the tangential. Relative velocity at the impeller discharge (W_2) makes a 50° angle with the radial. Volume flow rate is 2 m³/min and the pump has an efficiency of 94% with an input power of 2.12 hp.

    a. Sketch the exit velocity triangle.

    b. Find the blade velocity (U_2) at the exit (m/s).

    c. Determine the width (a) of the outlet (mm).

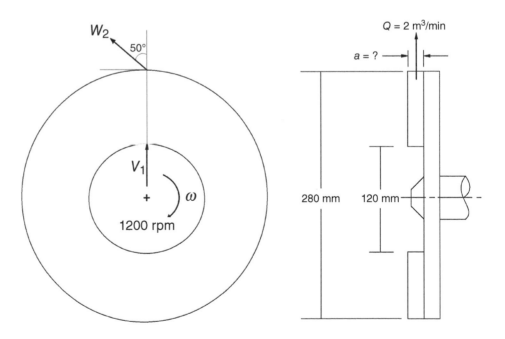

3.14 A free-piston Stirling cooler is used for the cryogenic cooling application of a superconducting magnetic energy storage (SMES) system. The cooler has a cooling capacity of 100 W and absorbs the heat through its cold head (40 mm diameter), as shown in the figure. A metal extension of the same diameter as the cold head is attached to the head for extending the cold surface area. If the inner surface temperature (T_0) of the 20 mm thick metal disk is $-52\,°C$, determine the outer surface temperature (T_1) of the disk if the material is

a. aluminum ($k = 205$ W/m K)

b. copper ($k = 390$ W/m K)

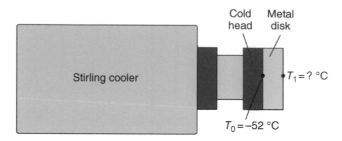

3.15 The conversion efficiency of a solar panel decreases with increasing temperature above its nominal operating cell temperature (NOTC). Therefore, air flow on a windy day can help enhance panel efficiency due to convective cooling. The surface temperature of a solar panel is measured to be $58\,°C$ on a day when the air temperature is $24\,°C$. The convective heat transfer coefficient, h, as a function of wind speed, U_∞, is given by the correlations below (Jürges, 1924):

$$h = 4.0\, U_\infty + 5.6 \quad U_\infty < 5\ \text{m/s}$$
$$h = 7.1\, U_\infty^{0.78} \quad\quad U_\infty > 5\ \text{m/s}$$

Determine the convective heat transfer rate from the solar panel to the surrounding air if the wind speed is

a. 3.7 m/s

b. 6.3 m/s

CHAPTER 4
Fossil Fuels

4.1 Introduction

Over millions of years, buried organisms such as prehistoric animals and plants have decomposed under pressure and heat from the Earth's crust. The pressure from the rocks and the heat from the crust transformed the decaying organic material into fossil fuels. This transformation resulted in three different kinds of fuels: oil (petroleum), coal, and natural gas. Fossil fuels are the world's primary energy source. Although the shares of oil and coal have been decreasing in recent years, fossil fuels make up more than 80% of global primary energy sources [1] (**Figure 4.1**).

Fossil fuels have served us well since the Industrial Revolution. They served us in our homes, in our industries and businesses, for our transportation, and in agriculture. They come with an environmental cost though. In 2018, fossil fuels were the source of approximately 75% of the total anthropogenic (human-caused) greenhouse gas emissions in the United States [2]. Despite the current dependence on fossil fuels in the United States and worldwide, international meetings, environmental necessities, and increasing awareness of climate change have been the catalyst for a transition towards two key elements: *energy efficiency* and *renewable energy*. A smart combination of these with improving technology and global awareness have already been proven to work, while there is still much to be done!

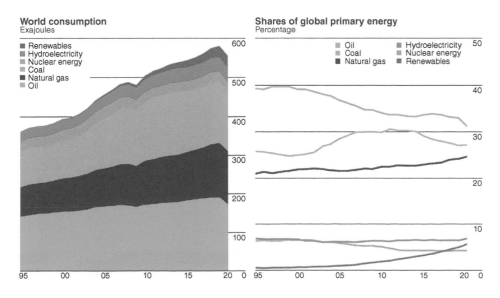

FIGURE 4.1 Shares of global primary energy sources. Source: BP Statistical Review of World Energy, 2021.

FURTHER LEARNING

OECD Energy Data
https://data.oecd.org/energy.htm

FURTHER LEARNING

IEA Data and Statistics
www.iea.org/data-and-statistics

FURTHER LEARNING

BP Statistical Review of World Energy
www.bp.com/en/global/corporate/
energy-economics/statistical-review-
of-world-energy.html

4.1.1 Types of Fossil Fuels

Fossil fuels including coal, oil, and natural gas are currently the primary energy source across the globe. In this section, each fuel is covered. There is an additional section on *shale gas*. Although shale gas is also natural gas, it is worth discussing it in a separate section due to the distinct extraction methods, controversies, and its impact on global gas trade.

4.1.1.1 COAL

Coal is a complex mixture of substances. It releases environmentally harmful by-products such as CO, Hg, SO_x, and NO_x when combusted. There are *clean coal* methods which are used to enhance the thermal efficiency of power plants and reduce the emission of harmful pollutants. Wet scrubbing with limestone and water reduces SO_x, gasification diminishes NO_x, and coal washing lessens the amount of S. Not all coal is the same in terms of energy output. **Table 4.1** lists four types of coal and their carbon content.

Coal is mostly burnt in thermal power plants and is still used for residential heating in some parts of the world (**Figure 4.2**). Due to China being the major consumer of coal, Asia-Pacific takes the lead in regional production in the global arena. During the *26th U.N. Climate Change Conference of the Parties* (COP26) in Glasgow in 2021, a pledge to

Table 4.1

Properties of different coal types [3]			
Coal type	% Carbon	Heat content	US deposits
Anthracite	92	High	Pennsylvania, New York
Bituminous	80	Medium	Appalachia, Midwest, Utah
Subbituminous	77	Medium	Rocky Mountains
Lignite	71	Low	Montana

FIGURE 4.2 Coal production and consumption by region. Source: BP Statistical Review of World Energy, 2021.

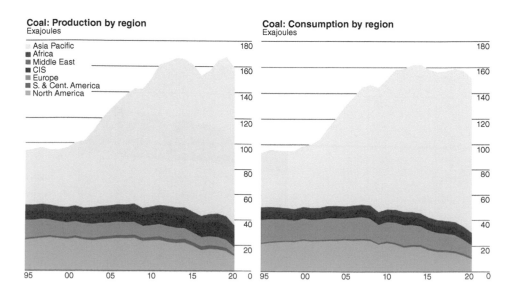

phase out coal was supported by many coal-using countries. The COP summits aim to achieve the target of limiting global temperature rise to 1.5 °C.

Example 4.1 Calculating Energy Equivalence

In 2021, China produced approximately 4 billion tonnes of coal. Calculate the amount of alternative fuels that could provide an equivalent amount of energy to replace coal that year. Use crude oil, natural gas, and gaseous LPG propane as alternative fuels.

Solution

First, approximate the energy content of coal and the alternative fuels mentioned based on their specific energy (GJ/tonnes) or energy density (MJ/m^3). The following values are obtained:

> 1 tonne of coal equivalent $\approx$ 24 GJ (assume local coal for electricity production)
> 1 tonne crude oil equivalent $\approx$ 45 GJ
> 1 m^3 natural gas equivalent $\approx$ 38.3 MJ
> 1 m^3 LPG propane gas equivalent $\approx$ 93.3 MJ

Next, the total energy content of coal needs to be determined:

Coal

$$E_{\text{total}} = (4 \times 10^9 \text{ tonnes}) \left| \frac{24 \text{ GJ}}{\text{tonne}} \right| = 96 \times 10^9 \text{ GJ}$$

Then, the amount of crude oil or natural gas required to provide the equivalent amount of total energy can be calculated:

Crude oil (mass in tonnes)

$$\text{LCOE} = \frac{\$141{,}126.7 + \$84{,}000}{5815 \text{ MWh}} = \frac{\$225{,}126.7}{5815 \text{ MWh}} \left| \frac{1 \text{ MWh}}{10^3 \text{ kWh}} \right| \left| \frac{100¢}{1\$} \right| = 3.87 \frac{¢}{\text{kWh}}$$

Natural gas (volume in bcm)

$$\forall_{\text{natural gas}} = \frac{96 \times 10^9 \text{ GJ}}{38.3 \dfrac{\text{MJ}}{\text{m}^3}} \left| \frac{10^3 \text{ MJ}}{1 \text{ GJ}} \right| \left| \frac{1 \text{ bcm}}{10^9 \text{ m}^3} \right| = 2506.5 \text{ bcm}$$

LPG propane gas (volume in bcm)

$$\forall_{\text{LPG propane gas}} = \frac{96 \times 10^9 \text{ GJ}}{93.3 \dfrac{\text{MJ}}{\text{m}^3}} \left| \frac{10^3 \text{ MJ}}{1 \text{ GJ}} \right| \left| \frac{1 \text{ bcm}}{10^9 \text{ m}^3} \right| = 1029 \text{ bcm}$$

Note: *The equivalencies depend on fuel composition and may vary for different contents. Amounts of fuel were calculated based on the equivalencies listed above. The quantity of natural gas required to replace the coal needs of China is about 2500 bcm, which is a significant amount. By way of comparison, the total natural gas consumption of Europe is about 460 bcm. As natural gas is a cleaner fossil fuel, transition to this fuel from coal could significantly reduce CO_2 emissions worldwide.*

4.1.1.2 OIL (PETROLEUM)

Oil (also called *crude oil* or *petroleum*) is a naturally existing fossil fuel that comes from dead organisms. It exists in liquid form in reservoirs beneath overburden. Oil deposits can be beneath land or beneath the seabed. Oil can be extracted by means of both *onshore* and *offshore drilling* depending on the location of the deposits. **Figure 4.3** depicts both of these drilling applications, as well as an oil refinery where the crude oil is transported after being extracted. A refinery is an industrial facility that processes crude oil to obtain derivatives of it such as liquefied petroleum gas (LPG), gasoline, diesel, jet fuel, kerosene, fuel oil, and asphalt. There are also other industrial facilities called *petrochemical plants*, which differ from refineries as these plants make products that are not burnt but are used in other applications such as production of olefins (e.g., ethylene and propylene) or aromatics (e.g., benzene and toluene). Olefins and aromatics are used in a wide range of applications including detergents, plastics, resins, lubricants, gels, and adhesives. While an oil refinery or petrochemical plant can be separate facilities, there are also integrated complexes which can perform both refining and petrochemical processes.

A *pump jack* (also known as a *nodding donkey*) is a mechanical system designed to extract crude oil from an oil deposit where the pressure in the well is not sufficient to push the oil up to the surface (**Figure 4.4**). Pump jacks operate in a similar manner as hand pumps used to deliver water from wells. A beam-type pump jack houses a crank shaft, a pitman arm, and a beam to convert rotary motion from the prime mover into a

FIGURE 4.3 Geological schematic of oil deposits. From left to right: an offshore drilling rig, a refinery, and a pump jack. Source: Youst/DigitalVision Vectors via Getty Images.

FIGURE 4.4 A pump jack in Alberta, Canada. Source: laughingmango/E + via Getty Images.

FIGURE 4.5 An offshore drilling rig in the Gulf of Thailand, Malaysia. Source: Mekdet/Moment via Getty Images.

FIGURE 4.6 A drillship in the Atlantic Ocean off West Africa. Source: Cavan Images/Cavan via Getty Images.

vertical reciprocating motion of sucker rods and a piston to transport the oil from the well.

An *offshore drilling rig* (also called an *offshore platform* or *oil platform*) is a floating structure built on water for exploring and extracting oil from deposits beneath the seabed (**Figure 4.5**). There are different types of platforms based on the location, construction, or operational conditions. Fixed platform rigs are attached to the seabed by their legs, which can be steel or concrete. *Mobile offshore drilling units* (MODUs), on the other hand, can be moved to different locations for exploration (**Figure 4.6**). These units can also be used for extracting oil, with adequate system modifications, once oil is found. Barges, jack-up rigs, semi-submersible platforms, and drillships are some examples of MODUs.

According to the BP Statistical Review 2021, the majority of the oil production in the world in 2019 came from three regions: Middle East (31.3%), North America (26.6%), and Commonwealth of Independent States (15.3%) [1] (**Figure 4.7**).

In the United States, the majority of the crude oil production comes from the Gulf of Mexico and five states: Texas, North Dakota, New Mexico, Oklahoma, and Colorado (**Figure 4.8**) [2].

The oil trade is one of the largest businesses worldwide. Just as in trading of any other asset internationally, grading is a necessity and a helpful tool for the oil market as well. The two commonly used grades are Brent North Sea Crude (*Brent crude*) and West Texas Intermediate (*WTI*). Brent crude's price is used as a benchmark for African, Middle Eastern, and European oil; WTI on the other hand is the benchmark crude for North America.

FIGURE 4.7 Oil production and consumption by region. Source: BP Statistical Review of World Energy, 2021.

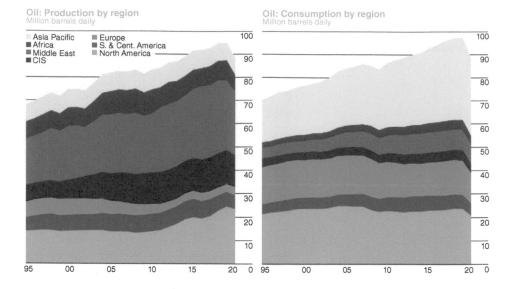

Oil: Production by region
Million barrels daily

Oil: Consumption by region
Million barrels daily

Asia Pacific Europe
Africa S. & Cent. America
Middle East North America
CIS

FIGURE 4.8 United States crude oil production by state or region by 2020. Source: U.S. Energy Information Administration.

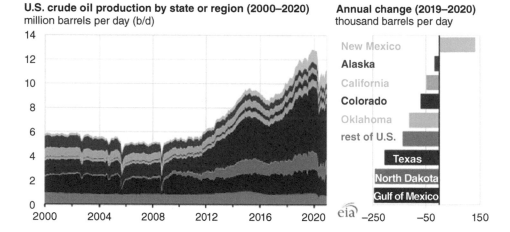

U.S. crude oil production by state or region (2000–2020)
million barrels per day (b/d)

Annual change (2019–2020)
thousand barrels per day

New Mexico
Alaska
California
Colorado
Oklahoma
rest of U.S.
Texas
North Dakota
Gulf of Mexico

FIGURE 4.9 1 barrel of oil = 42 gallons = 159 liters = 0.136 tonnes (metric).*
Source: jdillontoole/DigitalVision Vectors via Getty Images.
* Based on worldwide average gravity

4.1.1.3 NATURAL GAS

Natural gas is the cleanest of the fossil fuels in terms of carbon emissions. It contains many compounds, with methane (CH_4) being the dominant one. It can be categorized based on the geological conditions of the reserves.

- *Conventional natural gas*: found in large fractures and spaces between layers of overlying rock
- *Unconventional natural gas*: found in the small pores within formations of shale, sandstone, or other types of sedimentary rocks (*shale gas* or *tight gas*)
- *Associated natural gas*: found in the deposits of crude oil
- *Coalbed methane*: found in coal deposits

The bulk portion of global natural gas production in 2020 was from three regions: North America (28.3%), Commonwealth of Independent States (21.2%), and Middle East (17.4%) (**Figure 4.10**). The United States produced the most (23.1%), followed by Russia (17.0%) and Iran (6.1%) [1].

Once natural gas has been extracted from the gas or oil wells, it goes through a preliminary process of separation and then is sent to the gas processing plant. It is then sent either to the main line sales, to an underground storage reservoir, or to utility companies for distribution to consumers (**Figure 4.11**). For consumer use, methanethiol (also called *mercaptan*), a chemical that smells like rotten eggs, is added as an odorant to

FIGURE 4.10 Natural gas production and consumption by region. Source: BP Statistical Review of World Energy, 2021.

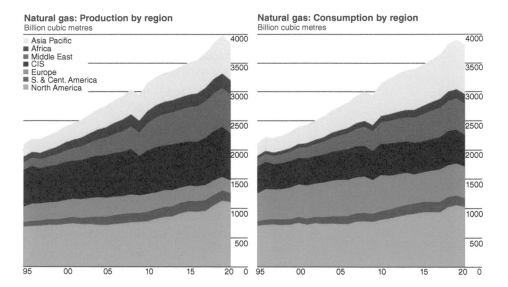

Natural gas: Production by region
Billion cubic metres

- Asia Pacific
- Africa
- Middle East
- CIS
- Europe
- S. & Cent. America
- North America

Natural gas: Consumption by region
Billion cubic metres

FIGURE 4.11 Stages of natural gas production and delivery. Source: U.S. Energy Information Administration.

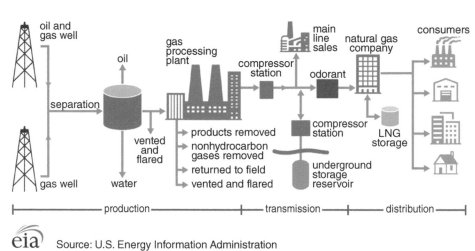

Source: U.S. Energy Information Administration

FIGURE 4.12 Blue methane flames coming out of a gas stove burner head. As natural gas has no color, odor, or taste, a chemical called *mercaptan* is added to the gas for easier detection in case of a leakage. Source: Rosmarie Wirz/Moment via Getty Images.

the gas. As natural gas has no color, odor, or taste, the addition of an odorant aims to help detection of gas leakages by consumers (**Figure 4.12**).

4.1.1.4 SHALE GAS

Not all oil or gas can be easily retrieved based on the depth the reservoir is at or the surrounding structures. Some of the oil or gas is trapped in deposits of shale, a type of sedimentary rock formation. This formation does not have a porous nature, therefore extracting the oil or gas through the shale layers requires different techniques.

Shale gas is natural gas that is trapped in these shale formations. Because shale has low permeability, extracting the gas requires fractures to provide gas flow. While the gas has been produced for a very long time from shale basins with natural fractures, implementation of *hydraulic fracturing* (alternatively called *hydrofracking, hydrofracturing,* or *fracking*) is more recent.

The hydrofracking method was first used for commercial purposes by the Standard Oil Company in 1947. The technique involved injection of a mixture of oil and acid into the well to stimulate shale oil production. Since then, various chemicals have been tried to increase the effectiveness of the hydrofracking fluid. In the 1950s, water was introduced as the working fluid.

After the 2008 recession, the shale gas industry and use of hydrofracking increased substantially. This noticeable change is known as the *shale boom*. This not only influenced investment and employment in the oil and gas industry in the United States, but also made a big impact, directly or indirectly, on the international oil and gas trades encompassing many countries.

Various methods of drilling exist for hydrofracking. Vertical drilling is the conventional application and does not require extensive amounts of fluid and components. It provides access to a limited lateral source for extracting the gas. Horizontal drilling has enabled the industry to produce more gas without having to bore more holes into the basin. The lateral surfaces on the horizontal section of the bore yield a high-volume hydraulic fracturing advantage. These horizontal wells can stretch up to two miles (3.2 km) along a shale deposit [4]. This method uses a significant amount of water compared to the vertical drilling application.

While hydrofracking has enabled leveraging the trapped gas and oil, supporting the economy with added investments and employment, it has also been a controversial topic due to the risk of methane contamination into aquifers and to the atmosphere. One important distinction to make is that natural gas and methane are not the same. Methane (CH_4) is one typical composition of natural gas. Natural gas being mostly methane (70–90%) can sometimes yield this energy jargon confusion. Other forms of natural gas exist, and we are familiar with these through a variety of daily living needs. Some of these forms are ethane (CH_6), propane (C_3H_8), and butane (C_4H_{10}). Methane is not toxic and does not cause health problems in drinking water. However, if the amount of contamination is high, gas may escape from water and seep into occupied spaces where an explosion can occur if the concentration reaches sufficient levels for the given volume of space. The major concern with methane is leakage into the atmosphere. It is a much more powerful greenhouse gas than carbon dioxide and can contribute to global warming at much higher potential. Its *global warming potential* (GWP) for a 100-year time horizon relative to CO_2 is 28, according to the *Fifth Assessment Report of the Intergovernmental Panel on Climate Change* (IPCC), which is the United Nations body for assessing the science related to climate change [5].

Lawrence Berkeley National Laboratory (LBNL) conducted a simulation study in collaboration with the EPA on six possible scenarios of upward fluid migration from a hydrofracking zone to the water aquifers. These scenarios are [6]:

- Scenario A: Defective well construction along with high pressures during hydraulic fracturing can result in damage to well integrity. This can cause formation of a migration pathway through which fluids could penetrate and move upward through the cement cylinder or clearance near the wellbore into the aquifers closer to surface. In this scenario, there are not necessarily fracture formations in the overburden.
- Scenario B1: Due to inadequate design of the hydraulic fracturing process, fracturing of the overburden can be observed. This can allow direct or indirect methanol migration into the aquifers above them. Indirect migration would be seen if fractures intercept a permeable formation between the shale gas formation and the water-table.
- Scenario B2: Similar to Scenario B1, fracturing of the overburden can allow indirect migration of fluids between the shale gas reservoir and the aquifers after

FIGURE 4.13 Scenario D1: Fracturing of the overburden opening pathways for the hydrocarbons and other contaminants into offset wells. Source: U.S. Environmental Protection Agency (EPA).

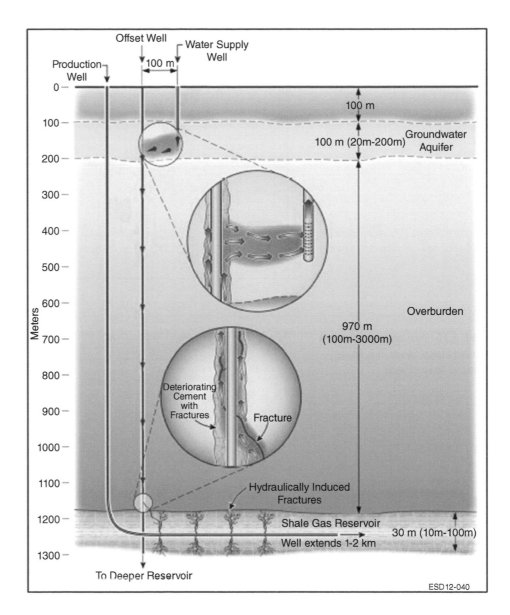

intercepting conventional hydrocarbon reservoirs, which can result in dual sources of contamination for the water-table.

- Scenario C: Dormant fractures and faults can be activated during the hydraulic fracturing operation, resulting in formation of pathways for upward migration of hydrocarbons and other contaminants.
- Scenario D1: Fracturing of the overburden can form pathways for migration of hydrocarbons and other contaminants into offset wells or their vicinity in conventional reservoirs with degradation of cement casing. The offset wells can intersect with aquifers, resulting in contamination. This scenario is illustrated in **Figure 4.13**.
- Scenario D2: Similar to Scenario D1, fracturing of the overburden can result in migration of hydrocarbons and other contaminants into improperly closed offset wells or their vicinity with vulnerable casings. The offset well can provide a low-resistance migration path to the contaminants from the shale gas reservoir to the aquifer.

FIGURE 4.14 Global shale basins. Source: U.S. Energy Information Administration.

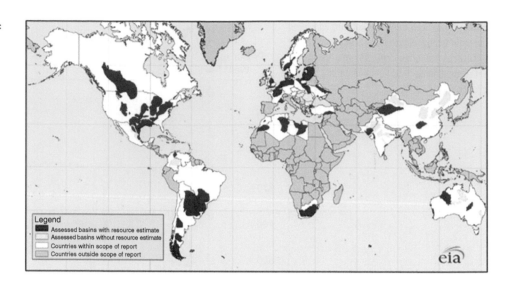

FIGURE 4.15 Shale plays and basins in the United States. Source: U.S. Energy Information Administration.

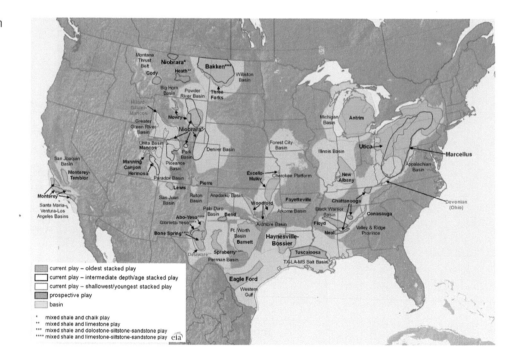

In this study, LBNL developed computational codes to simulate the migration of the gas, water, and dissolved chemicals to have some idea about the risk of methane and the chemicals making their way to the groundwater aquifer. It should be noted that this is a complicated multi-phase flow problem with a complex set of input parameters including the stratigraphy of the site, rock properties, gas and pore water compositions, hydraulic fracturing pressure and other thermal, physical, and geomechanical properties and conditions.

Global and US shale basins can be seen in **Figures 4.14** and **4.15**, respectively. The shale boom in the United States has been a game changer both regionally and globally. The increase in gas production in the United States has influenced the global natural gas

FIGURE 4.16 Hydrofracking process at a shale rig. The process is still a highly controversial topic. While its proponents endorse the economic benefits and argue that natural gas is cleaner than coal, the opponents are concerned with the potential groundwater and surface water contamination in case of chemicals migrating through the overburden. Source: roccomontoya/DigitalVision Vectors via Getty Images.

CLASSROOM DEBATE

Is shale gas a viable solution or not?

FURTHER LEARNING

Global Climate Change: Vital Signs of the Planet
https://climate.nasa.gov/causes/

FURTHER LEARNING

Greenhouse Gas Equivalencies Calculator
www.epa.gov/energy/greenhouse-gas-equivalencies-calculator

and liquefied natural gas (LNG) trends, which has had environmental, political, and technological influences, as well as economic impacts.

4.1.2 Environmental Impacts of Coal, Oil, and Gas

4.1.2.1 GREENHOUSE EFFECT

Electromagnetic radiation received by the Earth is transmitted, reflected, or absorbed. Some range of the radiation can pass through the atmosphere. This range is mostly visible light. Some radiation, however, is absorbed by gases such as ozone, carbon dioxide, and water vapor. These gases do not absorb the visible light, they let it pass. They absorb the infrared radiation, but a fraction of the infrared radiation emitted by the Earth is absorbed by these gases and partially emitted back to Earth again. This is what causes extra heating of the atmosphere. When we think of a greenhouse, the visible light is transmitted through the roof, reaching the plants. The plants emit infrared radiation, which is absorbed by the glass roof above them, and the roof emits the radiation back to the plants, causing the space to be warmer. This is called the *greenhouse effect*. The gases that can absorb infrared radiation and emit back to the Earth are called the *greenhouse gases*. Some of the most abundant greenhouse gases in Earth's atmosphere are:

- Water vapor (H_2O)
- Carbon dioxide (CO_2)
- Methane (CH_4)
- Ozone (O_3)
- Nitrous oxide (N_2O)
- Chlorofluorocarbons (CFCs)
- Hydrofluorocarbons (HCFCs and HFCs)

The environmental impact of fossil fuels is that the combustion process produces CO_2, which increases the greenhouse effect, which in turn increases the global temperature.

4.1.2.2 AIR POLLUTION

Air pollution refers to the low quality of atmospheric air rather than indoor air in HVAC applications. The EPA has identified six major pollutants as "criteria" air pollutants, as it uses these in assessing air quality and identifying permissible levels in consideration of human health-based and environmental-based criteria [7]. These pollutants are:

- Carbon monoxide
- Lead
- Nitrogen oxides
- Ozone
- Particulate matter
- Sulfur dioxide

Each of these pollutants can cause serious health effects, whether acting solely or in combination. Some of the health problems associated with these pollutants can be listed as dizziness, confusion, asthma, bronchitis, heart disease, and malfunctions of the nervous system, kidney function, immune system, and reproductive and developmental systems [7].

4.1.2.3 ACID RAIN

When sulfur dioxide (SO_2) and nitrogen oxides (NO_x) are released to the atmosphere, they are carried by air currents and winds. When the right conditions exist, they react with water, oxygen, and other chemicals to form sulfuric acid (H_2SO_4) and nitric acid (HNO_3). They then mix with water and other compounds and fall to the ground mixed with rain, snow, fog, or even hail. This unwelcome form of precipitation is called *acid rain*. The solution to address this problem is clear. Sulfur dioxide and nitrogen oxide emissions into the atmosphere should be reduced. The major sources of SO_2 and NO_x in the atmosphere are [8]:

- Fossil fuel combustion (two-thirds of SO_2 and one-fourth of NO_x in the atmosphere come from power plants used to generate electricity)
- Transportation (all uses including personal, commercial, industrial, etc.)
- Oil refineries, manufacturing facilities, and other industries

It is important to note that sulfur dioxide and nitrogen oxides can be carried by air currents for long distances. This can cause acid rain across borders, or in areas that do not host environmentally harmful industries. Therefore, this environmental issue is a problem for all countries, industrialized or developing.

4.1.2.4 SMOG

The term "smog" comes from the words *smoke* and *fog*. It was first used in the mid-twentieth century in London to describe the pollution. The common perception of smog rises from our optical senses, seeing the lower segment of the atmosphere closer to the ground as a yellow-brown haze (**Figure 4.17**). It is a mix of harmful pollutants including NO_x, SO_2, particulate matter (PM), and volatile organic compounds (VOCs). Smog can be observed mostly in highly populated cities, and especially is associated with transportation. Many drivers all around the world living in such cities have witnessed the not so pleasing haze during rush hour, driving to or from work. According to the EPA, the transportation sector in the United States is responsible for [9]:

- Over 55% of NO_x total emissions
- Less than 10% of VOC emissions
- Less than 10% of $PM_{2.5}$ and PM_{10} emissions

Smog can have serious health effects such as eye, nose, and throat irritation, and provoking existing respiratory or heart diseases.

FIGURE 4.17 Smog is a common problem of urban areas and gives a yellow-brown haze to the lower atmosphere. Source: Kiszon Pascal/Moment via Getty Images.

 **FURTHER LEARNING**

Carbon Footprint Calculator
www3.epa.gov/carbon-footprint-calculator

4.2 Theory

Fossil fuels are mostly used for generating electricity (power plants), transportation (vehicles), and in buildings mainly for heating and cooking purposes. In all these applications, a *combustion* process is taking place. This can be the combustion of coal at a thermal power plant, it can be combustion of diesel or gasoline in the engine of our vehicles, or it can be the combustion of natural gas in our furnaces for heating, or stoves for cooking.

Most fossil fuel combustion takes place in the power generation industry. Theoretical aspects of power plants in the fields of electrodynamics, chemical energy conversion, thermodynamics, fluid mechanics, and heat transfer were discussed in

Chapter 3. As a reminder for ourselves, let us recall the fundamental topics pertaining to theoretical analysis of power plants:

Electrodynamics: Electromotive force, Faraday's law, and Maxwell's equations
Chemistry: Combustion processes and energy conversion efficiency
Thermodynamics: Power generation executing Rankine or Brayton cycles
Fluid mechanics: Flow of fluid through control volumes and turbomachinery
Heat transfer: Conduction, convection, radiation, and combined heat transfer

It is important to remember the big picture in theoretical analysis and to be able to make the connection between different fields of physics in conducting the analyses. We should be able to zoom in on a section of the physical problem, solve it, and then zoom out and wrap things up, making sure the main task is accomplished and/or we are able to see the big picture and make interpretations on the outcome of the analysis. *We should not get lost in the forest while observing a single tree when the task is to examine the whole forest.* In this analogy, the trees are:

- Power generation (what we want to get)
- Combustion (what needs to be done for what we want to get)
- Carbon dioxide production (what we need to deal with after getting what we want)

4.2.1 Power Generation

Rankine (steam turbines) and Brayton (gas turbines) cycles are the two major thermodynamic cycles utilized in power plants for electricity generation. Four important things to know about these cycles for an energy engineer or an energy professional are:

- Schematic of the power plants with key components (i.e., boiler, turbine, pump etc.)
- *T–S* diagram of the cycle
- Determining enthalpy values at the inlet and exit of each control volume (the components)
- Calculating net work output and thermal efficiency of the plant

The schematic and *T–S* diagram of a Rankine cycle are illustrated in **Figure 4.18**. Thermal efficiency of a fossil fuel power station can be determined by:

$$\eta_{\text{th}} = \frac{\dot{W}_{\text{net}}}{\dot{Q}_{\text{in}}} = \frac{\dot{W}_{\text{turb}} - \dot{W}_{\text{pump}}}{\dot{Q}_{\text{boiler}}} \tag{4.1}$$

For actual systems, internal irreversibilities within the components of the power station need to be taken into account. To address this, isentropic efficiencies are utilized in the analysis. Isentropic turbine and pump efficiencies are given in **Equations 4.2 and 4.3**, respectively:

$$\eta_{\text{t}} = \frac{h_3 - h_4}{h_3 - h_{4s}} \tag{4.2}$$

FIGURE 4.18 Rankine vapor cycle with superheat and irreversibilities.

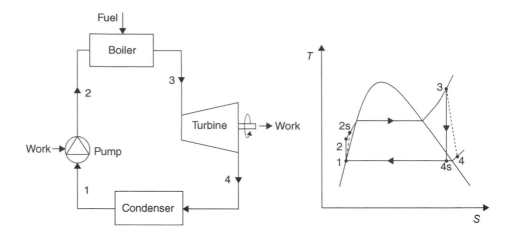

FIGURE 4.19 Air-standard Brayton cycle with irreversibilities.

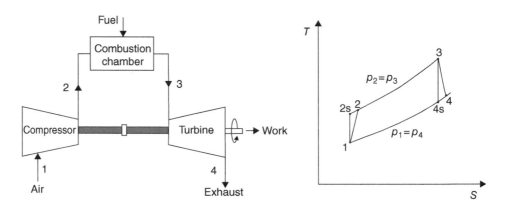

$$\eta_p = \frac{h_{2s} - h_1}{h_2 - h_1} \tag{4.3}$$

In the case of executing a Brayton cycle, the schematic and T–S diagram of a gas turbine system can be seen in **Figure 4.19**. Thermal efficiency, isentropic efficiencies of the compressor and the turbine, and the back work ratio (bwr) derived using air-standard analysis are given in **Equations 4.4–4.7**:

$$\eta_{th} = 1 - \frac{T_1}{T_2} = 1 - \left(\frac{p_1}{p_2}\right)^{\frac{k-1}{k}} \tag{4.4}$$

$$\eta_t = \frac{h_3 - h_4}{h_3 - h_{4s}} \tag{4.5}$$

$$\eta_c = \frac{h_{2s} - h_1}{h_2 - h_1} \tag{4.6}$$

$$\text{bwr} = \frac{\dot{W}_{comp}/\dot{m}}{\dot{W}_{turb}/\dot{m}} = \frac{h_2 - h_1}{h_3 - h_4} \tag{4.7}$$

In **Equation 4.4**, (p_1/p_2) is the compressor pressure ratio, and k is the specific heat ratio (c_p/c_v).

Example 4.2 Conversion of a Coal-Fired Power Plant

A coal-fired power plant executing a Rankine cycle has the following specifications. The plant owners are considering two options for converting the power plant into a more efficient facility with lower CO_2 emissions. The alternative options are biomass cofired with coal or natural gas-fired systems. Values for the existing and alternative plant configurations are given below. Determine the thermal efficiencies of all systems.

Fuel	Heat addition (MW)	Turbine output (MW)	Pump input (MW)
Coal	1637.5	550	6.4
Biomass–coal	1676.0	588	6.4
Natural gas	1860.8	790	6.6

Solution

The thermal efficiency of each cycle can be calculated by using **Equation 4.1**:

Coal-fired power plant

$$\eta_{\text{th,coal}} = \frac{\dot{W}_{\text{net}}}{\dot{Q}_{\text{in}}} = \frac{\dot{W}_{\text{turb}} - \dot{W}_{\text{pump}}}{\dot{Q}_{\text{boiler}}} = \frac{(550 - 6.4)\,\text{MW}}{1637.5\,\text{MW}} = 0.332\,(33.2\%)$$

Biomass–coal cofired power plant

$$\eta_{\text{th,cofired}} = \frac{\dot{W}_{\text{net}}}{\dot{Q}_{\text{in}}} = \frac{\dot{W}_{\text{turb}} - \dot{W}_{\text{pump}}}{\dot{Q}_{\text{boiler}}} = \frac{(588 - 6.4)\,\text{MW}}{1676\,\text{MW}} = 0.347\,(34.7\%)$$

Natural gas-fired power plant

$$\eta_{\text{th,NG}} = \frac{\dot{W}_{\text{net}}}{\dot{Q}_{\text{in}}} = \frac{\dot{W}_{\text{turb}} - \dot{W}_{\text{pump}}}{\dot{Q}_{\text{boiler}}} = \frac{(790 - 6.6)\,\text{MW}}{1860.7\,\text{MW}} = 0.421\,(42.1\%)$$

Thermal efficiency of the power plant can increase from 33.2% to 34.7% if the plant boiler is adjusted to be suitable for biomass cofiring. Efficiency would increase to 42.1% with conversion from coal to natural gas for this specific power plant. Either of the alternatives have environmental benefits as CO_2 emissions would be lower than the emissions from the existing coal-fired plant. There are, however, other challenges such as technical or economic factors. Ash deposition and disposal in biomass cofired boilers is one of the technical challenges. From an economic point of view, the feasibility of converting the plant into either a biomass cofired plant or a natural gas-fired plant depends on the investment cost, interest rate, and additional O&M costs.

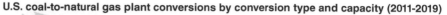

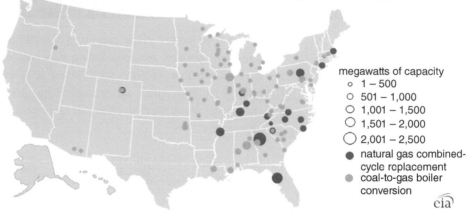

U.S. coal-to-natural gas plant conversions by conversion type and capacity (2011-2019)

megawatts of capacity
- 1 – 500
- 501 – 1,000
- 1,001 – 1,500
- 1,501 – 2,000
- 2,001 – 2,500
- natural gas combined-cycle replacement
- coal-to-gas boiler conversion

eia

Coal to natural gas conversion is a viable way to reduce CO_2 emissions. Source: U.S. Energy Information Administration, Annual Electric Generator Report and Preliminary Monthly Electric Generator Inventory.

4.2.2 Combustion

In this section, we will focus on the combustion process and the carbon dioxide emissions. Combustion of all fossil fuels yields a similar reaction:

$$\text{Fuel} + \text{Oxygen} \rightarrow \text{Carbon dioxide} + \text{Water} (+\text{Energy!})$$

The general equation for combustion of a hydrocarbon in oxygen is given as:

$$C_xH_y + \left(x + \frac{y}{4}\right)O_2 \rightarrow xCO_2 + \frac{y}{2}H_2O \tag{4.8}$$

Examples of combustion equations for four different fuels – methane, benzene, gasoline, and diesel – are derived below based on **Equation 4.8**:

$$\text{Methane} \quad CH_4 + 2O_2 \rightarrow CO_2 + 2H_2O$$

$$\text{Benzene} \quad C_6H_6 + \frac{15}{2}O_2 \rightarrow 6CO_2 + 3H_2O$$

$$\text{Gasoline} \quad C_8H_{18} + \frac{25}{2}O_2 \rightarrow 8CO_2 + 9H_2O$$

$$\text{Diesel} \quad C_{15}H_{28} + 22O_2 \rightarrow 15CO_2 + 14H_2O$$

Each of these combustion processes releases energy in addition to carbon dioxide and water production. The amount of energy released by burning a unit mass of fuel indicates its specific energy (MJ/kg or Btu/lb). This is also referred to as the heating value or calorific value. Heating values of some selected fuels are listed in **Table 4.2**.

Another useful parameter in energy analysis of fuels is energy density, which is the amount of energy released in the combustion process of a unit volume of fuel (MJ/L or Btu/gal).

Table 4.2

Heating values of selected fuels [10]	
Fuel	**Heating value (MJ/kg)**
Dry firewood[1]	16
Hard black coal	25
Diesel fuel	42–46
Petrol/gasoline	44–46
Natural gas	42–55
Natural uranium[2]	500,000

[1] Heating value of wood, peat, or coal decreases with increasing moisture content.
[2] In LWR (light water reactor)

4.2.3 Carbon Dioxide Production

The major source of greenhouse gases from human activities is combustion of fossil fuels for heating, generating electricity, and transportation. Carbon dioxide emissions is one of the critical concerns associated with greenhouse gas emissions. The amount of CO_2 released to the atmosphere during a combustion process depends on the chemical reaction of the fossil fuel with oxygen to deliver energy. During combustion (or burning), the species reacting with oxygen is oxidized. Fossil fuels are mainly composed of hydrocarbons, which contain carbon–hydrogen bonds. When the combustion reaction takes place, these bonds are broken in the reactants while new ones are formed in the products. Breaking of the bonds requires energy and forming of the bonds releases energy. Fossil fuels that have a higher hydrogen-to-carbon ratio have comparatively more bonds. These fuels also weigh less as the weight of hydrogen (1.00784 u) is almost 1/12 of the weight of carbon. Note that it is not exactly 1/12, although hydrogen has only one proton, and carbon (^{12}C) has six protons and six neutrons. The reason for this difference is the nuclear binding energy which keeps all protons together without repelling each other. There is more discussion of nuclear binding energy in Chapter 5. Going back again to comparison of fossil fuels with different hydrogen-to-carbon ratios, the higher the ratio, the more bonds are involved with less mass, and hence the amount of CO_2 released per unit energy delivered is less. Let's look at three examples below to illustrate this fact:

$$\text{Coal} \qquad C + O_2 \rightarrow CO_2 + 33 \text{ MJ/kg}$$

$$\text{Gasoline} \qquad C_8H_{18} + \frac{25}{2}O_2 \rightarrow 8CO_2 + 9H_2O + 48.4 \text{ MJ/kg}$$

$$\text{Natural gas} \qquad CH_4 + 2O_2 \rightarrow CO_2 + 2H_2O + 54.4 \text{ MJ/kg}$$

Gasoline has a hydrogen-to-carbon ratio of 2.25 and the amount of energy released per kg is approximately 48.4 MJ. Natural gas has a hydrogen-to-carbon ratio of 4 and the amount of energy released per kg is approximately 54.4 MJ. This shows that the combustion of natural gas is certainly cleaner than the combustion of coal or gasoline

Table 4.3

Molecular masses of combustion elements and compounds					
Element	Symbol	Molecular mass (kg/kmol)	Compound	Formula	Molecular mass (kg/kmol)
Carbon	C	12	Carbon dioxide	CO_2	44
Hydrogen	H	1	Water	H_2O	18
Oxygen	O	16	Oxygen	O_2	32

Table 4.4

Carbon dioxide produced for combustion of unit mass of selected fuels				
Fuel	Formula	kg C	Molar mass of fuel (kg/kmol)	kg CO_2 from 1 kg fuel
Methane	CH_4	12	16	2.75
Benzene	C_6H_6	72	78	3.38
Gasoline	C_8H_{18}	96	114	3.09
Diesel	$C_{15}H_{28}$	180	208	3.17

(or other derivatives of crude oil) based on the amount of CO_2 produced per unit amount of energy delivered.

Using the chemical equations discussed in **Section 4.2.2**, one can calculate the amount of CO_2 produced per unit mass of fuel burnt. For every 12 kg of carbon combusted, 44 kg of CO_2 is produced. Hence, for every kilogram of carbon burnt, the amount of CO_2 in the exhaust is 3.67 kg. **Table 4.3** gives the molecular masses of combustion elements and compounds. **Table 4.4** lists the four fuels we discussed in terms of the amount of CO_2 released within the exhaust per one kilogram of each fuel when combusted.

Example 4.3 Energy Released per Unit Amount of CO_2 Production

Consider the combustion equations of coal, gasoline, and natural gas discussed in this section. Determine the amount of energy delivered per kilogram of CO_2 released from burning 1 kg of fuel for all three fuel types and compare.

Solution

Coal $C + O_2 \rightarrow CO_2 + 33$ MJ/kg
Amount of CO_2 from 1 kg of coal = 3.67 kg CO_2 / kg fuel

Energy delivered per kilogram of CO_2 released from burning 1 kg of fuel

$$= \frac{33 \, \dfrac{MJ}{kg \, fuel}}{3.67 \, \dfrac{kg \, CO_2}{kg \, fuel}} = 9.0 \, \frac{MJ}{kg \, CO_2}$$

Gasoline $C_8H_{18} + \frac{25}{2}O_2 \rightarrow 8CO_2 + 9H_2O + 48.4 \, MJ/kg$

Amount of CO_2 from 1 kg of gasoline = 3.09 kg CO_2/kg fuel

Energy delivered per kilogram of CO_2 released from burning 1 kg of fuel

$$= \frac{48.4 \, \dfrac{MJ}{kg \, fuel}}{3.09 \, \dfrac{kg \, CO_2}{kg \, fuel}} = 15.7 \, \frac{MJ}{kg \, CO_2}$$

Natural gas $CH_4 + 2O_2 \rightarrow CO_2 + 2H_2O + 54.4 \, MJ/kg$

Amount of CO_2 from 1 kg of natural gas = 2.75 kg CO_2/kg fuel

Energy delivered per kilogram of CO_2 released from burning 1 kg of fuel

$$= \frac{54.4 \, \dfrac{MJ}{kg \, fuel}}{2.75 \, \dfrac{kg \, CO_2}{kg \, fuel}} = 19.8 \, \frac{MJ}{kg \, CO_2}$$

Comparison: As can be seen from calculations, natural gas produces more energy per unit amount of CO_2 released than coal and gasoline. While the unit value for gasoline can still be considered to be close, coal has by far the lowest value. This is because coal does not possess the hydrogen–carbon bonds that gasoline or natural gas have.

4.3 Applications and Case Study

In this section, we will observe real-world examples via two applications and a case study. The applications include a coal-fired power plant in Japan, and a gas-fired power plant in France. The case study is about the national dependence of India on fossil fuels.

Application 1 Coal-Fired Power Plant (Japan)

Figure 4.20 depicts a schematic of a typical power plant fueled by coal. The plant has the four key components needed to execute the Rankine cycle for power generation: boiler, turbine, condenser, and a pump. The working fluid is water which is heated to high-temperature, high-pressure steam that flows into the turbine. As the steam hits the turbine blades that are attached to a shaft, it forces the shaft to rotate, generating electricity by means of electromagnetic induction. Steam does work here on the turbine shaft, hence it leaves the turbine with a lower enthalpy value due to the conservation of energy. It then enters the condenser

FIGURE 4.20 Schematic of a coal-fired power plant. Source: Wikimedia commons, https://commons.wikimedia.org/wiki/File:Coal_fired_power_plant_diagram.svg.

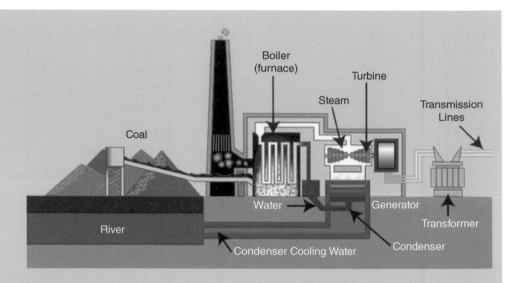

where it is condensed back to the liquid phase to be pumped into the boiler. The cycle is complete. There is a secondary fluid in this cycle, which according to the figure is river water. River water acts as the *heat sink*, absorbing heat from the steam to condense it. The diagram illustrates an open-loop cooling system. Not all thermal power plants are constructed next to a heat sink such as an ocean, sea, lake, or river. Some plants are built inland, far from natural heat sinks. Heat absorption from the condensers of these plants is performed by instrumenting a closed-loop of condenser cooling water. In such systems, a cooling tower is employed for heat rejection from the coolant to the atmosphere.

An application example of a coal-fired power plant is Isogo Power Plant in Japan (**Figure 4.21**). The plant has two units, with each unit having a capacity of 600 MW. Unit 2 of the power station utilizes HELE (high efficiency, low emission) technology. It has average emissions in single digits for SO_2 and NO_x, and <5 mg/m^3 for particulate matter. The Japanese coal fleet is known to be relatively younger than its peers and more modern, with the highest average efficiency (41.6% LHV, net) in the globe [11].

Isogo Power Plant
Location: Yokohama, Japan
Company: J-Power
Fuel: Coal
Capacity: 2 × 600 MW
Turbine: Siemens Steam Turbine
www.d-maps.com

FIGURE 4.21 Isogo thermal power plant in Japan. Source: DigiPub/Monet via Getty Images.

Application 2 Gas-Fired Power Plant (France)

Natural gas-fueled power plants have far lower emissions than coal-fueled plants. Gas-fired plants can also utilize combined-cycle gas turbines (CCGT), where electricity is generated using the gas turbine and the resulting waste heat is recovered to generate steam that drives the steam turbine to produce additional electricity, hence increasing the overall efficiency of the plant (**Figure 4.22**). Bouchain CCGT Power Plant in France has one of the largest and most efficient natural gas turbines with an efficiency of 62%. It was recognized by *Guinness World Records* as the most efficient CCGT plant in the world in 2016 (**Figure 4.23**).

FIGURE 4.22 Schematic of a combined-cycle gas/steam turbine power plant. Source: ENGIE, https://engie.com.au/home/engie-today/education/how-does-gas-fired-power-work/.

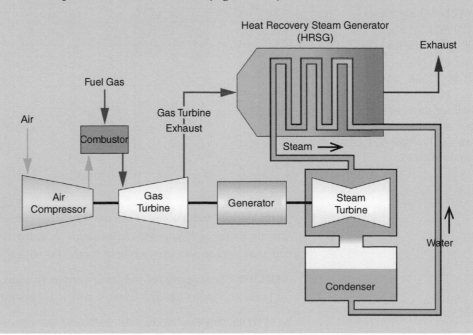

Bouchain Power Plant
Location: Bouchain, France
Company: EDF
Fuel: Natural gas
Capacity: 605 MW
Turbine: GE 9HA Gas Turbine
www.d-maps.com

FIGURE 4.23 Bouchain combined cycle power plant in northern France. Source: Wikimedia Commons author, Serge Ottaviani, https://commons .wikimedia.org/wiki/File:Centrale_ Thermique_de_Bouchain-004.JPG.

Case Study National Dependence on Fossil Fuels (India)

Population: 1,326,093,247
(2020 est.)
GDP: $9.474 trillion
(2017 est.)
GDP per capita: $7200
(2017 est.)
Fossil fuel consumption
world rank: 3rd
Data: CIA Factbook [12]
www.d-maps.com

INTRODUCTION India is one of the top fossil-fuel consuming nations. While this is due mainly to its high population, increases in industrialization and manufacturing in the past decade also play a role in its total fossil fuel consumption.

OBJECTIVE The objective of this case study is to learn about the national dependence of India on fossil fuels as a country that is becoming more industrialized.

METHOD This case study was conducted through researching articles and reports.

RESULTS As India takes steps towards modernization, this causes either more people to move to existing urban areas, or urbanization of the smaller local regions. As a result, there is a shift from biomass and waste use as sources of energy to fossil fuels. India's proven reserves makes up about 15% of the global total coal reserves. Its share of oil and natural gas in the global picture is below 1%. It is one of the largest coal producers in the world. Despite its high coal production capacity, it is still one of the largest coal importers, as well as ranking at top five in terms of oil and natural gas exports. Most of the coal is used for power generation in the country. While there is a significant amount of fossil fuel imports, India is also big in re-exporting petroleum products. Fossil fuel re-exports made up 15% of India's export value in 2018. Taxes, services, and other fees on fossil fuel production and consumption accounted for 17.8% of the general government revenue in 2017. Fossil fuel production subsidies in the same year were reported to be $0.3 billion (0.1% of general government revenue), while fossil fuel consumption subsidies summed up to a higher amount of $11 billion (1.9% of general government revenue) [13].

DISCUSSION National dependence on fossil fuels is one of the top topics for both developed and developing countries as it relates directly to energy security and economy. In addition to the reliance and environmental concerns about fossil fuels, subsidies that governments allocate for the fossil fuel sector can also be in significant amounts. Reduction of these subsidies can result in the redistribution of taxpayers' money for other sectors such as clean energy, education, and health.

RECOMMENDATIONS The dependence of a country on fossil fuels can be reduced significantly by taking several measures. These include formulating and adopting policies to improve energy efficiency across all consumption sectors, encouraging and supporting the renewable energy industry, and spreading the recycling and reuse culture in the country. Policies, incentives, subsidies, and education in schools are some of the key elements that can be utilized to implement the recommended measures.

Example 4.4 Retrieving and Plotting Historical Oil Prices Using Python

Historical oil prices can be retrieved from a variety of reliable sources. Plot the following for Europe Brent, obtaining the data between 1987 and 2021 from the EIA:

a. Historical daily percentage changes (%)
b. Historical daily level changes ($/bbl)
c. Oil price level ($/bbl) vs oil price change ($/bbl)
d. Oil price level ($/bbl) vs percentage oil price change (%)

Solution

The data can be retrieved from www.eia.gov/dnav/pet/hist_xls/RBRTEd.xls

```
import numpy as np
import matplotlib.pyplot as plt
import pandas as pd
```

a. First, assign the address for the spreadsheet to pull the data from:

```
prices=pd.read_excel("https://www.eia.gov/dnav/pet/hist_xls/
RBRTEd.xls", sheet_name="Data 1", skiprows=2)
```

Then, define the commands to retrieve the first and last two data cells:

```
prices.head(2)
```

Data	Date	Europe Brent Spot Price FOB ($ per barrel)
0	1987-05-20	18.63
1	1987-05-21	18.45

```
prices.tail(2)
```

Data	Date	Europe Brent Spot Price FOB ($ per barrel)
8725	2021-10-01	79.40
8726	2021-10-04	81.44

```
prices["Brent"]=prices
['Europe Brent Spot Price FOB (Dollars per Barrel)']
prices["Month"]=prices["Date"].dt.month
prices["Year"]=prices["Date"].dt.year
prices["Percent_Change"]=prices
['Europe Brent Spot Price FOB (Dollars per Barrel)'].pct_change()
prices["Price_Change"]=prices
['Europe Brent Spot Price FOB (Dollars per Barrel)'].diff()
plt.ylabel("% Change")
plt.title("Historical daily percent changes of Brent price")
plt.plot(prices.Date, 100*prices.Percent_Change)
```

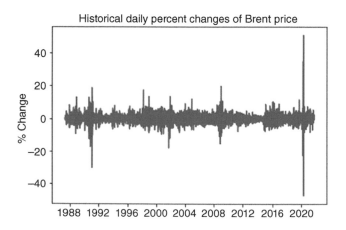

b. Historical daily level changes ($/bbl)

```
plt.ylabel("$/bbl Price change")
plt.plot(prices.Date, prices.Price_Change)
plt.title("Historical daily level (USD) changes of Brent price")
```

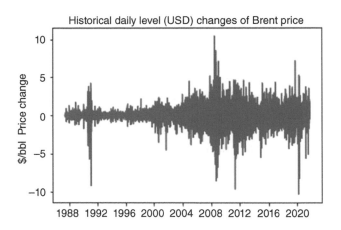

Historical daily level (USD) changes of Brent price

c. Oil price level ($/bbl) vs oil price change ($/bbl)

```
lt.ylabel("Change in oil prices ($/bbl)")
plt.xlabel("Oil price ($/bbl)")
plt.title("Oil price level vs oil price change in $/bbl")
plt.scatter(prices.Brent, prices.Price_Change)
```

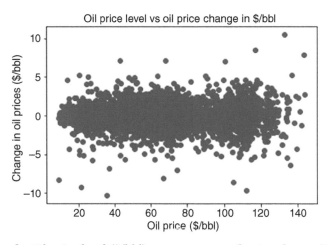

Oil price level vs oil price change in $/bbl

d. Oil price level ($/bbl) vs percentage oil price change (%)

```
plt.ylabel("% change in oil prices")
plt.xlabel("Oil price ($/bbl)")
plt.title("Oil price level ($/bbl) vs oil price change(%)")
plt.scatter(prices.Brent, 100*prices.Percent_Change)
```

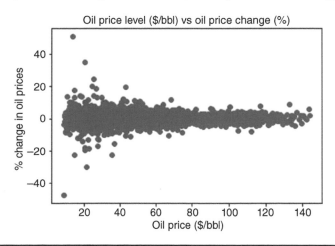

Oil price level ($/bbl) vs oil price change (%)

4.4 Economics

The economics of constructing and operating a power plant involves a variety of phases, each phase potentially having multiple cost items. From a business point of view, the economics of the power plant can be split in three sections:

Construction: Includes the cost of land, construction, and equipment

Operation and maintenance: Operating costs (i.e., fuel, utilities, wages and salaries, social security, unemployment insurance, workers' compensation, medical insurance, taxes) and maintenance costs (routine maintenance, troubleshooting, repairs)

Business: Marketing, sales, customer relations

Typical construction and operating costs for different types of fossil fuel power plants are listed in **Table 4.5** [14].

While this table provides an idea of the capital and operating costs, the most feasible plant option does not rely only on these numbers. The life of the power plant is significant in conducting comparative analysis. There is a parameter called *levelized cost of energy* or *levelized cost of electricity* (LCOE) which is a measure of net present cost of electricity generation for a power plant over its lifespan. This value allows comparison not only of different fossil fuel technologies to generate electricity, but also of renewable energy technologies such as solar or wind energy farms of unequal operating lifetimes, project sizes, and costs. The LCOE can be calculated by:

$$\text{LCOE} = \frac{I_\text{A} + \sum_{t=1}^{n} \dfrac{\text{OM}_t + F_t}{(1+r)^t}}{\sum_{t=1}^{n} \dfrac{E_t}{(1+r)^t}} \tag{4.9}$$

where

I_A: annualized investment costs

OM_t: operating and maintenance costs in year t

F_t: fuel costs in year t

E_t: electrical energy generated in year t

r: annual effective discount rate (amount of interest paid or earned as a percentage of the balance at the end of the annual period)

n: estimated lifespan of the power plant

Annualized investment cost can be calculated with known values of the total investment cost (I_T), annual discount rate (r), and anticipated lifespan (n) of the power plant:

Table 4.5

Approximate capital and operating costs for fossil fuel power plants [14]		
Technology	**Capital cost ($/kW)**	**Operating cost ($/kWh)**
Coal-fired combustion turbine	500–1000	0.02–0.04
Natural gas combustion turbine	400–800	0.04–0.10
Coal gasification combined cycle (IGCC)	1000–1500	0.04–0.08
Natural gas combined cycle	600–1200	0.04–0.10

FIGURE 4.24 Comparison of operating cost and operational flexibility for different power plant technologies. Source: S. Blumsack, www.e-education.psu.edu/eme801/node/530.

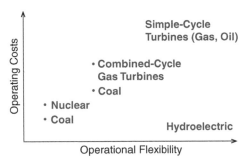

$$I_A = I_T \frac{r\left(1+r\right)^t}{\left(1+r\right)^t - 1} \tag{4.10}$$

In **Equation 4.10**, the multiplier of I_T on the right-hand side is called the *capacity recovery factor*, which is the ratio of constant annuity to the present value of owning that annuity over a selected period of time.

Another important parameter for power plants besides cost is the operational flexibility. Two significant quantities that are associated with operational flexibility are ramp time and minimum run time. Ramp time is the duration (h) it takes for the generator to start from rest and reach the lower operating limit, which is the minimum amount of power the plant can generate. Minimum run time is the shortest duration a plant can run after the system is on. **Figure 4.24** illustrates an approximate characteristic of different power generation technologies based on their operating costs and operational flexibilities.

Example 4.5 Determining LCOE for a Natural Gas Power Plant

A natural gas-fired power plant has an initial investment cost of $1,450,000. Fixed annual O&M expenditure and fuel costs together are reported to be $84,000. The plant produces 5815 MWh per year. Assume a 9% annual effective discount rate and 30-year lifespan for the plant. Calculate the LCOE for this power plant in ¢/kWh.

Solution

First, we obtain annualized initial costs employing **Equation 4.10**:

$$I_A = I_T \frac{r\left(1+r\right)^t}{\left(1+r\right)^t - 1} = (\$1,450,000)\frac{(0.09)(1+0.09)^{30}}{(1+0.09)^{30}-1} = \$141,126.7$$

Plugging the given values into **Equation 4.9**, we get:

$$\text{LCOE} = \frac{\$141,126.7 + \$84,000}{5815\ \text{MWh}} = \frac{\$225,126.7}{5815\ \text{MWh}}\left|\frac{1\ \text{MWh}}{10^3\ \text{kWh}}\right|\left|\frac{100\text{¢}}{1\$}\right| = 3.87\ \frac{\text{¢}}{\text{kWh}}$$

The LCOE for this plant is 3.87 ¢/kWh based on the simple method, assuming fixed O&M and fuel costs, and annual energy output. This is an approximate price at which electricity would need to be sold to "break even." If the company sells the electricity at a higher rate than this value, it will profit, and if the unit price of electricity is lower than the calculated value, the company will be in loss. An increase in the lifespan of the plant could result in reduced LCOE, although this may not be notable. Lowering the investment and/or O&M costs, as well as annual interest paid, can have a more significant reduction in the LCOE of the power plant.

4.5 Summary

Despite all the environmental concerns about fossil fuels and the progress made in renewable energy technologies, fossil fuels make up the majority of the primary energy sources globally. Looking at the demand, supply, and technological improvements, it can be said that we will be using fossil fuels for enough time to keep their environmental impact in mind. As of now, they account for more than 80% of global energy production. Fossil fuels are the source of the majority of total human-caused greenhouse gas emissions. Besides the greenhouse effect, they can cause air pollution, acid rain, and smog. These environmental concerns not only pose a threat to human health and lives, but also are obstacles in achieving the Sustainable Development Goals (SDGs).

There are examples of clean power generation plants with enhanced boiler, turbine, compressor, and generator technologies, which is encouraging. It is important to research and develop cleaner technologies to extract, transport, and more importantly to combust fossil fuels.

REFERENCES

[1] BP, "Statistical Review of World Energy," 2021. Accessed: Oct. 13, 2021 [Online]. Available: www.bp.com/statisticalreview.

[2] U.S. Energy Information Administration (EIA), "U.S. Crude Oil Production Fell by 8% in 2020, the Largest Annual Decrease on Record," EIA, Mar. 9, 2021. www.eia.gov/todayinenergy/detail.php?id=47056.

[3] LibreTexts Chemistry, "Fossil Fuels," Jun. 5, 2019. https://chem.libretexts.org/Courses/Prince_Georges_Community_College/Chemistry_2000%3A_Chemistry_for_Engineers_(Sinex)/Unit_6%3A_Thermo_and_Electrochemistry/Chapter_15%3A_First_Law_Thermochem/Chapter_15.7%3A_Fossil_Fuels (accessed Aug. 27, 2020).

[4] K. Glass, "Shale Gas and Oil Terminology Explained: Technology, Inputs and Operations," Environmental and Energy Study Institute, Dec. 2011. Accessed: Aug. 28, 2020 [Online].

[5] Core Writing Team, R. Pachauri, and L. Meyer (Eds.), *Fifth Assessment Report of the Intergovernmental Panel on Climate Change*, Geneva, Switzerland: IPCC, 2014.

[6] U.S. EPA, "Hydraulic Fracturing for Oil and Gas: Impacts from the Hydraulic Fracturing Water Cycle on Drinking Water Resources in the United States (Final Report)," Office of Research and Development, U.S. EPA, Washington, DC, Dec. 2016.

[7] U.S. EPA, "Lead Air Pollution," U.S. EPA, Feb. 25, 2016. www.epa.gov/lead-air-pollution (accessed Aug. 28, 2020).

[8] U.S. EPA, "Acid Rain," U.S. EPA, Jan. 28, 2019. www.epa.gov/acidrain (accessed Aug. 28, 2020).

[9] U.S. EPA, "Smog, Soot, and Other Air Pollution from Transportation," U.S. EPA, Sep. 10, 2015. www.epa.gov/transportation-air-pollution-and-climate-change/smog-soot-and-local-air-pollution#about (accessed Aug. 28, 2020).

[10] World Nuclear Association, "Heat Values of Various Fuels," 2018. www.world-nuclear.org/information-library/facts-and-figures/heat-values-of-various-fuels.aspx (accessed Aug. 29, 2020).

[11] M. Wiatros-Motyka, "An Overview of HELE Technology Deployment in the Coal Power Plant Fleets of China, EU, Japan and United States," IEA Clean Coal Centre,

Dec. 2016. https://usea.org/publication/overview-hele-technology-deployment-coal-power-plant-fleets-china-eu-japan-and-usa-ccc. Accessed: Aug. 29, 2020. [Online].

[12] CIA, "India: The World Factbook." www.cia.gov/the-world-factbook/countries/india/ (accessed Aug. 30, 2020).

[13] I. Gerasimchuk, *et al.*, "Beyond Fossil Fuels: Fiscal Transition in BRICS REPORT," International Institute for Sustainable Development, Nov. 2019. Accessed: Aug. 30, 2020 [Online]. Available: www.iisd.org/system/files/publications/beyond-fossil-fuels-brics.pdf.

[14] S. Blumsack, "Basic Economics of Power Generation, Transmission and Distribution," 2018. www.e-education.psu.edu/eme801/node/530 (accessed Aug. 31, 2020).

CHAPTER 4 EXERCISES

4.1 Carbon contents of various types of coal are listed below. Based on the values in the table, determine the amount of CO_2 produced in kilograms during the combustion process of 45 kg of each coal type. Assume all carbon in the coal is oxidized during all combustion cases.

Coal Type	% Carbon
Anthracite	92
Bituminous	80
Subbituminous	77
Lignite	71

4.2 Calculate the amount of water produced for every kilogram of the following fuels combusted:

Methane $CH_4 + 2O_2 \rightarrow CO_2 + 2H_2O$

Gasoline $C_8H_{18} + \dfrac{25}{2}O_2 \rightarrow 8CO_2 + 9H_2O$

Diesel $C_{15}H_{28} + 22O_2 \rightarrow 15CO_2 + 14H_2O$

4.3 706,309,263 short tons of coal were produced in the United States in 2019, according to the 2020 Annual Coal Report of the U.S. Energy Information Administration. Calculate the amount of energy that would be released if all of this coal was combusted. Assume an average energy density of 24 MJ/kg for all types of coal extracted. Remember to convert short tons (US ton) to tonnes (metric ton).

4.4 The total electricity generation of Poland was 163.9 TWh in 2019, according to the U.S. Energy Information Administration. Determine the amount of:

a. coal needed in tonnes (assume bituminous coal with specific energy of 27 MJ/kg) if all the electricity was to be generated by coal-fired power plants with an average overall thermal efficiency of 36%;

b. natural gas needed in bcm (assume an energy density of 38 MJ/m^3) if all the electricity was to be generated by natural gas-fired power plants with an average overall thermal efficiency of 40%;

c. CO_2 produced in tons when coal is used (assume bituminous coal to have 80% carbon);

d. CO_2 produced in tons when natural gas is used.

4.5   Shale gas at a basin in Texas has an energy density value of 37 MJ/m^3. Determine the amount of crude oil that would yield the equivalent amount of energy that can be harvested from 1 bcm natural gas extracted from this shale play.

A shale rig in a corn field. Source: grandriver/E+ via Getty Images.

4.6   A petroleum company in Baghdad, Iraq, processes crude oil into gasoline at a cost of \$38 per barrel. The company sells gasoline at its gas stations to end-users at a price of \$0.52 per liter. What is the profit (%) of the oil company?

4.7   Considering the specific energy values of the fuels listed below, calculate the amount of energy yielded by combustion of each fuel for 1 kg of CO_2 released.

   a.   Gasoline, C_8H_{18} (46.4 MJ/kg)

   b.   Natural gas, CH_4 (53.6 MJ/kg)

   c.   LPG propane, C_3H_8 (49.6 MJ/kg)

   d.   LPG butane, C_4H_{10} (49.1 MJ/kg)

4.8   One part per million (ppm) of CO_2 concentration in the atmosphere is approximately equivalent to 7.82 billion metric tonnes of CO_2 released to the atmosphere. Based on the NOAA data provided below, the CO_2 concentration in 1980 was about 340 ppm. In 2020, it was approximately 418 ppm.

   a.   Calculate the amount of CO_2 released to the atmosphere between 1980 and 2020.

   b.   Determine the total amount of CO_2 present in the atmosphere as of 2020.

   c.   Compare the global CO_2 amount until 1980 to the total amount released in the following 40 years only. Comment on your findings.

Global monthly mean CO_2 concentrations. Source: www.esrl.noaa.gov/

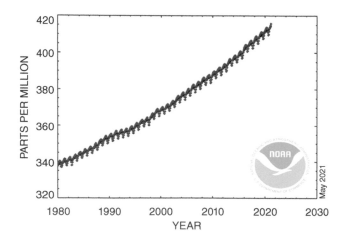

4.9 The turbine of a coal-fired power plant operating based on the Rankine cycle receives steam at 12 MPa and 600 °C. The steam expands to 20 kPa pressure as it enters the condenser. The working fluid leaves the condenser as saturated liquid. Isentropic efficiencies of the turbine and the pump are 83% and 86%, respectively.

    a. Sketch the *T–S* diagram for the cycle.

    b. Determine the thermal efficiency of the power plant.

    c. Calculate the rate of heat addition from the coal to the working fluid if the mass flow rate of steam is 65 kg/s. Assume negligible heat loss from boiler to the surroundings.

4.10 A natural gas-fired power generation unit operating based on the ideal air-standard Brayton cycle has a pressure ratio of 8. Air enters the compressor at 100 kPa and 27 °C at a flow rate of 280 m³/min. As it runs through the combustion chamber, it is heated to 1027 °C during the isobaric process. Determine:

    a. mass flow rate of air (kg/s)

    b. power generation capacity of the unit (MW)

    c. thermal efficiency of the unit (%)

CHAPTER 5
Nuclear Energy

History of Nuclear Energy [1, 2]

1942. The first self-sustaining nuclear chain reaction achieved at the University of Chicago.

1945. Atomic bombs dropped on Hiroshima and Nagasaki in Japan.

1946. The U.S. Atomic Energy Act of 1946 formed the Atomic Energy Commission (AEC) to control nuclear energy development.

1954. First nuclear-powered grid: 5 MW graphite moderated, light water cooled, enriched uranium reactor at Obninsk, USSR.

1954. First nuclear-powered submarine, USS Nautilus, was commissioned.

5.1 Introduction

Nuclear energy is one of the key players in the energy world due to its immense potential to deliver energy and it being free of greenhouse gas emissions. It is not an intermittent energy source. While the construction cost is high, operating costs are lower than its rivals such as coal and natural gas. It does, however, come with the obvious question of whether or not it is safe, especially after several notable accidents of the past 40 years, which we will examine in this chapter. The other concern about nuclear energy is nuclear waste, which is harmful to the environment and takes a long time to degrade. Following the last significant nuclear trauma, the Fukushima accident, there has been increasing motivation, research, and investment in nuclear energy. Improvement of the technology requires progress in various fields including exploration of new fuel resources, development of advanced reactor designs, more economic, safe, and sustainable waste management methods, and enhancement in the use of nuclear fusion for energy generation.

In this chapter, the history of nuclear energy and its global overview are covered, followed by fission and fusion. The challenges of fusion are discussed. The basic components of nuclear power plants and reactor types including advanced reactor technologies are reviewed. A theoretical section covers discussion of how much energy can be harvested from nuclear fuels, electricity generation executing a thermodynamic cycle, the amount of nuclear waste produced, and comparison of nuclear fuels with other fuels. In the last section, global examples of two applications with different reactor types and a case study are discussed, followed by a glance at the economics of nuclear power plants.

5.1.1 Global Overview

The input-equivalent of total global nuclear energy consumption in 2019 in was 24.9 EJ with the United States ranking first with a 7.6 EJ share. Regionally North America and Europe made up the majority, each region having shares of 34.5% and 33.2%, respectively, followed by the Asia Pacific region with a share of 23.2%. France was the major consumer in Europe, and China consumed the highest amount in the Asia Pacific region [3]. **Figure 5.2** shows the regional distribution of nuclear energy consumption.

In the first quarter of 2020, nuclear power generation was recorded to be 3% less than that for the first quarter of 2019. While this was expected due to planned maintenance, the unexpected COVID-19 crisis added to the drop due to decreased electricity demands [4].

When we look at nuclear energy production on a country basis, we see that the United States is by far the leading country, followed by France, China, and the Russian

1957. UN formed the International Atomic Energy Agency (IAEA) in Vienna, Austria, to promote the peaceful use of nuclear energy and stop the spread of nuclear weapons throughout the globe.

1979. Three Mile Island accident, Middletown, Pennsylvania, United States.

1984. Nuclear energy share surpasses natural gas and hydropower, becoming the second largest energy source preceded by coal.

1986. Chernobyl accident, Chernobyl, Ukraine (former USSR).

1996. First Generation III reactor (ABWR) in use at Kashiwasaki–Kariwa Nuclear Power Plant in Japan.

2011. Fukushima accident, Fukushima, Japan.

2013. Voyager 1, powered by radioisotope thermoelectric generators using plutonium-238 fuel, enters interstellar space.

FIGURE 5.1 President Harry S Truman signs the Atomic Energy Act of 1946 establishing the U.S. Atomic Energy Commission. Source: Department of Energy, Office of History and Heritage.

Federation. Besides the amount of nuclear energy generation, the share of nuclear energy in the total electricity production of countries is a helpful parameter in determining the significance of nuclear energy; strategically, economically, and environmentally. In this regard, France is at the top of the list with a 71.7% share, followed by Slovakia (55%) and Ukraine (53%). **Figure 5.3** lists the countries in terms of their total nuclear energy production, and the share of nuclear energy in total electricity generation [5].

The IAEA report on electricity and nuclear power projections up to year 2050 provides estimates on nuclear electrical generating capacity and production values [5]. Projections are reported based on two different scenarios, namely the low case and the high case. While both of these cases assume the same economic and electricity demand growth based on current expectations, the scenarios differ, based mainly on policies and regulations. The low case makes predictions based on the assumption that the current market, technology, and resource trends are maintained with a few changes in laws, regulations, and policies. The high case, on the other hand, assumes a more rigorous and ambitious approach, including country policies towards climate change. The IAEA projections on world nuclear electricity generating capacity are as follows [5]:

- The world nuclear electrical generating capacity is projected to increase to 496 GW(e) by 2030, and to 715 GW(e) by 2050 in the *high case.*
- In the *low case*, the world nuclear electrical generating capacity is projected to decrease gradually until 2040, and then bounce back to 371 GW(e) by 2050.
- The world total electrical generating capacity is expected to increase from 7188 GW(e) in 2018 to 9782 GW(e) by 2030, and to 13633 GW(e) by 2050.
- The share of nuclear electrical generating capacity in the world total electrical capacity is estimated to be roughly 3% in the *low case*, and around 5% in the *high case* by 2050.

The IAEA projects the following for global nuclear generating power:

- The total nuclear electricity production in the world is expected to continue increasing until 2050.
- In the *high case*, nuclear electricity production will increase by 50% from the 2018 level of 2563 TWh by 2030, and a further increase of 50% will occur over the following 20 years. This would mean the production in 2050 will be about 2.25 times the production level in 2018.
- In the *low case*, despite nuclear electrical generating capacity declining from the present level until 2040 and then going up again, nuclear electricity production will increase by about 11% by 2030, and about 16% by 2050. This would translate to the production in 2050 being about 1.29 times the production level in 2018 in the *low case* scenario.
- The share of nuclear electricity in total electricity production in the world will decrease in the *low case* from about 10.2% in 2018, to 8.5% in 2030, and down to 6.1% in 2050.
- In the *high case* scenario, its share will increase from 10.2% in 2018, to 11.5% in 2030, and up to 11.7% in 2050.

Of course, these are only projections, and many factors could affect these estimates. Unexpected developments (economic, political, environmental, global health related) have the potential to change the estimated values. A nuclear incident (hope not) can always take these numbers closer to the low case scenario estimates, while technological

FIGURE 5.2 World nuclear electricity consumption by region. Source: BP Statistical Review of World Energy, 2021.

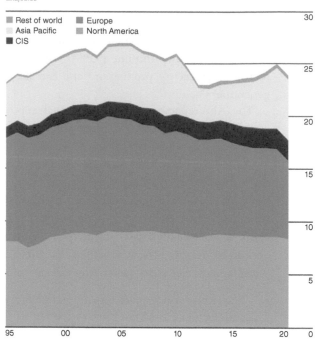

Nuclear energy consumption by region
Exajoules

FURTHER LEARNING

World Nuclear Association
https://world-nuclear.org/

FURTHER LEARNING

Nuclear Energy Fact Sheets
https://nei.org/resources/fact-sheets

FURTHER LEARNING

International Nuclear Information System (INIS)
www.iaea.org/resources/databases/inis

improvements (e.g., reactor technology, wide use of modular systems) can result in shares that are closer to or even higher than the high case estimates.

The *Power Reactor Information System* (PRIS) developed by the IAEA provides a comprehensive database on nuclear power plants all around the world. The database includes information not only on the reactors that are in operation, but also on reactors that are under construction, or ones that are being decommissioned. Information can be obtained on reactor specification data including status, location, operator, owner, suppliers, and milestone dates, as well as on technical design characteristics and performance data including energy production and energy loss data, outage, and operational event logs [6].

5.1.2 Nuclear Fission: Power from Splitting

The word *fission* means "splitting or breaking up into parts." Energy in the form of heat is released when atoms are split in a nuclear fission process during which the nucleus of an atom splits into smaller nuclei. The process takes place when an isotope is bombarded with high-speed particles such as neutrons. When the neutrons hit the nucleus of the isotope, they break the nucleus into two smaller and lighter nuclei. The resulting energy is utilized in obtaining high-temperature, high-pressure steam to be sent into a turbine to generate electricity.

5.1.3 Nuclear Fusion: Promises and Challenges

Fusion means "the process of merging two or more pieces together to form a single entity." Nuclear fusion refers to the joining of nuclei to form a heavier nucleus while yielding a significant amount of energy. Deuterium is the most abundant nuclear fuel. Atoms of deuterium and tritium collide and merge to form a helium atom and a neutron. All the stars, including the Sun, are powered by fusion. The amount of energy released

FIGURE 5.3 (a) World nuclear electricity production, and (b) share of nuclear in total electricity production in the world in 2018. Note: Nuclear electricity production in Taiwan, China was 26.7 TWh. Source: International Atomic Energy Agency.

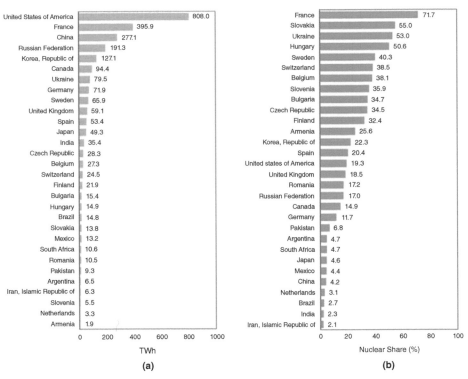

(a)

(b)

FIGURE 5.4 The word *tokamak* is an acronym for the technical term in Russian which translates to "toroidal chamber with an axial magnetic field." Inside its toroidal (donut-shaped) vacuum chamber, gaseous hydrogen fuel becomes plasma in which the fusion of hydrogen atoms takes place. Source: ITER Organization.

FURTHER LEARNING

Map: World's Nuclear Power Plants
www.carbonbrief.org/mapped-the-worlds-nuclear-power-plants

CLASSROOM DEBATE

Which form of nuclear energy production is better, fission or fusion?

during the *fusion* process is several times greater than it is during *fission*. Fusion also has the advantage in terms of byproducts. The process produces helium which does not pose any threat, unlike the highly radioactive waste of nuclear fission. Another advantage of fusion is that because it requires a high amount of energy to start, it is easy to control, or even to halt the process if necessary.

Despite these promising advantages, there are some challenges with the fusion process. It requires uniting atoms of lighter elements such as hydrogen. As the nucleus of each atom is charged positive, there is a repelling force between atoms which makes it difficult to bring them close enough together to undergo fusion. For the process to take place, extreme temperature and pressure levels are required to overcome the forces repelling the atoms. This requires a highly engineered design, which is the technological challenge. The other challenge is economic. The design also needs to be economically viable. Being able to generate electricity via fusion and to supply it to the grid is one thing; however, making the cost attractive for the producer and reasonable for the end-user is another thing.

In 1985, a collaborative international project on developing fusion energy for peaceful purposes was proposed at the Geneva Superpower Summit. In 1986, an agreement was reached by the EU, Japan, USSR, and United States to work together on a large fusion facility called the *International Thermonuclear Experimental Reactor* (ITER). In 2006, an agreement was signed by the seven parties: China, EU (Euratom), India, Japan, Russia, South Korea, and the United States to work jointly on the project. After Brexit, the United Kingdom expressed its interest in continuing to be a part of the project; this has been granted until a new agreement is reached.

The ITER will be the largest tokamak in the world (**Figure 5.4**). This giant reactor will have a weight of 23,000 tons, a 150 million °C plasma temperature, and an output power of 500 MW from fusion energy.

5.1.4 Basic Components of a Nuclear Power Plant

A nuclear power plant mainly consists of a reactor, steam turbine, condenser, and a feedwater pump. These are the components that enable the power plant to generate electricity by executing the Rankine thermodynamic cycle. Except for the reactor part of the power plant, the rest of the system is similar to a coal-fired or partially similar to a gas-fired power plant. The major difference of a nuclear power plant is the reactor which in itself is associated with critical components in regard to energy generation and safety measures.

The main components of a nuclear power plant that make up the reactor are fuel, nuclear (fuel) moderator, control rods, reactor coolant, pressure vessel, and containment.

Fuel

Uranium is the most broadly used fuel for nuclear energy generation. Naturally existing uranium has two isotopes, uranium-238 and uranium-235. More than 99% of it is U-238 and about 0.7% is the U-235 isotope, which is the one that can undergo fission as its atoms can be split apart easily. For some reactor designs, a minimum of 2% U-235 is required to accomplish the reaction. Hence this *fissile* portion is first enriched, and then used to fabricate the fuel pellets to go into the fuel rods. These pellets are stacked inside the rods like woodpeckers place acorns into trees, except that the pellets are arranged in series. Fuel rods are bundled together to form fuel assemblies. These fuel assemblies are then placed inside the reactor core. In terms of bringing pieces together, this is similar to solar cells coming together to form solar panels, and panels being put together to form solar arrays.

Nuclear Moderator

Nuclear or fuel moderators are placed in the reactor core to slow down the nuclear reaction. Collisions occur more efficiently when the neutrons produced by the nuclear reaction do not move as fast. The moderator medium can be graphite, light water (H_2O), heavy water (D_2O), or carbon dioxide (CO_2).

Control Rods

The fuel rods are bundled together with *control rods* which are made of neutron-absorbing materials such as cadmium (Cd), hafnium (Hf), or boron (B). Control rods are either inserted or pulled from the reactor core to control the rate of the nuclear reaction.

Reactor Coolant

Nuclear reactor coolant is the primary working fluid in the nuclear power plant. It absorbs the heat that is released during fission, and carries it either to an external steam generator or to the turbine, depending on the reactor design. The most common coolant is light water. Other substances used as coolants include heavy water, molten salt, and molten sodium.

Pressure Vessel

Reactor pressure vessels are usually steel vessels that contain the reactor core, moderating medium, and the coolant.

Containment

Containment is a steel and concrete structure to protect the operating personnel and the power plant from radiation and the working fluid which is at high temperature and high pressure. It is designed to contain any possible leak of radioactive steam. The personnel at the power plant working in the reactor area still wear a protective shield to reduce the amount of occupational radiation exposure.

FIGURE 5.5 Stages of uranium production and use. Source: U.S. Department of Energy.

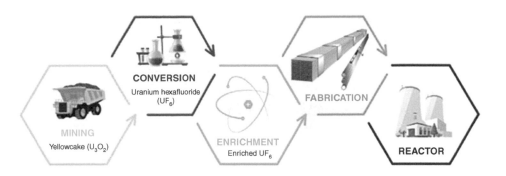

5.1.5 Reactor Types

Fission reactors come in different designs depending on parameters such as the coolant they use, moderating media, heat exchange designs and conditions, and safety measures. There are technologies that have already been in use for a long time and have been undergoing continuous improvement. These are mostly light water reactors (LWRs). There are also designs utilizing heavy water reactors (HWRs) or other coolants.

5.1.5.1 PRESSURIZED WATER REACTOR (PWR)

Pressurized water reactors are the most common type being utilized for power generation. The primary working fluid (coolant) reaches higher pressures without boiling before being sent into the steam generator. Heat exchange occurs between the primary and secondary working fluids here. The steam produced within the secondary loop goes into the turbine to generate electricity (**Figure 5.6**). Having a secondary loop keeps radioactive particles from the turbine, hence making turbine maintenance easier and safer.

5.1.5.2 BOILING WATER REACTOR (BWR)

Boiling water reactors are the second most common type used in nuclear energy production. Unlike the PWR design, BWRs have only a single fluid loop in which the

FIGURE 5.6 Schematic of a pressurized water reactor (PWR). Source: Graphic by Sarah Harman, U.S. Department of Energy.

PRESSURIZED WATER REACTOR (PWR)

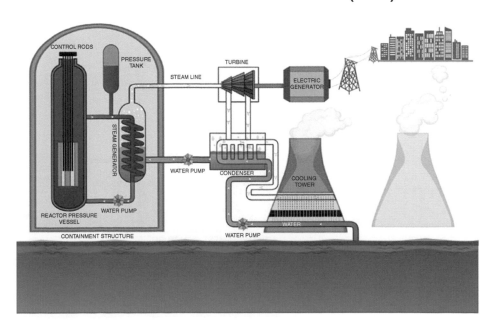

FIGURE 5.7 The pressure cooker was invented by a French physicist, Denis Papin, in 1679 and was originally known as a steam digester. It allows pressure to build up inside the pot, increasing the boiling point of water, thus cooking the food faster. Source: Abhisek Kain/EyeEm via Getty Images.

CLASSROOM DEBATE

Which reactor is more advantageous, BWR or PWR?

FURTHER LEARNING

Nuclear Reactor Simulators
www.iaea.org/topics/nuclear-power-reactors/nuclear-reactor-simulators-for-education-and-training

FIGURE 5.8 Schematic of a boiling water reactor (BWR). Source: Graphic by Sarah Harman, U.S. Department of Energy.

steam generated within the reactor pressure vessel flows directly into the turbine. It is then condensed and directed back into the pressure vessel to absorb heat from the fuel assemblies and complete the loop (**Figure 5.8**). These systems have the advantage of having a simpler plumbing design compared to PWRs. However, having a direct flow from the pressure vessel into the turbine increases the risk of radioactive contamination within the turbine should there be a leak from the fuel rods.

5.1.5.3 CANADA DEUTERIUM–URANIUM REACTOR (CANDU)

These reactors take the name from where they were first developed, and the moderator and fuel they use. CANDU reactors use pressurized heavy water, deuterium oxide (D_2O), as the moderator fluid. The fuel used is natural uranium. This can be considered an advantage of these systems as uranium enrichment is not required. However, they have the disadvantage of leakage of radioactive tritium, which is produced due to neutron absorption in deuterium.

5.1.5.4 HIGH-POWER CHANNEL REACTOR (RBMK)

The Soviet-designed RBMK reactor takes its acronym from the Russian name (*Reaktor Bolshoy Moshchnosty Kanalny*) which means high-power channel reactor. It employs a graphite moderator and light water as the fluid. Energy flow into the turbine section is similar to that in BWRs. Steam produced by the heat from the fuel rods goes directly into the turbine. The containment of these reactors is not as secure as other designs. The reactor core sits in a reinforced concrete volume that acts as a radiation barrier.

5.1.6 Advanced Reactor Technologies

There is ongoing research on advanced reactor technologies to achieve systems that are more efficient, more economic, and safer. As in any technology undergoing R&D, these

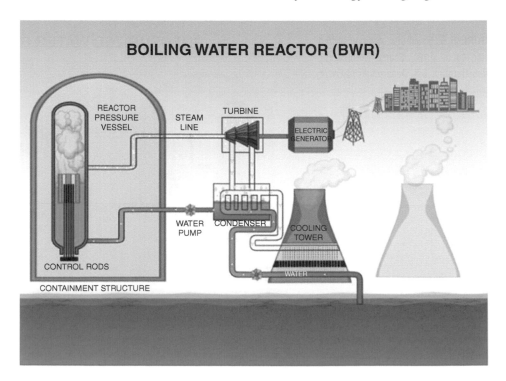

BOILING WATER REACTOR (BWR)

Table 5.1

Classification of advanced reactor designs based on coolant and neutron spectrum [7]		
Coolant	**Thermal neutron spectrum**	**Fast neutron spectrum**
Water	Small modular reactor (SMR)	—
Helium	High-temperature gas-cooled reactor (HTGR), Very High Temperature Reactor (VHTR)	Gas-cooled fast reactor (GFR)
Liquid metal	—	Sodium-cooled fast reactor (SFR), lead-cooled fast reactor (LFR)
Molten salt	Fluoride-cooled high-temperature reactor (FHR), Molten Salt Reactor - Fluoride (MSR-fluoride)	Molten salt reactor - chloride (MSR-chloride)

systems also come with technological challenges in addition to the advantages they can provide. Advanced reactor technologies can be categorized based on the coolant they utilize and their neutron spectrum. **Table 5.1** lists these technologies [7]. Neutron spectrum depends on the energy of the free neutrons. There are different groups consisting of a range of neutron energy such as cold neutron, thermal neutron, slow neutron, and fast neutron. Thermal neutron identifies a free neutron having a kinetic energy of about 0.025 eV. Fast neutrons have a much higher energy, in the range of 1–20 MeV.

5.1.6.1 SMALL MODULAR REACTOR (SMR)

Small modular reactors provide lower capacities, as small as tens of megawatts to no more than several hundred megawatts of generation capacity per unit. They can operate both with light water or with other coolants such as gas, molten salt, or liquid metal. They have several advantages, the first of which is their compact size. Other strengths include siting flexibility, lower cost, improved efficiency, and higher safety.

5.1.6.2 HIGH-TEMPERATURE GAS-COOLED REACTOR (HTGR)

The HTGRs employ a graphite moderator and helium coolant. They have lower power density, meaning the amount of energy generation per unit volume of reactor is lower than other designs. Hence, these designs are more fit to be used for purposes other than energy generation, such as transportation, industrial, and residential use. The fuel pellets they use have their own containment (e.g., coated pebbles) which makes them safer due to the added shield between the radioactive material and the surroundings. Control of gas flow in these reactors is critical as inadequate coolant flow can cause a rapid temperature increase in the reactors. There are also modular HTGR designs. Multiples of these modules can be staggered to achieve higher capacities. They come with the safety and flexibility advantages of any modular reactor design.

5.1.6.3 MOLTEN SALT REACTOR (MSR)

Molten salt reactors utilize fluid fuel in the form of high-temperature fluoride salt or chloride salt. The thermal designs use fluoride salt (MSR-fluoride), and the fast neutron types have chloride salt within the fluid fuel (MSR-chloride). Since the fuel is liquid, it can be used as the coolant as well. Molten salt reactors have the advantages of being

Accidents

Fukushima Daiichi
Location: Fukushima, Japan
Date: March 11, 2011
Capacity: 5306 MW
Reactor type: BWR
Incident: External power loss due to earthquake, backup diesel generator malfunction due to tsunami, hydrogen explosions due to overheating, fuel in three reactor cores melted

FIGURE 5.9A Fukushima Daiichi.
Source: Taro Hama/Moment, Getty Images.

Chernobyl
Location: Chernobyl, Ukraine (former USSR)
Date: April 26, 1986
Capacity: 4000 MW
Reactor type: RBMK-1000
Incident: Testing how long turbines would continue operating in case of loss of main electric power preceding routine maintenance shutdown, disabling automatic shutdown mechanism, sudden power surge resulting in explosions

FIGURE 5.9B Chernobyl power plant.
Source: Edward Neyburg/Moment, Getty Images.

able to breed new fuel and enabling use of thorium, which is an alternative to uranium. One major concern about these systems is the risk of corrosion due to the presence of salt. Corrosion not only poses a threat of leakage and contamination, but also from a heat transfer point of view it reduces the convective heat transfer coefficient within the pipes. This results in decreasing power plant efficiency.

5.1.6.4 GAS-COOLED FAST REACTOR (GFR)

A gas-cooled fast reactor is a type of gas-cooled reactor in which the fission chain reaction is sustained by fast neutrons. Hence these are fast neutron reactors that are cooled by high-pressure helium. They do not require a moderating medium. There are other types of gas-cooled reactors besides the fast neutron designs, such as high-temperature gas-cooled reactors (HTGRs), pebble bed reactors (PBRs), and modular designs of them.

5.1.6.5 SODIUM-COOLED FAST REACTOR (SFR)

Similar to the GFR, the SFR is also a fast-neutron reactor, the difference being in the coolant it employs. Instead of a gas, this reactor uses liquid sodium metal as a coolant. Sodium is heavier than hydrogen, and neutrons can move faster in it, yielding the fast reactor. The SFRs run at a high power density due to the non-appearance of a moderating medium and the greater heat absorption capacity of liquid sodium.

5.2 Theory

Theoretical analysis of nuclear energy and nuclear power plants involves six important topics which we discuss in this section:

- Nuclear reactions
- Nuclear binding energy
- Radiation
- Radioactive decay
- Nuclear electricity generation
- Cost of nuclear energy production

5.2.1 Nuclear Reactions

Nuclear changes can be classified as a variety of nuclear reactions. The two important types of nuclear reactions are *fission* and *fusion*.

Fission
Uranium-235 is the most common fuel used in fission during which the nucleus is split into two nuclei when bombarded by a neutron. The splitting of the isotope releases two neutrons and energy. The energy is then used in heating a working fluid, then generating electricity through a turbine-generator coupling. As for the neutrons released, some of them are absorbed by other atoms of U-235. The chain reaction occurring uses a thermal neutron whose energy is approximated by:

$$E = 0.5 \, kT \tag{5.1}$$

where k is the Boltzmann constant (8.617×10^{-5} eV/K) and T is the absolute temperature (K) within the core.

A typical reaction U-235 undergoes is:

$$^{235}_{92}U + ^{1}_{0}n \rightarrow ^{236}_{92}U \rightarrow ^{144}_{56}Ba + ^{89}_{36}Kr + 3^{1}_{0}n + 177 \text{ MeV} \qquad (5.2)$$

Formation of barium (Ba) and krypton (Kr) is an example fission reaction. Different byproducts can also be seen.

As mentioned previously, uranium in nature contains mostly the U-238 isotope, which makes up more than 99% of it. Unlike U-235, which only is about 0.7% of the uranium, U-238 is not fissile. That is the major reason for the necessity for U-235 enrichment, as the fissionable portion of uranium is unfortunately a very small percentage of it. The abundant part of it though, U-238, can be used to *breed* another fissile fuel, plutonium-239. The series of reactions yielding Pu-239 along with release of beta particles (beta decay) is shown in **Equations 5.3–5.5**:

$$^{238}_{92}U + ^{1}_{0}n \rightarrow ^{239}_{92}U \qquad (5.3)$$

$$^{239}_{92}U \rightarrow ^{239}_{93}Np + ^{0}_{-1}\beta \qquad (5.4)$$

$$^{239}_{93}Np \rightarrow ^{239}_{94}Pu + ^{0}_{-1}\beta \qquad (5.5)$$

Fusion

Unlike in fission, merging of nuclei is the basic principle in harvesting energy in fusion. While this technology is still in the R&D phase, it can provide significant amounts of energy once the process is achieved. An example fusion reaction to generate electricity is between two deuterium (^{2}H) atoms:

$$^{2}_{1}H + ^{2}_{1}H \xrightarrow{\text{input energy}} ^{3}_{2}He + ^{1}_{0}n + 3.27 \text{ MeV} \qquad (5.6)$$

Another example is a fusion reaction between two different isotopes of hydrogen, deuterium (^{2}H) and tritium (^{3}H):

$$^{2}_{1}H + ^{3}_{1}H \xrightarrow{\text{input energy}} ^{4}_{2}He + ^{1}_{0}n + 17.6 \text{ MeV} \qquad (5.7)$$

5.2.2 Nuclear Binding Energy

Nuclear binding energy is the minimum amount of energy required to break the nucleus into its parts. In terms of nuclear energy, both splitting (fission) and uniting (fusion) of nuclei can result in the release of nuclear binding energy, as shown in **Equations 5.2, 5.6, and 5.7**. The amount of energy released can be calculated by determining the difference in the mass of the atom and the sum of the masses of its constituent particles. This relation between mass and energy originates from the famous *theory of special relativity* of Albert Einstein:

$$E = mc^2 \qquad (5.8)$$

The changes in the masses of nuclei during the nuclear reaction are tiny. However, considering that the tiny number is multiplied by the square of the speed of light, which is approximately 300,000 km/s, the resulting energy that is released can be

Three Mile Island
Location: Middletown, PA, USA
Date: March 28, 1979
Capacity: 819 MW
Reactor type: PWR
Incident: Malfunction in the main feedwater pumps sending coolant to steam generator, turbine and rector automatic shutdown, pressure relief valve opened and got stuck, loss of coolant through the valve, overheating of the reactor

FIGURE 5.9C Three Mile Island power plant.Source: Robert J. Polett/ Design Pics, Getty Images.

significant, for instance, sufficient to heat a working fluid to generate steam for electricity production.

$$\Delta m = \left(N m_n + Z m_p\right) - A \qquad (5.9)$$

$$BE = \left[\left(N m_n + Z m_p\right) - A\right]c^2 \qquad (5.10)$$

where

BE: binding energy

N: number of neutrons

Z: number of protons

m_n: mass of a neutron in u (atomic mass unit) at rest

m_p: mass of a proton in u at rest

A: atomic mass number (or *nucleon* number)

Nuclear mass can be expressed in terms of atomic mass unit, u (also denoted amu). 1 u in terms of kilograms is:

$$1 \, u = 1.66054 \times 10^{-27} \, kg = 931.5 \, MeV/c^2 \qquad (5.11)$$

A relation between the mass and energy in terms of the speed of light can be seen in **Equation 5.11**, relating it to Einstein's special theory of relativity as given in **Equation 5.8**.

Example 5.1 Determining Binding Energy

What is the approximate binding energy of U-235 per each of its nucleons?

Solution

Binding energy for U-235 can be calculated by **Equation 5.1**:

$$BE = \left[\left(N m_n + Z m_p\right) - A\right]c^2$$

The mass of a neutron (m_n) is 1.675×10^{-27} kg, which is equivalent to 1.0087 u. Similarly, the mass of a proton (m_p) is 1.673×10^{-27} kg which is 1.0075 u. U-235 has 143 neutrons ($N = 143$) and 92 protons ($Z = 92$). Then:

$$BE = \left[(143 \times 1.0087 \, u) + (92 \times 1.0075 \, u) - 235 \, u\right]\left(3 \times 10^8 \, \frac{m}{s}\right)^2$$

$$= (1.934 \, u)\left(3 \times 10^8 \, \frac{m}{s}\right)^2 \left|\frac{1.66054 \times 10^{-27} \, kg}{1 \, u}\right|\left|\frac{1 \, J}{1 \, kg \frac{m^2}{s^2}}\right|\left|\frac{1 \, MeV}{1.60 \times 10^{-13} \, J}\right|$$

$$= 1806.46 \, MeV$$

Then, the binding energy per nucleon can be calculated by dividing the binding energy by the number of nucleons:

$$\frac{BE}{A} = \frac{1806.46}{235} \cong 7.7 \, MeV/u$$

FIGURE 5.10 Energy equivalents of one uranium fuel pellet. Source: American Nuclear Society (ANS).

Example 5.2 Energy Released in a Fission Reaction

Determine the energy released in the fission reaction given below:

$$^{235}_{92}\text{U} + ^{1}_{0}\text{n} \rightarrow ^{96}_{37}\text{Rb} + ^{138}_{55}\text{Cs} + 2\,^{1}_{0}\text{n}$$

Solution

First, we write the masses of both sides of the nuclear reaction (reactants and the products) to see the change in mass which is released as energy:

Reactants

$$\left.\begin{array}{l} \text{Mass of } ^{235}_{92}\text{U} = 235.043929 \text{ u} \\ \text{Mass of } ^{1}_{0}\text{n} = 1.008665 \text{ u} \end{array}\right\} \quad \text{Total } m_{\text{reactants}} = 236.052594 \text{ u}$$

Products

$$\left.\begin{array}{l} \text{Mass of } ^{96}_{37}\text{Rb} = 95.93427 \text{ u} \\ \text{Mass of } ^{138}_{55}\text{Cs} = 137.911017 \text{ u} \\ \text{Mass of } 2\,^{1}_{0}\text{n} = 2.01733 \text{ u} \end{array}\right\} \quad \text{Total } m_{\text{products}} = 235.862617 \text{ u}$$

The difference in mass is $\Delta m = 236.052594 - 235.862617 = 0.189977$ u

Then, the equivalent energy would be:

$$\Delta E = \Delta mc^2 = (0.189977 \text{ u})(c^2)\left|\dfrac{931.5\,\dfrac{\text{MeV}}{c^2}}{1 \text{ u}}\right| \cong 177 \text{ MeV} \equiv 2.836 \times 10^{-11} \text{ J}$$

5.2.3 Radiation

Radiation can be measured and assessed in different ways and for different purposes. One might be looking at the amount of radioactivity in drinking water, in outside air, in the soil in an agricultural area, or in the fish caught close to a nuclear power plant. If a person has been exposed to a radioactive source, the dose received by the individual would be what is measured. As there can be different purposes and domains, knowing the terminology and related units of measurement is important.

Radioactivity is associated with the amount of ionizing radiation released by a substance. Ionizing radiation is a means of attenuation of energy released by atoms that

Table 5.2

Radiation unit conversions			
Quantity	**SI unit**	**US unit**	**Conversion factor**
Radioactivity	becquerel (Bq)	curie (Ci)	$1 \text{ Bq} = 2.7 \times 10^{-11} \text{ Ci}$
Exposure	coulomb per kg (C/kg)	roentgen (R)	$1 \text{ C/kg} = 3876 \text{ R}$
Absorbed dose	gray (Gy)	rad	$1 \text{ Gy} = 100 \text{ rad}$
Effective dose	sievert (Sv)	rem	$1 \text{ Sv} = 100 \text{ rem}$

FURTHER LEARNING

Radiation Dose Calculator
www.ans.org/nuclear/dosechart/

travels in the form of electromagnetic waves (gamma rays or X-rays), or particles (alpha or beta particles, or neutrons). The radioactivity of a substance identifies how many atoms in it decay over a time period. The units of radioactivity are becquerel (Bq, SI) or curie (Ci, US) (**Table 5.2**).

Exposure defines the airborne radiation level. It has the units of coulomb per kilogram (C/kg, SI) or roentgen (R, US).

Absorbed dose is the amount of radiation absorbed by a substance or a person. It has the units of gray (Gy, SI) or rad (US).

Effective dose (or *dose equivalent*) is a combined measure of the amount of radiation absorbed by an individual and the associated health impacts of the radiation on the individual's organs. Units for effective dose are sievert (Sv, SI) or rem (US).

5.2.4 Radioactive Decay

Unstable nuclei decay into different forms in time to become more stable. This phenomenon is called *nuclear decay* (or *radioactive decay*) and the material that contains unstable nuclei is defined to be radioactive. Nuclear decay can be classified into four types based on the radiation produced.

Alpha decay is the emission of an alpha particle, which is a helium nucleus. An example of alpha decay is:

$$^{218}_{84}\text{Po} \rightarrow \ ^{214}_{82}\text{Pb} + \ ^{4}_{2}\text{He} \tag{5.12}$$

Beta decay is the emission of an electron (e^-) from the nucleus. An example of beta decay can be:

$$^{235}_{92}\text{U} \rightarrow \ ^{235}_{93}\text{Np} + \ ^{0}_{-1}e^- \tag{5.13}$$

Positron emission (or β^+ decay) is the emission of a positron (e^+) from the nucleus. A positron emission example is:

$$^{37}_{19}\text{K} \rightarrow \ ^{37}_{18}\text{Ar} + \ ^{0}_{+1}e^+ \tag{5.14}$$

Gamma decay is the emission of a gamma ray, where a high-energy photon is released from the nucleus. In gamma decay, there is no change in mass number or atomic number, unless another form of decay is occurring along with the gamma emission. An example of gamma decay is:

$$^{234}_{91}\text{Pa} \rightarrow \ ^{234}_{91}\text{Pa} + \ ^{0}_{0}\gamma \tag{5.15}$$

The left-hand side term that is decaying is the *parent* $\left(^{234}_{91}\text{Pa}\right)$, and the right-hand side terms are the *daughter* $\left(^{234}_{91}\text{Pa}\right)$ and the *gamma ray* $\left(^{0}_{0}\gamma\right)$.

If the number of nuclei of an isotope at time t is $N(t)$, then the change in the number over a time period of Δt is given by:

$$\Delta N = -\lambda N(t)\,\Delta t \tag{5.16}$$

where λ is the *decay constant*. The negative sign on the right-hand side indicates a decrease in $N(t)$ over the given period, Δt. Let the initial number of nuclei at time t be N_0. Then, with the known rate of change of the number of nuclei, **Equation 5.16** can be rearranged and manipulated as below:

$$\int_{N_0}^{N} \frac{dN}{N(t)} = -\int_{t}^{t+\Delta t} \lambda\,dt \tag{5.17}$$

Integrating **Equation 5.17** yields:

$$N(t) = N_0\,e^{-\lambda t} \tag{5.18}$$

The time it takes for the number of nuclei to reduce to half of its initial value is known as half-life, $t_{1/2}$, which can be determined from **Equation 5.18** by simply equating $N(t_{1/2})$ to $N_0/2$. Solving for the half-life gives:

$$t_{1/2} = \frac{\ln(2)}{\lambda} \tag{5.19}$$

The exponential characteristic of nuclear decay for an isotope can be seen in **Figure 5.11**. Half-lives and decay types of some isotopes are listed in **Table 5.3**.

FIGURE 5.11 Exponential nuclear decay of an isotope. Source: National Oceanic and Atmospheric Administration (NOAA) Ocean Explorer.

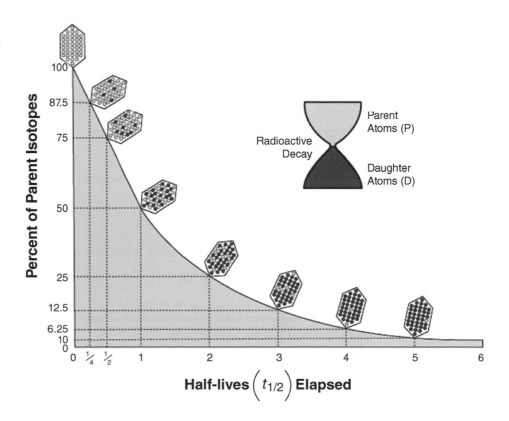

Table 5.3

Half-lives and decay types of some isotopes			
Parent isotope	**Daughter isotope**	**Half-life**	**Decay type**
Rubidium-87	Strontium-87	48.8 billion years	β^-
Beryllium-10	Boron-10	1.39 million years	β^-
Radon-222	Polonium-218	3.8 days	α
Carbon-10	Boron-10	19.3 seconds	β^+
Uranium-215	Thorium-211	2.24 milliseconds	α

Example 5.3 Calculating Nuclear Decay Time

Iodine-131 has a decay half-life of about 8 days. It experiences a beta decay into xenon-131. In how many days, would 80% of iodine-131 decay into xenon-131?

Solution

When 80% of iodine-131 decays, the ratio of current amount to initial amount is:

$$\frac{N(t)}{N_0} = \frac{100 - 80}{100} = 0.20$$

Then, using **Equation 5.18** we get:

$$\frac{N(t)}{N_0} = e^{-\lambda t} = 0.20$$

and rearranging with t on the left-hand side gives:

$$t = \frac{-\ln(0.20)}{\lambda}$$

The decay constant, λ, can be determined from **Equation 5.19** using the half-life, $t_{1/2}$:

$$\lambda = \frac{\ln(2)}{t_{1/2}} = \frac{0.693}{8 \text{ d}} = 0.0866 \text{ d}^{-1}$$

Then, plugging λ into the expression we obtained using **Equation 5.18** yields:

$$t = \frac{-\ln(0.20)}{\lambda} = \frac{-(-1.609)}{0.0866 \text{ d}^{-1}} = 18.58 \text{ days}$$

5.2.5 Nuclear Electricity Generation

Nuclear power plants do not differ from coal-fired or natural gas-fired power plants in terms of the execution of the thermodynamic cycle, whether they operate based on the Rankine cycle (steam turbine) or the Brayton cycle (gas turbine). The only difference is that

FIGURE 5.12 Ideal Rankine vapor cycle with superheat.

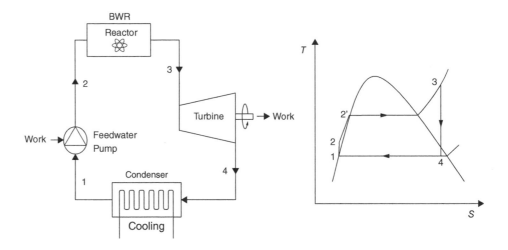

the boiler in a coal power plant, or the combustion chamber in a gas power plant is replaced with a reactor, where the reactor core acts as the heat source to generate high-temperature and high-pressure steam or gas, depending on the coolant. For instance, a light water reactor such as a BWR would be part of the Rankine cycle, while a gas-cooled reactor such as an HTGR can be the heat source component for a Brayton cycle.

A schematic and the T–S diagram of a nuclear power plant utilizing a BWR and executing an ideal Rankine cycle are illustrated in **Figure 5.12**. Thermal efficiency, work, and heat amounts for the components of the cycle with a coolant mass flow rate of $\dot{m}$ are given by:

$$\eta_{th} = \frac{\dot{W}_{net}}{\dot{Q}_{in}} = \frac{\dot{W}_{turb} - \dot{W}_{pump}}{\dot{Q}_{reactor}} \qquad (5.20)$$

$$\dot{W}_{turb} = \dot{m}\left(h_3 - h_4\right) \qquad (5.21)$$

$$\dot{W}_{pump} = v_f\left(p_2 - p_1\right) \qquad (5.22)$$

$$\dot{Q}_{reactor,BWR} = \dot{m}\left(h_3 - h_2\right) \qquad (5.23)$$

A schematic and the T–S diagram of a nuclear power plant utilizing a GCR and executing an ideal Brayton cycle are illustrated in **Figure 5.13**. Thermal efficiency of the cycle assuming cold-air standard analysis can be calculated by:

FIGURE 5.13 Ideal Brayton cycle.

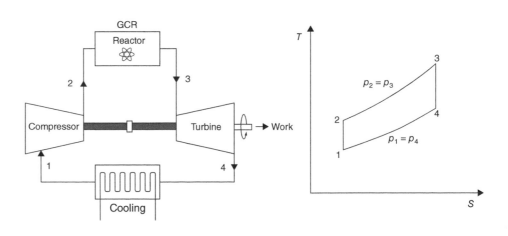

$$\eta_{th} = 1 - \frac{T_1}{T_2} = 1 - \left(\frac{p_1}{p_2}\right)^{\frac{k-1}{k}} \qquad (5.24)$$

$$\dot{W}_{turb} = \dot{m}\left(h_3 - h_4\right) \qquad (5.25)$$

$$\dot{W}_{comp} = \dot{m}\left(h_2 - h_1\right) \qquad (5.26)$$

$$\dot{Q}_{reactor,GCR} = \dot{m}\left(h_3 - h_2\right) \qquad (5.27)$$

Example 5.4 Thermal Analysis of a Nuclear Power Plant Employing a BWR

A nuclear power plant situated by a lake uses lake water as the heat sink for the condenser. The BWR within the containment building produces steam at 8 MPa and 480 °C with a mass flow rate of 6.5 kg/s. The steam exits the turbine at 8 kPa and condenses to saturated liquid within the condenser. The isentropic efficiencies of the turbine and the pump are 85% and 80%, respectively. Cooling water from the lake enters the condenser at 16 °C and leaves at 30 °C. Obtain:

a. the net power output of the power plant (kW)
b. the rate of heat rejection from the condenser (kW)
c. the mass flow rate of lake water running through the condenser (kg/s)
d. the thermal efficiency of the power plant (%)

Solution

First, all h values need to be determined:

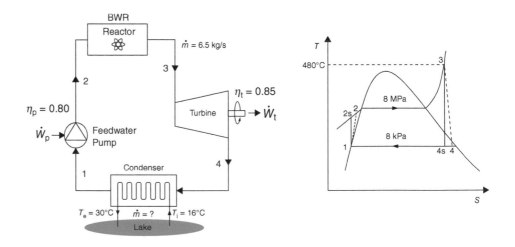

State 1

Saturated liquid ($x = 0$), $p_1 = 8$ kPa, then $h_1 = h_{f@8kPa} = 173.88$ kJ/kg **(Table B.2)**

State 2

From **Equation 3.21**:

$$h_{2s} = h_1 + \left[v_f(p_2 - p_1)\right]$$
$$= 173.88\,\frac{kJ}{kg} + \left[(1.008 \times 10^{-3})\,\frac{m^3}{kg}\,(80 - 0.08)\,bar\right]\left|\frac{10^5\,Nm^2}{1\,bar}\right|\left|\frac{1\,kJ}{10^3\,N\,m}\right| = 181.94\,\frac{kJ}{kg}$$

Then, using the isentropic pump efficiency information provided:

$$\eta_p = \frac{h_{2s} - h_1}{h_2 - h_1} = 0.80$$

Solving for actual pump exit enthalpy gives $h_2 = 183.96$ kJ/kg

State 3

$T_3 = 480\,°C$ and $p_3 = 80$ bar, then $h_3 = 3348.4$ kJ/kg (**Table B.3**)
$s_3 = 6.6586$ kJ/kg K (to be used for finding h_4 in the next step)

State 4

$s_{4s} = s_3 = 6.6586$ kJ/kg K, then quality of steam at the turbine exit will be:

$$x_{4s} = \frac{s_{4s} - s_f}{s_g - s_f} = \frac{6.6586 - 0.5926}{8.2287 - 0.5926} = 0.794\ \ (\text{or } 79.4\%)$$

$$h_{4s} = h_f + x_{4s}\left(h_g - h_f\right) = 173.88 + 0.794(2577 - 173.88) = 2081.96\,\frac{kJ}{kg}$$

Then, using the isentropic turbine efficiency value:

$$\eta_t = \frac{h_3 - h_4}{h_3 - h_{4s}} = 0.85$$

Solving for actual turbine exit enthalpy gives $h_4 = 2271.9$ kJ/kg
Now that we have found all h values, we can solve for each part of the problem.

a. Net power output of the plant is:

$$\dot{W}_{net} = \dot{W}_{turb} - \dot{W}_{pump} = \dot{m}[(h_3 - h_4) - (h_2 - h_1)]$$
$$= 6.5\,\frac{kg}{s}[(3348.4 - 2271.9) - (183.96 - 173.88)]\,\frac{kJ}{kg}\left|\frac{1\,kW}{1\,kJ/s}\right|$$
$$= 6931.7\,kW$$

b. The rate of heat rejection from the condenser is:

$$\dot{Q}_{out} = \dot{m}(h_4 - h_1) = 6.5\,\frac{kg}{s}(2271.9 - 173.88)\,\frac{kJ}{kg}\left|\frac{1\,kW}{1\,kJ/s}\right| = 13{,}637.1\,kW$$

c. The mass flow rate of the lake water absorbing the heat from the steam in the condenser can be calculated by:

$$\dot{Q}_{out} = \dot{Q}_{lake\ absorbed}$$

as heat given off by the steam will all be absorbed by the cooling water from the lake. Then:

$$\dot{m}_w c_{p,w}\,(T_e - T_i) = 13{,}637.1\,kW$$
$$\dot{m}_w = \frac{13{,}637.1\,kW}{c_{p,w}(T_e - T_i)} = \frac{13{,}637.1\,kW}{\left(4.18\,\frac{kJ}{kg\,°C}\right)(30 - 16)\,°C}\left|\frac{1\,kJ/s}{1\,kW}\right| = 233\,\frac{kg}{s}$$

d. Finally, thermal efficiency of the plant can be determined by:

$$\eta_{th} = \frac{\dot{W}_{net}}{\dot{Q}_{reactor}} = \frac{\dot{W}_{net}}{\dot{m}(h_3 - h_2)} = \frac{6931.7 \text{ kW}}{6.5 \, \frac{\text{kg}}{\text{s}} \, (3348.4 - 183.96) \frac{\text{kJ}}{\text{kg}} \left| \frac{1 \text{ kW}}{1 \text{ kJ/s}} \right|}$$

$$= \frac{6931.7 \text{ kW}}{20{,}568.9 \text{ kW}} = 0.337$$

Hence the thermal efficiency of the power plant is calculated as 33.7%.

5.2.6 Cost of Nuclear Energy Production

Calculation of the cost of energy production at a nuclear power plant involves the capital, fuel, waste management, and operating and maintenance costs over the lifetime of the plant. This is the levelized cost of energy (LCOE) of the plant. Calculation of LCOE is discussed in Chapter 4 and the calculation of it can be performed using **Equation 4.9**. A more detailed approach can also be used to determine the LCOE for a nuclear power plant [7]:

$$\text{LCOE} = \frac{1}{8766 \, L} \left[\Phi \frac{I}{K} + \frac{O}{K} \right] + \text{FCC} \tag{5.28}$$

where

    LCOE: levelized cost of electricity [$/kWh]
    L: annual capacity factor (actual kWh/rated kWh)
    Φ: levelized fixed charge rate, accounting for both taxes and depreciation [yr^{-1}]
    I: capital cost of the power plant [$]
    K: power plant size [kW]
    O: annual operating and maintenance costs [$/yr]
    FCC: fuel cycle cost [$/kW-days-thermal]

Levelized fixed charge rate, Φ, can be calculated using:

$$\Phi = \frac{(A/P, \chi, N)}{(1 - \tau)} - \frac{\tau}{(1 - \tau)} \left(\frac{1}{N} \right) \tag{5.29}$$

where $(A/P, \chi, N)$ is the capital recovery factor, which is given by:

$$A/P, \chi, N = \frac{\left[\chi (1 + \chi)^N \right]}{\left[(1 + \chi)^N - 1 \right]} \tag{5.30}$$

and

$$\chi = f_s r_s + f_b f_b (1 - \tau) \tag{5.31}$$

In **Equations 5.30 and 5.31**,

    χ: discount rate
    N: estimated plant economic life [yr]
    f_s: fraction debt
    f_b: fraction equity
    r_s: rate on equity (stock) [%]
    r_b: rate on debts (bonds) [%]
    τ: composite tax rate
    1/N: straight-line depreciation fraction

FURTHER LEARNING

Levelized Cost of Electricity
Calculator
https://cnpce.ne.anl.gov/cgi-bin/
qnecost?select=home

The very last term in **Equation 5.28** is the fuel cycle cost, which accounts for the enrichment, conversion, fabrication, and disposal costs of the nuclear fuel. The FCC can be determined by:

$$\text{FCC} = \frac{C_f}{24\,\eta B} \tag{5.32}$$

where:

C_f: total fuel cycle cost per kg of fuel [\$/kg]

B: fuel burnup at discharge [kW-days-thermal/kg fuel]

η: power plant thermal efficiency [kW$_e$/kW$_{th}$]

Normalized LCOE values of different energy sources in various countries are compared in **Table 5.4** [7]. The costs for natural gas and coal are presented for cases of both excluding and including the carbon costs.

Besides **Equations 4.9** and **5.28,** which give the LCOE for a power plant, there is another cost analysis that excludes the capital cost and helps determine the cost of energy per kWh production of the nuclear power plant. This cost can be calculated by:

$$\text{Total cost per kWh} \left(\frac{\$}{\text{kWh}}\right) = \frac{\text{Total annual cost} \left(\dfrac{\$}{\text{yr}}\right)}{\text{Total annual output} \left(\dfrac{\text{kWh}}{\text{yr}}\right)} \tag{5.33}$$

The numerator of **Equation 5.33** is the total annual cost of the fuel and operating and maintenance costs. The denominator needs to account for fuel spent, energy released, and thermal efficiency of the power plant:

$$\text{Total annual output} \left(\frac{\text{kWh}}{\text{yr}}\right) = \eta_{\text{plant}} m_{\text{fuel}} e_{\text{fuel}} \tag{5.34}$$

where m_{fuel} (g/s) is the amount of fuel consumption, and e_{fuel} (GJ/g) is the amount of energy released per one gram of fuel fissioned.

The efficiency of the power plant is:

$$\eta_{\text{plant}} = \left[\frac{(\Delta h)_{\text{turbine}}}{(\Delta h)_{\text{reactor}}}\right] \eta_{\text{turbine}} \eta_{\text{pump}} \eta_{\text{generator}} \tag{5.35}$$

Table 5.4

Normalized LCOE values of natural gas and coal with reference to nuclear energy for selected countries [7]					
Country	Natural gas		Coal		Nuclear
	LCOE	LCOE + CO$_2$ cost[*]	LCOE	LCOE + CO$_2$ cost	LWR
United States	0.67	0.85	0.88	1.21	1.0
South Korea	1.54–2.69	1.78–2.93	1.40	1.99	1.0
Japan	0.92–1.46	1.05–1.58	0.94	1.23	1.0
China	0.74–1.72	0.97–1.95	1.03	1.63	1.0
France	0.58–1.05	0.71–1.18	—	—	1.0

[*] Carbon cost is assumed as \$30/tonne of CO$_2$

In **Equation 5.35**, η_{turbine} and η_{pump} are the isentropic efficiencies of the turbine and pump, respectively, and $\eta_{\text{generator}}$ is the generator efficiency.

Substituting **Equations 5.34** and **5.35** into **Equation 5.33** gives the total cost of electricity production per year for the power plant:

$$\text{Total cost per kWh}\left(\frac{\$}{\text{kWh}}\right) = \frac{[\text{Fuel cost} + \text{O\&M cost}]\left(\frac{\$}{\text{yr}}\right)}{\left[\frac{(\Delta h)_{\text{turbine}}}{(\Delta h)_{\text{reactor}}}\right]\eta_{\text{turbine}}\eta_{\text{pump}}\eta_{\text{generator}} m_{\text{fuel}} e_{\text{fuel}}\left(\frac{\text{kWh}}{\text{yr}}\right)}$$

(5.36)

Example 5.5 Electricity Production Cost

Consider the nuclear power plant in **Example 5.4**. The plant operates with 144 metric tons of uranium, which contains 3.4% U-235. Every year, a quarter of the uranium in the reactor is renewed, which costs \$30 million. In addition to the fuel cost, the O&M cost of the power plant is \$153 million annually. Calculate the cost per kWh of electricity production for this nuclear power plant.

Solution

Implementing **Equation 5.34** into **5.33**, we get:

$$\text{Total cost per kWh}\left(\frac{\$}{\text{kWh}}\right) = \frac{\text{Total annual cost}\left(\frac{\$}{\text{yr}}\right)}{\text{Total annual output}\left(\frac{\text{kWh}}{\text{yr}}\right)}$$

$$= \frac{[\text{Fuel cost} + \text{O\&M cost}]\left(\frac{\$}{\text{yr}}\right)}{\left[\eta_{\text{plant}} m_{\text{fuel}} e_{\text{fuel}}\right]\left(\frac{\text{kWh}}{\text{yr}}\right)}$$

Thermal efficiency of the plant was calculated as 33.7% in **Example 5.4**. The annual fuel consumption is:

$$m_{\text{fuel}} = (0.25\ \text{yr}^{-1})(144\ \text{ton})(0.034)\left|\frac{10^6\ \text{g}}{1\ \text{ton}}\right|\left|\frac{1\ \text{yr}}{3.154 \times 10^7\ \text{s}}\right| = 0.039\ \frac{\text{g}}{\text{s}}$$

Assuming approximately 82 GJ/g of energy release from U-235, the total cost equation becomes:

$$\text{Cost per kWh} = \frac{(30 + 153) \times 10^6\ \dfrac{\$}{\text{yr}}}{\left[(0.337)\left(0.039\ \dfrac{\text{g}}{\text{s}}\right)\left(82\ \dfrac{\text{GJ}}{\text{g}}\right)\right]\left|\dfrac{3.154 \times 10^7\ \text{s}}{1\ \text{yr}}\right|\left|\dfrac{10^6\ \text{kJ}}{1\ \text{GJ}}\right|\left|\dfrac{1\ \text{kWh}}{3600\ \text{kJ}}\right|}$$

$$= 0.0194\ \frac{\$}{\text{kWh}}$$

Hence, the total cost per unit energy produced is \$0.0194/kWh (or 1.94 cents per kWh).

5.3 Applications and Case Study

In this section, we will observe real-world examples via two applications and a case study. The applications include a nuclear power plant in Ukraine utilizing a pressurized water reactor (PWR), and another plant in Switzerland having a boiling water reactor (BWR). The case study is about nuclear power developments in the United Arab Emirates.

Application 1 Pressurized Water Reactor (Ukraine)

Zaporizhzhia Nuclear Power Plant
Location: Enerhodar, Ukraine
Owner: Energoatom (State-run)
Reactor Type: PWR (VVER)
Capacity: 6 × 950 MW
Annual production:
   38,430 GWh (2019)
Data: www.npp.zp.ua [8]
www.d-maps.com

An application example for a nuclear power plant utilizing pressurized water reactors is the *Zaporizhzhia Nuclear Power Plant* in Ukraine. With an installed capacity of 5700 MW, Zaphorizhzhia is the largest NPP in Europe and one of the top 10 plants worldwide. The plant uses water–energetic reactors (WWER, or VVER in Russian), which are a type of PWR designed during the Soviet Union era. The plant is owned by the national nuclear energy generating company of Ukraine, Energoatom. It is situated by the Kakhovka water reservoir, which acts as the heat sink for the Rankine cycle executed.

Application 2 Boiling Water Reactor (Switzerland)

Leibstadt Nuclear Power Plant
Location: Leibstadt, Switzerland
Owner: Leibstadt AG (consortium)
Reactor Type: BWR
Capacity: 1 × 1220 MW
Annual production: 9600 GWh
Data: www.alpicq.com [9]
www.d-maps.com

Leibstadt NPP provides about one-sixth of the total electricity demand in Switzerland. The plant utilizes a boiling water reactor and has a capacity of 1220 MW. The cooling water for the condenser system is cooled within a closed loop circulating through a 144 m tall cooling tower, instead of having an open loop using water from the Rhine River (**Figure 5.14**). Considering the amount of heat dissipation, power plants have the potential to increase the temperature of the heat sink they are constructed by, if the heat sink is not large enough. Temperature increases of the water in rivers or lakes can be a threat to the habitat. Therefore, use of a closed-loop system instrumenting a cooling tower is a considerate step from an environmental point of view.

FIGURE 5.14 Leibstadt Nuclear Power Plant. Source: Tambako the Jaguar via Getty Images.

Case Study IAEA Milestone Approach for Nuclear Infrastructure (United Arab Emirates)

United Arab Emirates
Population: 9,992,083 (2020 est.)
GDP: $696 billion (2017 est.)
GDP per capita: $68,600 (2017 est.)
CO_2 emission (energy) world rank: 24th
Data: CIA Factbook [10]
www.d-maps.com

INTRODUCTION Many countries in the world contemplate adding nuclear power to their energy generation mix to diversify their sources. The International Atomic Energy Agency (IAEA) has developed a "Milestone Approach," which is a nuclear energy program to help such countries build their infrastructure.

OBJECTIVE The objective of this case study is to learn about the implementation of a nuclear energy program in a country.

METHOD This case study was conducted through researching articles and reports.

RESULTS Based on the recommendation of the IAEA, the United Arab Emirates (UAE) published its Nuclear Energy Policy in 2008. In 2009, the Emirates Nuclear Energy Corporation (ENEC) was built to implement the nuclear energy program of the UAE. In the same year, an agreement was made on the construction of four APR-1400 units with a consortium led by the Korean Electric Power Company (KEPCO). An APR-1400 is a type of advanced PWR which is a Generation III reactor. Total capacity of the Barakah Nuclear Energy Power Plant, which is close to the Saudi Arabia border, is 5600 MW. After being granted the Construction License from the Federal Authority for Nuclear Regulation (FANR) in 2012, construction of the first unit at the Barakah Nuclear Energy Power Plant started. By 2015, all four units were under construction. In February 2020, Unit 1 received the Operating

License. The NPP is owned by ENEC and is operated as a joint venture between ENEC and KEPCO. The plant uses uranium as the fuel. As per the non-proliferation agreement, the UAE will not enrich the fuel. Uranium enrichment is performed in South Korea. Once all four units are operational, the nuclear power plant is estimated to provide 25% of the total electricity demand in the UAE. Up until the nuclear energy generation, the energy supply of the UAE was mainly from natural gas (~88%), oil (~8%), coal (~3%), and the remainder from other energy sources such as wind and solar [11].

DISCUSSION Nuclear energy comes with the benefits of reduced reliance on imported fossil fuels, having low greenhouse gas emissions, generating more jobs, and supporting national security. Implementation of a nuclear energy program for power generation and peaceful use of nuclear technologies is essential for developing countries to ensure a safer and cleaner growth in terms of industry, transportation, and economy.

RECOMMENDATIONS Some of the challenges for nuclear energy are construction and operating costs, public perception, and nuclear waste management. These various challenges can be addressed by different methods such as improved technologies with lower capital costs, education and promoting public awareness, and safer waste transportation, storage, and disposal solutions.

5.4 Economics

The total cost of any type of power plant, including nuclear energy plants, is made up of three main line items:

Capital cost: This is the total cost of construction of the entire power plant. It has two components, the overnight cost and the interest cost. Overnight cost is a term used in the power generation sector and is associated with the cost of the land, construction, equipment, and labor. This would have been the total capital cost in a world where the interest rate was zero. That being said, the second component of the capital cost is the cost of interest for the loan.

Operating and maintenance cost: Operating costs include utilities, wages and salaries, social security, unemployment insurance, workers' compensation, medical insurance, and taxes. Maintenance costs include routine maintenance, troubleshooting, and repair costs.

Fuel cost: This includes enrichment, conversion, fabrication, and disposal costs of the fuel utilized to generate electricity.

Depending on the country and location, reactor technology, other major equipment such as turbine and generator, cooling system, staff on site, taxes, and other line items, the share of costs within the total cost can vary in different NPP applications. As an example, a breakdown of costs for plants having five different light water reactors (LWRs) is presented in **Table 5.5**, retrieved from an MIT report titled *The Future of Nuclear Energy in a Carbon-Constrained World* [7].

In **Table 5.5**, the nuclear island equipment consists of the reactor core, reactor coolant pumps, pressure vessel, and steam generators (for PWRs). The turbine-generator equipment line consists of the conventional island components of the plant. The yard, cooling, and installation item includes all costs for the site, heat sink for the power plant, and installation of equipment. The last two items on the table consist of indirect and direct engineering costs, supervisory costs, fees, permits, taxes, and costs for spare parts and

Table 5.5

Share of various costs for NPPs utilizing different light water reactors [7]					
	Cost breakdown (% of total cost)				
Budget item	**Generic AP1000**	**Historic US LWR (median case)**	**Historic US LWR (best case)**	**South Korean APR1400**	**EPR**
Nuclear island equipment	12.6	9.9	16.5	21.9	18.0
Turbine-generator equipment	4.9	7.0	11.9	5.6	6.3
Yard, cooling, and installation	47.5	46.3	49.3	45.5	49.7
Engineering, procurement, and construction cost	15.9	17.6	7.7	20.0	15.3
Owner's cost	19.1	19.2	14.6	7.0	10.7

commissioning. As can be seen in the table, almost half of the total cost of NPPs comes from the site, cooling system, and installation. The cost of a nuclear island varies between 10% and 22%, depending on the reactor technology and accompanying components.

Example 5.6 Retrieving and Plotting Nuclear Generation Data Using Python

Historical electricity net generation data from various energy sources can be retrieved from the EIA website. Plot the following for US electricity generation from nuclear obtaining the data between 1973 and 2021 from EIA[1]:

a. Monthly electricity generation from nuclear (GWh/month)
b. Comparison of monthly electricity generation from nuclear, coal, and petroleum (GWh/month)

Solution

The data can be retrieved from: www.eia.gov/totalenergy/data/browser/xls.php?tbl=T07.02A&freq=m

```
import numpy as np
import matplotlib.pyplot as plt
import pandas as pd
```

First, assign the address for the spreadsheet to pull the data from:

```
gen=pd.read_excel("https://www.eia.gov/totalenergy/data/browser/
xls.php?tbl=T07.02A&freq=m",sheet_name="Monthly Data",
skiprows=10)

gen = gen.drop(labels=0, axis=0)
```

Then, define the commands to retrieve the first two data cells for generation from each source listed in the spreadsheet:

```
gen.head(2)
```

[1] U.S. Energy Information Administration, Monthly Energy Review September 2021.

Month		Electricity net generation from coal, all sectors	Electricity net generation from petroleum, all sectors	Electricity net generation from natural gas, all sectors	Electricity net generation from other gases, all sectors	Electricity net generation from nuclear electric power, all sectors	Electricity net generation from hydroelectric pumped storage, all sectors	Electricity net generation from conventional hydroelectric power, all sectors	Electricity net generation from wood, all sectors	Electricity net generation from waste, all sectors	Electricity net generation from geothermal, all sectors	Electricity net generation from solar, all sectors	Electricity net generation from wind, all sectors	Electricity net generation total (including from sources not shown), all sectors
1	1973-01-01	75,190.1	31,183.6	21,185	N/A	N/A	6246.25	N/A	26,249.2	5.157	15.152	143.49	N/A	160,218
2	1973-02-01	67,797.9	26,087	20,252.5	N/A	N/A	5928.07	N/A	23,313	15.138	13.874	131.171	N/A	143,539

a. Monthly electricity generation from nuclear (GWh/month)

```
plt.title("Monthly U.S. Nuclear Generation")
plt.ylabel("GWh/month")
plt.plot(gen.Month,gen.iloc[:,5])
```

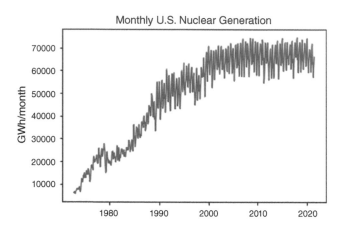

b. Comparison of monthly electricity generation from nuclear, coal, and petroleum (GWh/month)

As "month" is column 0, then each energy source by column number can be expressed as:

1: Coal, 2: Petroleum, 3: Natural gas, 4: Other gases, and 5: Nuclear

Then, we can plot the data for the corresponding columns:

```
plt.ylabel("GWh/month")
plt.plot(gen.Month,gen.iloc[:,1], label="Coal")
plt.plot(gen.Month,gen.iloc[:,2], label="Petroleum")
plt.plot(gen.Month,gen.iloc[:,5], label="Nuclear")
plt.legend()
plt.title ("U.S. Nuclear Generation Comparison"
```

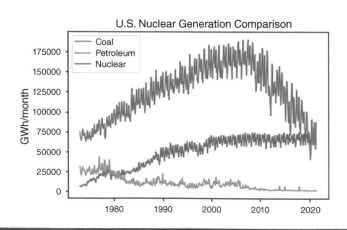

5.5 Summary

Electricity generation through nuclear energy has the advantages of zero CO_2 emissions and low cost of energy production. As for the concerns, besides what the accidents in history have taught us, there is still a necessity to find permanent solutions for nuclear waste. The risk of proliferation of nuclear weapons and materials is another major global concern.

From an engineering point of view, there is a significant amount of research on enhancement of current reactors, and development of advanced reactor technologies. Despite the ability to generate electricity at low cost per kWh, capital costs are still high. Development of safer, more efficient, and more economic systems is an on-going challenge in the nuclear industry. Besides the developments in fission technologies, there has been progress towards nuclear fusion since the initial steps taken at an international level in Geneva in 1985. Collaboration at an international level not only enhances the research capacity and abilities, but also strengthens the motivation for a globally peaceful nuclear industry.

REFERENCES

[1]   U.S. Department of Energy, Office of Nuclear Energy, Science and Technology, "The History of Nuclear Energy," 2006. Accessed: Jul. 14, 2020 [Online]. Available: www.energy.gov/sites/prod/files/The%20History%20of%20Nuclear%20Energy_0.pdf.

[2]   International Atomic Energy Agency, *Fifty Years of Nuclear Power: The Next Fifty Years,"* Vienna: IAEA, 2004.

[3]   BP, "Statistical Review of World Energy," 2021. Accessed: Oct. 13, 2021 [Online]. Available: www.bp.com/statisticalreview.

[4]   International Energy Agency, "Global Energy Review 2020," IEA, Paris, 2020. Accessed: Jul. 14, 2020 [Online]. Available: www.iea.org/reports/global-energy-review-2020.

[5]   International Atomic Energy Agency, *Energy, Electricity and Nuclear Power Estimates for the Period up to 2050*, Reference Data Series No. 1, Vienna: IAEA, 2019.

[6]   International Atomic Energy Agency, "Power Reactor Information System (PRIS)," IAEA, Apr. 2, 2019. www.iaea.org/resources/databases/power-reactor-information-system-pris (accessed Jul. 15, 2020).

[7]   Massachusetts Institute of Technology, "The Future of Nuclear Energy in a Carbon-Constrained World: An Interdisciplinary MIT Study," MIT, 2018. Accessed: Jul. 15, 2020 [Online]. Available: https://energy.mit.edu/wp-content/uploads/2018/09/The-Future-of-Nuclear-Energy-in-a-Carbon-Constrained-World.pdf.

[8]   112 Ukraine, "EU Delivered Equipment for Ukrainian Nuclear Power Plant," 112. international, Oct. 21, 2015. https://112.international/ukraine-and-eu/eu-delivered-equipment-for-ukrainian-nuclear-power-plant-1215.html (accessed Mar. 4, 2021).

[9]   Alpiq, "Leibstadt Nuclear Power Plant." www.alpiq.com/power-generation/thermal-power-plants/nuclear-power/leibstadt (accessed Jul. 17, 2020).

[10] CIA, "The World Factbook : United Arab Emirates." www.cia.gov/the-world-factbook/countries/united-arab-emirates/ (accessed Jul. 17, 2020).

[11] International Energy Agency, "Data & Statistics," IEA, 2020. www.iea.org/data-and-statistics?country=UAE&fuel=Energy%20supply&indicator=TPESbySource (accessed Jul. 21, 2020).

**CHAPTER 5
EXERCISES**

5.1 Compare the energy that can be harvested from 1 kg U-235 with energy from

    a. 1 barrel of oil

    b. 1 kg of coal

    c. 1 m^3 of natural gas

5.2 Calculate the binding energy of the copper isotope ^{63}Cu, in atomic mass unit (u).

5.3 Determine the binding energy of the helium isotope ^{4}He, in atomic mass unit (u). Calculate it in terms of MeV and J as well.

5.4 Match the pieces below to give a correct statement:

Positron emission is affiliated with ____.	**a.** conversion of a proton to a neutron
	b. atomic number ↑ by 2, and mass number ↑ by 4
Beta emission is affiliated with ____.	**c.** electron decay
	d. no change in either atomic or mass number
Alpha emission is affiliated with ____.	**e.** conversion of a neutron to a proton
Gamma emission is affiliated with ____.	**f.** atomic number ↓ by 2, and mass number ↓ by 4

5.5 Determine the daughter isotopes and indicate the decay type for the listed parent isotopes and the decay they experience:

    a. $^{60}_{27}$Co → ? + $^{0}_{-1}$e$^-$

    b. $^{232}_{90}$Th → ? + $^{0}_{+1}$e$^+$

    c. $^{149}_{64}$Gd → ? + $^4_2$He

    d. $^{137}_{56}$Ba → ? + $^{137}_{56}$Ba

    e. $^{235}_{92}$U → ? + $^4_2$He

5.6 ^{238}U decays into ^{234}Th with a half-life of 4.47 billion years. How many years would it take for 87.5% of ^{238}U to decay?

5.7 Answer the following questions for ^{19}Ne:

    a. What type of decay would it go through to form ^{19}F?

    b. What is its half-life?

    c. How long will it take for 70% of ^{19}Ne to decay into ^{19}F?

    d. How much ^{19}Ne was initially used if there is 4.50 mg left after 86 s?

5.8 Research and find a nuclear power plant example for each reactor type listed below. Include the name, location, and capacity of the power plant.

    a. Pressurized water reactor (PWR)

    b. Boiling water reactor (BWR)

    c. Canada deuterium–uranium reactor (CANDU)

    d. High-power channel reactor (RBMK)

5.9 A nuclear power plant with a PWR produces superheated steam at 18 MPa and 540 °C. The steam exits the turbine at 10 kPa and condenses to saturated liquid within the condenser. The isentropic efficiencies of the turbine and the pump are 80% and 76%, respectively. Determine:

a. the net power output of the power plant per unit mass flow rate of steam (kJ/kg)

b. the rate of heat transfer to the working fluid per unit mass flow rate of fluid through the steam generator (kJ/kg)

c. the rate of heat transfer from the working fluid per unit mass flow rate of fluid through the condenser (kJ/kg)

d. the thermal efficiency of the power plant (%)

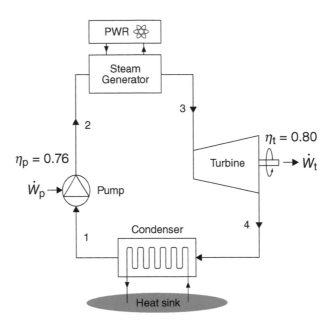

5.10 Reconsider **Example 5.5**. Calculate the total cost per kWh if the thermal efficiency of the power plant is increased to 34.5% with improvements in the Rankine cycle components and the annual O&M costs are reduced by 5%. The amount of fuel replaced annually remains the same.

CHAPTER 6
Renewable Energy Resources

LEARNING OBJECTIVES

After reading this chapter, you should have acquired knowledge on:

- Renewable energy resources and associated technologies
- Global overview of renewable energy production
- Costs, benefits, and challenges of renewable energy applications
- Technical and ethical roles of engineers in promoting renewable energy

6.1 Introduction

Renewable energy is the energy that can be harvested from natural resources that are not exhaustible as they are naturally replenishing, such as sunlight or wind. These resources are inexhaustible over time; however, the amount of energy they can provide within a unit amount of time is limited and differs depending on the type of renewable energy resource. Renewable energy is an environmentally friendly alternative to the conventional energy that relies on fossil fuels. It is a key instrument in mitigating climate change due to its little or no greenhouse gas emissions.

The major types of renewable energy resources are:

- Biomass
- Geothermal
- Hydropower
- Ocean
- Solar
- Wind

A variety of renewable energy technologies from different resources and their end-use applications are listed in **Table 6.1**. Their applications are electricity, industry, buildings, and transport. While electricity generated by means of renewable energy technologies can still be used by the last three applications, these technologies can also be utilized directly in industry, buildings, and transport in a variety of applications such as space heating, hot water, and daylighting.

In addition to the engineering, business, and marketing layers of renewable energy systems, development of policies addressing both technological and socio-economic aspects of these systems is important for a complete picture. A collaborative report by IRENA, IEA, and REN21 classified the policies into three categories: *direct policies*, *integrating policies*, and *enabling policies* [2]. *Direct policies* help support development and deployment of renewable energy technologies via a variety of mandates such as electricity quotas or biofuel use, and opportunities such as tax incentives, subsidies, and grants. *Integrating policies* help in steering the implementation of infrastructure to provide system functionality and flexibility, such as development of charging stations for EVs, or infrastructure for district heating. *Enabling policies* act at a "macro" level for bringing renewable energy technologies into effective action. These policies can be regional or national policies, or policies that pertain to market, industry, land-use, education, and other fields to promote renewable energy use and development. **Table 6.2** illustrates the broad spectrum of policies that have been categorized in the IRENA, IEA, and REN21 report.

Table 6.1

Renewable energy technologies and end-use applications [1]					
		End-use application			
Resource	**Technology**	**Electricity[1]**	**Industry**	**Buildings**	**Transportation**
Solar	Photovoltaics – flat plate	✓			
	Photovoltaics – concentrator	✓			
	Solar thermal parabolic trough	✓	✓		
	Solar thermal dish/Stirling	✓			
	Solar thermal central receiver	✓	✓		
	Solar ponds	✓	✓	✓	
	Passive heating			✓	
	Active heating			✓	
	Daylighting			✓	
Wind	Horizontal axis turbine	✓			
	Vertical axis turbine	✓			
Biomass	Direct combustion	✓	✓	✓	
	Gasification/pyrolysis	✓	✓		✓
	Anaerobic digestion	✓	✓	✓	
	Fermentation				✓
Geothermal	Dry steam	✓			
	Flash steam	✓			
	Binary cycle	✓			
	Heat pump			✓	
	Direct use		✓	✓	
Hydropower	Conventional	✓			
	Pumped storage	✓			
	Micro-hydro	✓			
Ocean	Tidal energy	✓			
	Thermal energy conversion	✓			

[1] Electricity generated by any of these technologies can be used in the other end-use applications listed above. The table lists the end-use applications that can be achieved directly via these technologies.

Table 6.2

Classification of renewable energy policies [2]				
Policies to achieve the energy transition		Deployment (installation and generation) of renewables in the general context	Deployment (installation and generation) of renewables in the access context (including energy services)	Maximization of socio-economic development from renewable energy deployment
Direct policies	Push	• Binding targets for use of renewable energy • Electricity quotas and obligations • Building codes • Mandates (e.g., solar water heaters, renewables in district heating) • Blending mandates		
	Pull	• Regulatory and pricing policies (e.g., feed-in tariffs and premiums, auctions) • Tradable certificates • Instruments for self-consumption (e.g., net billing and net metering) • Measures to support voluntary programs		
	Fiscal and financial	• Tax incentives (e.g., investment and production tax credits, accelerated depreciation, tax reductions) • Subsidies • Grants		
Integrating policies		• Measures to enhance system flexibility (e.g., promotion of flexible resources such as storage, dispatchable supply, load shaping)	• Policies for integration of off-grid systems with main grid • Policies for mini-grids and smart distributed energy systems • Coupling renewable energy policies with efficient appliances and energy services	
		• Policies to ensure the presence of needed infrastructure (e.g., transmission and distribution networks, electric vehicles charging stations, district heating infrastructure, road access) • Policies for sector coupling • RD&D support for technology development (e.g., storage)		
		• Better alignment of energy efficiency and renewable energy policies • Incorporation of decarbonization objectives into national energy plans • Adaptation measures of socio-economic structure to the energy transition		
Enabling policies		• Policies to level the playing field (e.g., fossil fuel subsidy reforms, carbon pricing policies) • Measures to adapt design of energy markets (e.g., flexible short-term trading, long-term price signal) • Policies to ensure the reliability of technology (e.g., quality and technical standards, certificates)	• Industrial policy (e.g., leveraging local capacity) • Trade policies (e.g., trade agreements, export promotion) • Environmental and climate policies (e.g., environmental regulations)	

Table 6.2 (cont.)

Policies to achieve the energy transition	Deployment (installation and generation) of renewables in the general context	Deployment (installation and generation) of renewables in the access context (including energy services)	Maximization of socio-economic development from renewable energy deployment
	• National renewable energy policy (e.g., objectives, targets) • Policies to facilitate access to affordable financing for all stakeholders • Education policies (e.g., inclusion of renewable energy in curricula, coordination of education and training with assessments of actual and needed skills • Labor policies (e.g., labor-market policies, training and retraining programs)		
	• Land-use policies • RD&D and innovation policies (e.g., grants and funds, partnerships, facilitation of entrepreneurship, industry cluster formation) • Urban policies (e.g., local mandates on fuel use) • Public health policies		
Enabling and integrating policies	• Supportive governance and institutional architecture (e.g., streamlined permitting procedures, dedicated institutions for renewables) • Awareness programs on the importance and urgency of the energy transition geared towards awareness and behavioral change • Social protection policies to address disruptions • Measures for integrated resource management (e.g., the nexus of energy, food and water)		

This table is based on "Renewable Energy Policies in a Time of Transition" developed by IRENA, OECD/IEA and REN21 (2018) but the resulting work has been prepared by the author and does not necessarily reflect the views of IRENA, OECD/IEA, or REN21. Neither IRENA, OECD/IEA, nor REN21 accepts any responsibility or liability for this work/translation.

6.2 Global Overview

Renewable energy technologies leaped from niche to mainstream over the past few decades with growing global interest. Renewable energy generation was 3147 TWh by 2020, including hydropower, wind, solar, geothermal, biomass, wave, and tidal [3]. Traditional biomass is not included in these figures.

According to U.S. Energy Information's International Energy Outlook 2021, global energy consumption will increase by approximately 50% by 2050. It is also projected that renewable energy sources will have the largest share in world primary energy consumption by fuel type in 2050 [4]. **Figure 6.1** shows the history and future projection of shares of fuel types. If we look at the shares of fuel types at a global scale for net electricity generation only, the contribution of renewable energy is further amplified, as seen in **Figure 6.2**.

Renewable energy generation by global regions is presented in **Figure 6.3**. Existing data and future projections (2030 and 2040) are provided for two different IEA scenarios [5]. The projected amount of energy generation via renewable energy

FIGURE 6.1 Global primary energy consumption by energy source. Source: U.S. Energy Information Administration, International Energy Outlook 2021.

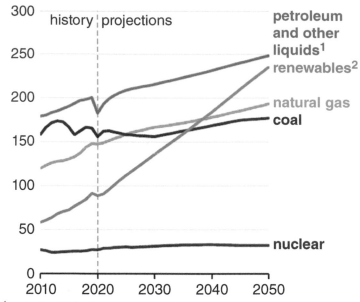

Primary energy consumption by energy source, world
quadrillion British thermal units

[1] Includes biofuels
[2] Electricity generation from renewable sources is converted to Btu at a rate of 8124 Btu/kWh

FIGURE 6.2 Global net electricity generation by source. Source: U.S. Energy Information Administration, International Energy Outlook 2021.

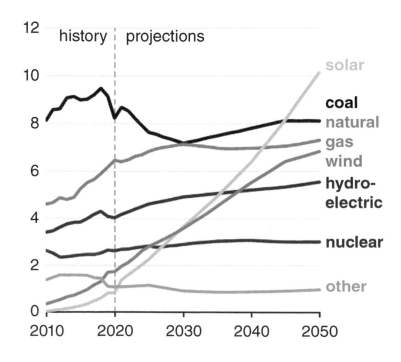

World net electricity generation by source
trillion kilowatthours

FIGURE 6.3 Renewable electricity generation by region for two scenarios, Stated Policies Scenario and Sustainable Development Scenario. Source: International Energy Agency, World Energy Outlook 2019.

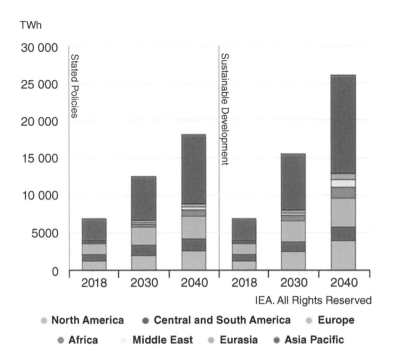

● North America ● Central and South America ● Europe ● Africa ● Middle East ● Eurasia ● Asia Pacific

FURTHER LEARNING

Charts on Global Renewable Energy
https://ourworldindata.org/renewable-energy

FURTHER LEARNING

Surfing for Renewable Energy
U.S. Energy Information Administration (EIA),
www.eia.gov/
National Renewable Energy Laboratory (NREL),
www.nrel.gov/
International Energy Agency (IEA),
www.iea.org/
International Renewable Energy Agency (IRENA),
www.irena.org/
REN21,
www.ren21.net/

technologies is 25% higher by 2030, and 45% higher by 2040 if the measures taken are in line with the Sustainable Development Scenario (SDS), instead of the Stated Policies Scenario (STEPS). Naturally, these enhancements translate into significant reductions in CO_2 emissions. According to the IEA, STEPS reflects the outcome of existing policies. Although there was an increasing global sensitivity and efforts to reduce CO_2 emissions after the Paris Agreement, this motivation did not last long at the desired level. Data obtained and projections for the future showed that the world is not on course to achieve the UN Sustainable Development Goals (SDGs) that are pertinent to energy. The IEA came up with a more ambitious and rigorous plan, the SDS, to ensure the SDGs associated with energy can be delivered in a more realistic manner while also being cost-effective. This scenario is developed by means of a retrograde analysis, working backwards from the goals that are to be achieved to the steps that need to be taken to reach those goals.

Towards 2040, renewables are expected to make up remarkable portions of capacity additions worldwide. **Figure 6.4** illustrates the renewable shares of capacity additions for various countries and regions according to both the STEPS and SDS cases [5]. More than 70% of the total capacity additions are anticipated to be from renewables in all listed countries and regions, if SDS actions are in place.

A regional analysis of renewable capacities can be helpful to interpret renewable energy potential, investments, and R&D interests around the globe. Renewable power capacities of the world, BRICS, EU, and the top six countries are listed in **Table 6.3** [6].

Energy consumption from renewable and non-renewable sources in the United States estimated for 2020 is illustrated in **Figure 6.5**. As can be seen in the flow chart, solar, hydro, and wind contribute to electricity generation in modest amounts with a small amount of geothermal. Biomass is mostly used for industrial and transportation energy needs.

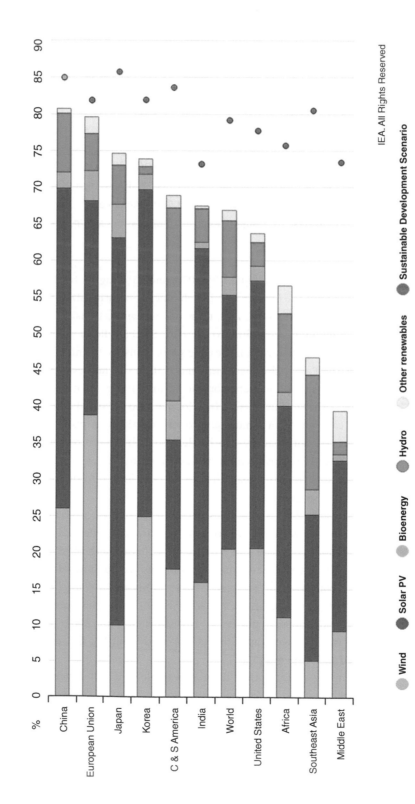

FIGURE 6.4 Renewables share in capacity additions by region in Stated Policies Scenario, and according to the Sustainable Development Scenario, between 2019 and 2040. Source: International Energy Agency, World Energy Outlook 2019.

Table 6.3

Renewable power capacities: World, regions, and top six countries,[1] 2019 [6]

Technology	World	BRICS[2]	EU-28	China	United States	India	Germany	Japan	United Kingdom
	GW	GW					GW		
Biomass	139	48	44	22.5	16.0	10.8	8.9	4.3	7.9
Geothermal	13.9	0.1	0.9	~0	2.5	0	~0	0.6	0
Hydropower	1150	530	131	326	80	45	5.6	22	1.9
Ocean	0.5	0	0.2	0	0	0	0	0	~0
Solar PV[3]	627	256	132	205	76	43	49	63	13.4
Concentrating solar thermal power (CSP)	6.2	1.1	2.3	0.4	1.7	0.2	0	0	0
Wind	651	292	192	236	106	38	61	3.9	24
Total renewable power capacity (including hydropower)	2588	1127	502	790	282	137	124	94	47
Total renewable power capacity (**not** including hydropower)	1438	597	371	464	202	92	119	72	45
Per capita capacity (kW per inhabitant, not including hydropower)	0.2	0.2	0.7	0.3	0.6	0.1	1.4	0.6	0.7

[1] Table lists the top six countries by total renewable power capacity not including hydropower. If hydropower were included, countries and rankings would change and the top six countries would be China, the United States, Brazil, India, Germany, and Canada.

[2] The five BRICS countries are Brazil, the Russian Federation, India, China, and South Africa.

[3] Solar PV data are in direct current (DC).

Note: Where totals do not add up, the difference is due to rounding. Capacities that are less than 50 MW (including pilot projects) are designated by "~0".

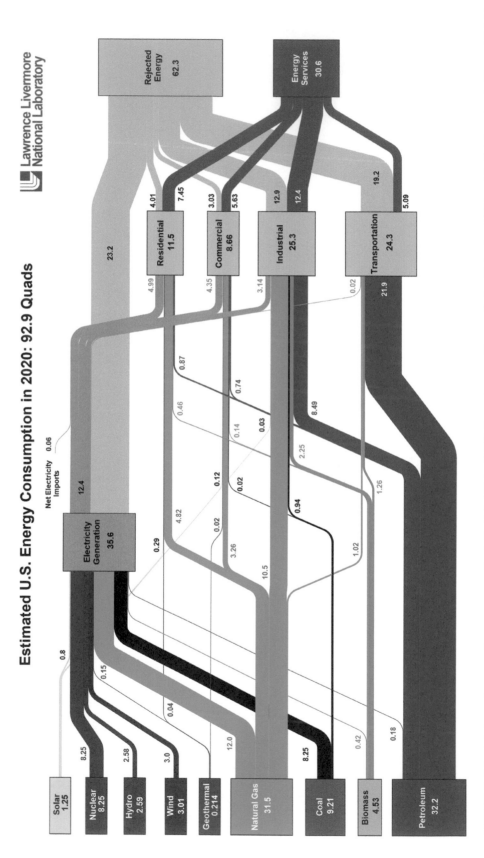

FIGURE 6.5 Estimated shares of resources in energy consumption in the United States (2020). Source: Lawrence Livermore National Laboratory (LLNL) and the U.S. Department of Energy (DOE), March 2021.

6.3 Costs, Benefits, and Challenges of Renewable Energy Technologies

6.3.1 Costs

Investment in any new electricity generation facility requires a thorough feasibility study as these projects also mean a long-term financial commitment along with climate impacts. Having lower initial costs does not necessarily make a power plant or "farm" that generates electricity a more cost-effective facility. Its capacity factor, lifetime, O&M costs, and all other costs to keep it functioning through its lifespan are all important factors in determining if it is a better investment than an alternative energy technology. If you are looking for a piano, scuba diving gear, or a car, the least expensive one will not always mean it is the optimum one to buy. How long will you be using it? How much will you have to spend to maintain it? How functional will it be? All of these parameters should be factored in when making a decision.

As discussed in previous chapters, levelized cost of energy (LCOE) is a useful tool in assessing the average lifetime cost of unit amount of electricity for a power generation facility. As such, this tool is helpful in determining renewable energy investments for individuals, the private sector, or governments. Calculation of LCOE for different renewable technologies enables comparison of long-term outcomes showing life cycle costs and identification of opportunities to develop projects at different scales ranging from residential to commercial. There are sources that provide up-to-date LCOE values of renewable energy technologies, including the IEA, IRENA, EIA, and NREL. For user-defined input values, interactive LCOE calculators also exist. The NREL in partnership with the U.S. Department of Energy (DOE) developed an LCOE calculator that provides metrics on the costs of different energy generation methods. These costs include capital costs, O&M costs, and fuel costs. In **Figure 6.6**, a sample illustration is given for a

FURTHER LEARNING

Local Renewable Energy Benefits and Resources
www.epa.gov/statelocalenergy/local-renewable-energy-benefits-and-resources

FIGURE 6.6 LCOE values ($/MWh) for various renewable and conventional electricity generating technologies for moderate scenario, market & policies case, 30-year recovery period, 2020. Source: National Renewable Energy Laboratory (NREL).

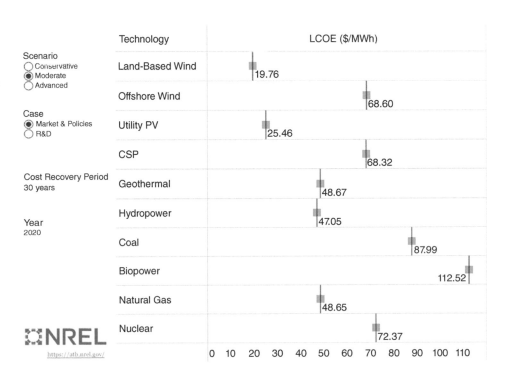

moderate scenario for 2020. Another example is the LCOE calculator developed by the Danish Energy Agency. This calculator accounts for climate externalities, air pollution, system costs, and heat revenue in addition to the capital, O&M, and fuel costs. There are a number of other online LCOE calculators with various input and output quantities that can be helpful for specific projects.

6.3.2 Benefits

There are four major benefits that renewable energy resources offer:

- Environmental
- Economic
- Diversity
- Public health

Environmental

Renewable energy technologies generate little to no greenhouse gas emissions. This includes the life cycle emissions of these technologies, accounting for manufacturing, installation, operation, and decommissioning. Among all renewables, energy harvesting from biomass can yield emissions that depend on the resource and the method by which it is processed for generating energy.

Economic

One significant benefit of renewable energy is providing opportunities for economic development and new jobs in manufacturing, installation, operation, and related technical and non-technical fields. **Figure 6.7** depicts global average jobs opened per million dollars of investment, and cost of CO_2 emissions reductions for a variety of renewable energy technologies, as well as some environmentally friendly energy measures such as improving efficiency or use of EVs.

Besides contributing to the job market and lowering the cost of CO_2 emissions reductions, renewable energy systems help individuals, businesses, or communities to reduce their utility bills. As for the utility companies, the short-sighted ones may try to block the growth of renewable energy use in their territories while the provident

FIGURE 6.7 Global average job opportunities yielded and cost-effectiveness of CO_2 emissions reductions for selected energy measures. Source: International Energy Agency (IEA).

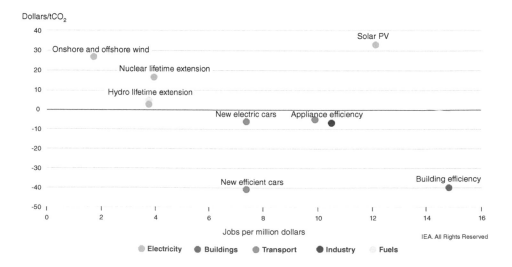

ones have already started investing in renewable energy for two main reasons: to be able to keep up with consumer demand with more economic infrastructural solutions, and to be ready for future distributed generation.

Diversity

Diversification of energy supplies helps reduce dependence on fossil fuels. This is helpful for the environment. In addition, if the fossil fuels are being imported, less dependence on them provides more energy security, especially in times of recession, catastrophes, or conflicts between the trading nations.

Public Health

Decreasing the emissions of harmful gases including CO_2, SO_2, and NO_x reduces the environmental impacts of fossil fuels such as air pollution, smog, and acid rain. Some of the health problems associated with these impacts include asthma, bronchitis, heart disease, and malfunction of the nervous system or immune system. A transition to renewable energy technologies will mitigate the emissions, which will result in improved public health.

The benefits listed above have interdependence as well. For instance, improving public health reduces a significant budget load from governments, which can be directed to other good uses such as improving health systems, education, or energy infrastructure. Also, each benefit listed above serves for one or more of the 17 SDGs that were discussed in **Chapter 2**. Taking public health as an example again, this benefit would be directly helpful in achieving SDG 3: *Good health and well-being*.

FIGURE 6.8 Decreasing emissions of harmful gases through sustainable energy promotes public health, which serves in achieving SDG 3 of the United Nations: *Good health and well-being*. Source: Hill Street Studios/DigitalVision Vectors via Getty Images.

6.3.3 Challenges

There are five major challenges that renewable energy faces:

- Public perception
- Economic and financial challenges
- Intermittency and grid issues
- Regulatory challenges
- Technological challenges

Public Perception

Transitioning into anything new comes with a potential risk of public resistance. Switching to renewable energy technologies has encountered opposition for differing reasons. The idea of renewables not being capable of supplying the total demand is one of the reasons for resistance. The belief that the number of jobs renewables would cut due to closing conventional power plants can be greater than the jobs they bring in can be another social challenge to be addressed. Lack of information can catalyze resistance based on such assumptions. Some people, although in favor of renewables, can still oppose the idea of having them in their close proximity. This is known as the *not in my backyard* (nimby) response, where the residents in a given area oppose proposed developments in their neighborhood, thinking that these developments can be detrimental to their quality of life or lower the value of their property. Another reason for public opposition can be due to the need for comparatively larger land area, which can have an impact on its use for other purposes such as agriculture.

Economic and Financial Challenges

Despite the economic benefits that renewable energy provides, it can itself have some economic and financial challenges. Depending on the application, project size, location, and other parameters affecting the cost, initial costs and finding investors can be a challenge. Another obstacle to growth of renewables is the cost of their competitor, fossil fuels. When fossil fuel prices are low, governments and industry can choose to lean towards traditional electricity generation methods. Limited subsidies and incentives play a role in slowing down the expansion of renewable energy applications worldwide.

Intermittency and Grid Issues

Intermittency is the *Achilles heel* of some of the renewable energy technologies such as solar and wind. The magnitude of this problem would be smaller if there were cost-effective and reliable ways of storing energy. With the ongoing research on energy storage systems, the intermittency issue of renewable energy sources can be lessened. Another challenge of renewable energy generation is grid reliability. The grid is the network stretching from the power generation plants or farms to transmission and distribution lines, and to the end-users. The existing grids in most of the developed countries are already aged and with increasing population and demand, the problem is not getting any better. With or without renewables, grid issues should be addressed for reliable energy supply. The equation gets more complex with the addition of renewables, not only because it is one extra term in the equation, but also because they are intermittent.

Regulatory Challenges

Regulatory challenges include policies that are not in support of renewables even if not against them, excessive paperwork for permits and certifications, insufficient incentives to encourage transitions, and lack of communication between the policymakers and the community. Other regulatory issues include licensing and siting, cost recovery, evaluating investment incentives, and overseeing purchase power agreements. A vast number of successful examples from around the world exist in terms of policies made, bureaucracy minimized, and effective communication with the investors, utility companies, businesses, or end-users established.

Technological Challenges

If we think about renewable energy generation and use with a *source–path–receiver* concept, as in acoustic analyses, intermittency would be a *source* problem and grid issue would be associated with *path*. Within the same analogy, public perception can be matched with the *receiver*. In addition to these problems, renewable energy needs to overcome technological challenges at every step from the source to the receiver. This is mainly the R&D challenge for system improvement for more efficient energy conversion, transmission, distribution, storage, and use. This is the challenge for engineers to take.

6.4 Role of Engineers: Technical and Ethical

In the previous section, the technological challenges of renewable energy were discussed and we alluded to engineers. The role of engineers in improving renewable energy technologies has two components: *technical* and *ethical*.

The technical role of engineers involves the R&D process, which can be broken down into six steps. These steps are: fostering ideas, filtering, analysis and

FURTHER LEARNING

Renewable Energy Challenges
www.esrl.noaa.gov/gsd/renewable/
challenges.html

CLASSROOM DEBATE

Do you think renewable energy can replace fossil fuels by the end of the century? Consider all technological developments, as well as population projections and changes in people's energy consumption habits.

FURTHER LEARNING

NREL System Advisor Model (SAM)
https://sam.nrel.gov/

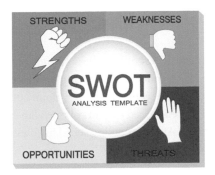

FIGURE 6.9 *SWOT analysis* is a technique for assessing four aspects (**S**trengths, **W**eaknesses, **O**pportunities, and **T**hreats) of a project. This method can be employed during the brainstorming phase of a renewable energy project to ensure feasibility and sustainable development economy. Source: Maria Stavreva/DigitalVision Vectors via Getty Images.

FIGURE 6.10 Engineers and energy professionals are expected to present the highest standards of honesty and integrity in their work, while being committed to protecting public health, safety, and welfare. This encompasses environmental responsibilities to meet sustainability, as well. Source: RapidEye/E+ via Getty Images.

development, prototyping and testing, marketing, and launching a product. Of these six steps, the first four fall into the engineers' basket.

Fostering the ideas: A list of ideas is collected through brainstorming. Different brainstorming techniques such as mind mapping, step ladder, SWOT analysis (**Figure 6.9**), or any other method that fits the nature of the study can be used. This step is the *divergent thinking* phase of R&D. Any idea, even though it may seem unrealistic, is welcome and in fact encouraged at this step.

Filtering: This is the *convergent thinking* phase of ideation. The team evaluates the pool of ideas generated and then filters them to focus on the ones with potential.

Analysis: In this step the ideas are further analyzed, including non-engineering aspects such as market readiness analysis through market surveys. An idea to start working on is selected.

Prototyping and testing: This is where a physical model is designed, fabricated, and tested. This is an iterative process to achieve a marketable product that is functional, efficient, reliable, and cost-effective in terms of mass production.

These four steps portray how engineers can either come up with new renewable energy technology ideas or improve existing systems. While the last two steps, marketing and launching, are not directly engineering related, teams in those departments are in close communication with the engineering teams for product and strategy development. which is an ongoing process.

The other component of the role of engineers in improving the technology is non-technical. It is about ethical responsibilities. As stated by the six fundamental canons of the National Society of Professional Engineers (NSPE) Code of Ethics [7], engineers in the fulfillment of their professional duties shall:

1. *Hold paramount the safety, health, and welfare of the public.*
2. *Perform services only in areas of their competence.*
3. *Issue public statements only in an objective and truthful manner.*
4. *Act for each employer or client as faithful agents or trustees.*
5. *Avoid deceptive acts.*
6. *Conduct themselves honorably, responsibly, ethically, and lawfully so as to enhance the honor, reputation, and usefulness of the profession.*

These canons address a broad spectrum of ethical responsibilities which applies to any profession. According to the first canon, engineers working on improving renewable energy technologies are to hold paramount the safety, health, and welfare of the public. This canon aligns with SDG 3 of the United Nations. Engineers should also look to develop solutions that pose no or minimal harm to the environment and the ecosystem.

6.5 Summary

Renewable energy technologies are pivotal in mitigating climate change due to their low or no greenhouse gas emissions. A variety of renewable energy resources such as biomass, geothermal, hydropower, ocean, solar, and wind, and technologies that are derivative outcomes from each resource come with their own set of benefits and challenges. Improvement of benefits and alleviation of challenges associated with renewables is a must for sustainable development. Renewable energy directly addresses 4 of the 17 SDGs:

- SDG 3: Good Health and Well-being
- SDG 7: Affordable and Clean Energy

- SDG 11: Sustainable Cities and Communities
- SDG 13: Climate Action

Environmental, economic, diversity, and health benefits need not only to be amplified for specific regions but should also be distributed globally for broader impact. Being aware of the challenges is the initial step. It has been proven already that technological development alone will not be adequate in achieving the widespread and enhanced benefits of renewable energy. Education and clear communication are instrumental in overcoming social hurdles. The role of policymakers is critical in bettering regulatory issues.

REFERENCES

[1] J. A. Sathaye and S. Meyers, *Greenhouse Gas Mitigation Assessment: A Guidebook.* Dordrecht; London: Springer, 2011.

[2] IRENA, OECD/IEA, and REN21, "Renewable Energy Policies in a Time of Transition," 2018. Accessed: Sep. 02, 2020 [Online]Available: www.irena.org/-/media/Files/IRENA/Agency/Publication/2018/Apr/IRENA_IEA_REN21_Policies_2018.pdf.

[3] BP, "Statistical Review of World Energy," 2021. Accessed: Jan. 15, 2022 [Online]. Available: www.bp.com/statisticalreview.

[4] U.S. Energy Information Administration, "International Energy Outlook 2021," EIA, Washington, DC, Oct. 2021. Accessed: Jan. 15, 2022 [Online]. Available: www.eia.gov/outlooks/ieo/.

[5] International Energy Agency, "World Energy Outlook 2019," IEA, Paris, Nov. 2019. Accessed: Sep. 18, 2020 [Online]. Available: www.iea.org/reports/world-energy-outlook-2019.

[6] REN21, "Renewables 2020 Global Status Report," Paris: REN21 Secretariat, 2020. Accessed: Nov. 8, 2020 [Online]. Available: www.ren21.net/wp-content/uploads/2019/05/gsr_2020_full_report_en.pdf.

[7] National Society of Professional Engineers, "Code of Ethics," 2013. www.nspe.org/resources/ethics/code-ethics (accessed Sep. 21, 2020).

CHAPTER 6 EXERCISES

6.1 Fill in the chart below for the six renewable energy sources listed. Briefly describe how each resource is used for harvesting energy. Share a conventional and an advanced (or still in R&D phase) application from around the world. List two advantages and two disadvantages for each technology.

Energy resource	Working principle	Application example (conventional)	Application example (advanced or in R&D phase)	List two advantages	List two disadvantages
Biomass					
Geothermal					
Hydropower					
Ocean					
Solar					
Wind					

6.2 Using Table 6.1, list the technologies used with each energy resource to generate electricity. Discuss why electricity is not generated using some of the technologies listed in the table.

6.3 Enabling policies for renewable energy technologies are categorized in Table 6.2. Research an example policy for each item listed below. Policy examples can be from the United States, Europe, or other parts of the world.
  a. Land-use policies
  b. RD&D and innovation policies (e.g., grants and funds, partnerships, facilitation of entrepreneurship, industry cluster formation)
  c. Urban policies (e.g., local mandates on fuel use)
  d. Public health policies

6.4 Compare renewable energy production among Germany, France, and the United Kingdom in 2021. Draw a conclusion about the possible reasons for the differences.

6.5 Fill in the concept map below for the given renewable energy sources with corresponding technologies associated with each source.

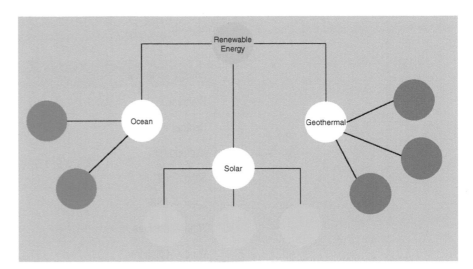

6.6 Using the information provided in Figure 6.5, determine the percentage shares of each energy resource within the sectors listed below:
  a. Solar energy in residential sector
  b. Wind energy in electricity generation
  c. Geothermal energy in commercial sector
  d. Biomass in transportation sector
  e. Hydropower in electricity generation

6.7 Using the information provided in Figure 6.5, plot pie charts for both residential and commercial energy consumption showing the share of each energy source. Treat electricity as a single source in itself without breaking it into each energy source used in generating electricity.

6.8 Apply the SWOT analysis for the proposed projects below:
  a. A hydroelectric power plant on the Põltsamaa River in Estonia
  b. A biomass power plant in Battambang, Cambodia

A rice field in the countryside of Battambang, Cambodia. Rice fields can be a good source of biomass energy as the rice husks are abundant and easy to collect. Source: Fabien Astre/Moment via Getty Images.

6.9 Research the renewable energy efforts in Tanzania and include which SDGs of the UN have been addressed with these developments.

6.10 Obtain 2021 global weighted average levelized cost of electricity (LCOE) values for the renewable energy technologies listed below. Include the reference of your research.

Energy resource	LCOE ($/MWh)
Biomass	
Geothermal	
Hydropower	
Solar – PV	
Solar – CSP	
Wind – Onshore	
Wind – Offshore	

CHAPTER 7
Biomass

LEARNING OBJECTIVES

After reading this chapter, you should have acquired knowledge on:

- Global biofuel production
- Biomass resources and energy conversion
- Different types of biofuels
- Combustion, pyrolysis, gasification, transesterification, fermentation, and anaerobic digestion processes
- Municipal solid waste management
- Combined heat and power (CHP) plants for electricity generation
- Economics of biomass energy

7.1 Introduction

Biomass is organic material that comes from plants and animals. It can be defined as organisms that are living or have recently died, and the byproducts of these organisms. It is a renewable source, and it is known to be the oldest source of energy for human beings, after the Sun. Biomass also gets its energy from the Sun. For instance, during photosynthesis, sunlight gives plants the energy they need to convert water and carbon dioxide into oxygen and sugars. These sugars are carbohydrates, and they supply the plants with energy. They are a good source of energy for animals and humans as well. Photosynthesis is the process by which plants and some other organisms convert sunlight into chemical energy. The chemical energy is stored in carbohydrate molecules, such as sugars. The reaction is in the form:

$$\text{Water} + \text{Carbon dioxide} + \text{Sunlight} \rightarrow \text{Glucose} + \text{Oxygen}$$

The energy stored in biomass can be used to generate process heat or electricity. A variety of biofuels obtained from different biomass sources can be utilized in harvesting bioenergy. These sources and biofuels are discussed in later sections.

While biomass has the advantages of being renewable, producing less greenhouse gas (GHG) emissions, and contributing to waste and landfill reduction, it can come with some disadvantages such as land use, water supply and quality, and potential impact on wildlife and the ecosystem. The environmental impacts of biomass energy can be minimized through a number of practices such as more sustainable land and water use measures, well-planned replanting schedules, and enhanced technological innovations.

7.1.1 Global Overview

Bioenergy accounts for approximately 10% of total global primary energy supply. Energy from biomass is harvested by means of traditional and modern methods. Traditional use of biomass entails cooking and heating purposes, which are still seen in developing countries and remote areas of developed countries. Modern bioenergy on the other hand is an important renewable energy, which on the global scale supplies five times more energy than wind and solar photovoltaic (PV) combined [1].

Global biofuel production has been increasing steadily over the past decade. In 2010, total global production was equivalent to 59 Mtoe. In 2019, it was recorded as 96 Mtoe. According to the Sustainable Development Scenario (SDS) of the IEA, the targeted global value is 298 Mtoe by 2030 [1]. While it does not look realistic to achieve the SDS goal with current developments, more supportive policies and regulations, enhancement of technology, scaling up biofuel consumption, and adoption of it by other transportation means (aviation and marine) can help in closing the gaps between the achieved and targeted values.

FIGURE 7.1 Global biofuel production in different regions of the world. Source: BP Statistical Review of World Energy 2021.

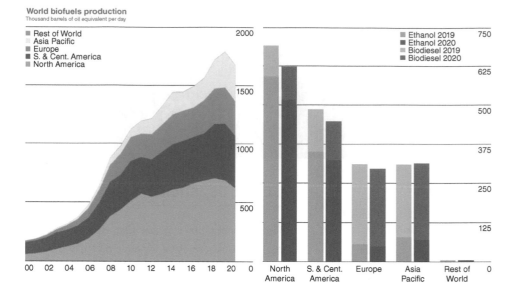

FIGURE 7.2 Country rankings based on biofuel production in 2020. Source: Graph generated based on data from BP Statistical Review of World Energy 2021.

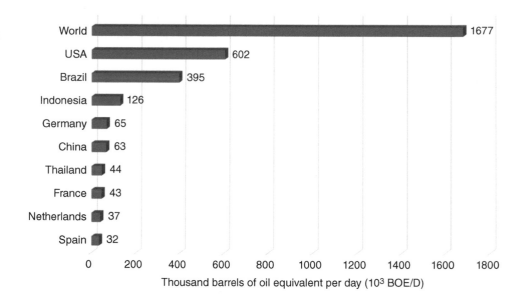

As for global biomass electricity generation, the progress is more encouraging. Total global bioenergy power generation was 310 TWh in 2010, and 589 TWh in 2019. Target production by 2030 is 1168 TWh per the SDS [1]. The historic trend, owing to the policy and market developments in emerging economies, indicates that reaching this goal is more pragmatic than theoretical.

In terms of biofuel production, the United States, Brazil, and Indonesia were the top three countries in 2019. As for bioenergy production, China was ranked first, followed by Brazil, and the United States in 2020. **Figures 7.1** and **7.2** show global biofuel production on a regional basis and by country rankings, respectively.

7.1.2 Biomass Resources

Biomass resources can be used directly as a fuel for heating and electricity generation, or as a feedstock to be converted into another form of fuel or energy source

FIGURE 7.3 Biomass sources.

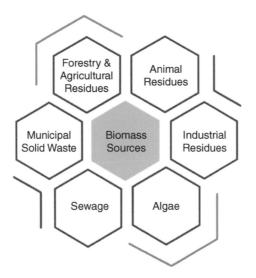

(e.g., transportation). These resources include *forestry residues* (limbs, branches, culled trees), *agricultural residues* (corn stover, wheat straw, oat straw), *animal residues* (solid excreta), *industrial residues* (sludge, whey), *algae* (microalgae, macroalgae), *sewage* (municipal wastewater), and *municipal solid waste* (trash) (**Figure 7.3**).

7.1.3 Biofuels

As discussed in the previous section, biomass resources can be utilized directly as a fuel for heating and electricity generation, or as a feedstock to be converted into another form of fuel or energy source. Different forms of energy conversion for harvesting energy or producing biofuels are [2]:

- Direct combustion: to produce heat or electricity
- Thermochemical conversion: to produce solid, gaseous, and liquid fuels
- Chemical conversion: to produce liquid fuels
- Biological conversion: to produce liquid and gaseous fuels

Direct combustion is the most common method for harvesting energy from biomass. It helps produce bioenergy, as opposed to the other conversion techniques listed which are used for biofuel production. In direct combustion, biomass is burnt directly for residential heating, for process heat, or for producing steam to generate electricity.

Thermochemical conversion of biomass includes pyrolysis and gasification processes. Both are thermal decomposition processes in which biomass feedstocks are heated in gasifiers at high temperatures. The two processes differ in the process temperatures and amount of oxygen present during the conversion process.

Chemical conversion process is used for converting vegetable oils, animal fats, and greases into fatty acid methyl esters (FAME), which are used to produce biodiesel. This conversion process is called transesterification (**Figure 7.4**). A strong base or a strong acid can be used as a catalyst during the reaction.

Biological conversion of biomass can involve fermentation to convert biomass into ethanol, or anaerobic digestion to produce biomethane, which is a biogas also known as renewable natural gas (RNG). The most common ethanol production facilities use yeast to ferment the starch and sugars in corn, sugar cane, and sugar beets.

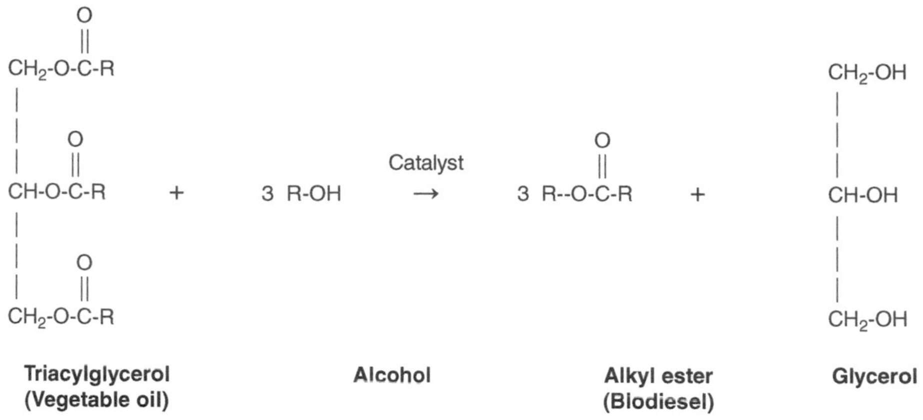

FIGURE 7.4 Transesterification process. Source: J. Van Gerpen, B. Shanks, R. Pruszko, D. Clements, and G. Knothe, "Biodiesel production technology," National Renewable Energy Laboratory, Golden, CO (2004). www.nrel.gov/docs/fy04osti/36244.pdf.

$$
\begin{array}{ccccccc}
\begin{array}{c}\text{O}\\ \| \\ \text{CH}_2\text{-O-C-R} \\ | \\ \quad\;\; \text{O} \\ | \quad\; \| \\ \text{CH-O-C-R} \\ | \\ \quad\;\; \text{O} \\ | \quad\; \| \\ \text{CH}_2\text{-O-C-R}\end{array} & + & 3\ \text{R-OH} & \xrightarrow{\text{Catalyst}} & \begin{array}{c}\text{O}\\ \| \\ 3\ \text{R--O-C-R}\end{array} & + & \begin{array}{c}\text{CH}_2\text{-OH} \\ | \\ \text{CH-OH} \\ | \\ \text{CH}_2\text{-OH}\end{array} \\[2em]
\textbf{Triacylglycerol} & & \textbf{Alcohol} & & \textbf{Alkyl ester} & & \textbf{Glycerol} \\
\textbf{(Vegetable oil)} & & & & \textbf{(Biodiesel)} & &
\end{array}
$$

7.1.3.1 ETHANOL

Conversion: Biological

Process: Fermentation

Ethanol (or ethyl alcohol) is an organic chemical compound with the formula C_2H_5OH. It can be produced by fermentation. Microorganisms such as bacteria or yeast metabolize plant sugars yielding ethanol. In the United States, which is the largest fuel ethanol producer in the world, corn is used for ethanol production due to its abundance and low price. Brazil, the world's second largest fuel ethanol producer, uses sugar cane as feedstock.

Ethanol is a renewable fuel and is used as a blending agent in gasoline to raise the octane level. Most of the gasoline sold in the United States has some ethanol content. This is mostly performed to meet the requirements of the 1990 Clean Air Act, and the Renewable Fuel Standard detailed in the 2007 Energy Independence and Security Act. There are three main types of ethanol blends: E10 (10% ethanol, 90% gasoline), E15 (15% ethanol, 85% gasoline), and E85 (51–83% ethanol, depending on the geography and season; **Figure 7.5**) [3]. There are *flexible fuel vehicles* (FFVs) which are capable of operating on gasoline and any type of ethanol blend up to 83%.

Ethanol can also be produced by breaking down cellulose in plant fibers. The product is *cellulosic ethanol* which is considered an advanced biofuel. Cellulosic feedstocks are non-food based and are composed of cellulose, hemicellulose, and lignin. Lignin is removed from the feedstocks for producing heat and electricity at biomass plants. Producing cellulosic ethanol involves a more complicated production process than fermentation. Commercial production of cellulosic ethanol is relatively small compared to ethanol.

7.1.3.2 METHANOL

Conversion: Thermochemical

Process: Gasification

FIGURE 7.5 E85 is a high-level ethanol–gasoline blend that can contain 51–83% ethanol depending on the geographic location and the season. This type of fuel can be used in *flexible fuel vehicles* (FFVs). Source: Peter Dazeley/The Image Bank via Getty Images.

Methanol (or methyl alcohol) has the chemical formula CH_3OH. It is common practice to denote it as MeOH. Its main feedstocks can be fossil fuels (natural gas, coal) or biomass (wood). Methanol from biomass can be produced by gasification, which is a thermochemical process. The substances are reacted at high temperatures without experiencing combustion. This process yields *syngas* (or synthesis gas), which itself is another

FIGURE 7.6 Biodiesel-powered trimaran, EarthRace, offshore from Mexico. The engine ran on an animal fat and vegetable oil mix biodiesel. The boat had other eco-friendly features such as use of vegetable oil lubricants and hemp composites. Source: Peter Reindl via Wikimedia Commons.

type of biofuel. The syngas is then reacted by use of a catalyst to form methanol. Similar to ethanol fuels, methanol fuels can also reduce net GHG emissions from vehicles. M85 (85% methanol, 15% gasoline) is a fuel like E85; however, it cannot be utilized by vehicles that are designed for E85. There are FFVs that can operate with M85, but the use of these vehicles has so far been limited.

7.1.3.3 BIODIESEL

Conversion: Chemical
Process: Transesterification

Biodiesel (or biomass-based diesel fuel) is produced through transesterification, which is a chemical process that converts fats and oils into fatty acid methyl esters. It can be produced by a variety of feedstocks that have sufficient free fatty acids. In most of the biodiesel production in the United States, raw vegetable oils, used cooking oils, yellow grease, and animal fats are used as feedstocks for the transesterification process. Some alternative feedstocks used in biodiesel production in different parts of the world include sunflower oil, palm oil, and rapeseed oil [4]. Just like ethanol and methanol, biodiesel also can be blended in different ratios. The most common blends are B2 (2% biodiesel, 98% petroleum diesel), B5 (5% biodiesel, 95% petroleum diesel), and B20 (20% biodiesel, 80% petroleum diesel). While pure biodiesel (B100) can also be used in different applications, it is mostly used as a blend stock and is rarely utilized as a transportation biofuel.

7.1.3.4 PYROLYSIS OIL

Conversion: Thermochemical
Process: Pyrolysis

Pyrolysis oil, also known as bio-oil, is a product of the pyrolysis process using biomass. The most common biomass used is wood. Other types of biomass including straw, olive pits, and sorghum can also be used although the efficiency and economic indicators may not be as high. The product is a synthetic fuel that is obtained by heating dried biomass in a chemical reactor in the absence of oxygen, followed by a cooling process. Two types of pyrolysis exist: slow (conventional) pyrolysis and fast pyrolysis. They differ based on the heating rate of the biomass. The byproducts are liquid pyrolysis oil, solid char, and some non-condensable gases. Pyrolysis oil comes out in the form of tar, which is a dark brown or black viscous liquid.

7.1.3.5 BIOGAS

Conversion: Biological
Process: Anaerobic digestion

Biogas is another renewable fuel which can be produced from a variety of organic waste including livestock manure, food processing wastes, and sewage sludge. It is also known as *gobar gas* in India, as the word *gobar* means "cow manure" in Hindi. Biogas is produced through biological means by anaerobic digestion of manure and other feedstocks. *Aerobic* means "with air," and *anaerobic* means "without air." *Anaerobic digestion* is a process through which bacteria break down organic matter such as manure in the absence of oxygen. The anaerobic digester is a sealed, air-tight vessel. Organic material is heated in the digester and the microorganisms that are homogeneously distributed within the digester by a mixer break down the waste, producing biogas.

7.1.3.6 PRODUCER GAS

Conversion: Thermochemical

Process: Gasification

Producer gas is a mixture of combustible (H_2, CH_4, and CO) and non-combustible (N_2, CO_2) gases. It is produced by thermochemical means through a gasification process. In the United States, producer gas can also take other names such as *wood gas* or *town gas*. In the United Kingdom, it is also known as *suction gas*, due to air being pulled in to flow through the fuel to generate carbon monoxide during the exothermic reaction:

$$2C + O_2 = 2CO$$

Producer gas can be burnt as a fuel gas for heating purposes, or in a boiler to produce steam for electricity generation. It can also serve both purposes at a combined heat and power (CHP) facility.

7.1.3.7 SYNTHESIS GAS

Conversion: Thermochemical

Process: Gasification

Synthesis gas or *syngas* is a mixture of CO and H_2 which are products of gasification of organic material to a gaseous product. It can be obtained from fossil fuels (coal) or biomass (wood, municipal solid waste). It follows the reactions:

$$C + O_2 = CO_2$$

$$CO_2 + C = 2CO$$

$$C + H_2O = CO + H_2$$

Syngas can be burnt in gas engines to produce methanol and hydrogen, or it can be converted into synthetic fuel (also known as *synfuel*) through a method called the *Fischer–Tropsch process*, which involves a series of chemical reactions to obtain liquid hydrocarbons.

The gasification process that is used in obtaining syngas and other biofuels such as producer gas or methanol is used to convert fossil fuels or biomass feedstocks into carbonaceous fuels. Gasification takes place in systems called *gasifiers* or *gasification reactors* (**Figure 7.7**). There are three types of gasifiers:

- Fixed-bed gasifiers
- Fluidized-bed gasifiers
- Entrained-bed gasifiers

Fixed-bed gasifiers can be *updraft* (counter-current) or *downdraft* (co-current) types. In *updraft gasifiers*, the fuel and air (as the gasification agent) move in opposite directions. Air enters the system from the bottom and the produced gas leaves the gasifier from the top. In the lower section, the combustion reaction occurs. Above this is where the

FIGURE 7.7 Gasifier (gasification reactor) types. Source: Adapted from Kopiersperre, Wikimedia Commons author, https://commons.wikimedia.org/wiki/File: Gasifier_types.svg.

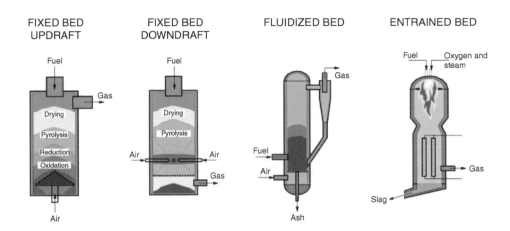

reduction reaction takes place, followed by sections for pyrolysis and drying. The major advantages of updraft gasifiers are that the system has a simple design, it can handle a diverse group of fuels from coal to biomass, and the high heat transfer coefficient inside the system enables lower exit temperatures for gas, hence increasing the efficiency. In *downdraft gasifiers* air and fuel flow downward in the same direction and the gas is retrieved from the bottom. Unlike in updraft designs, air enters downdraft gasifiers at or above the combustion zone, hence lowering the chance for tar entrainment, which can be a disadvantage of updraft designs. While the tar levels are lower with these designs, efficiency is also lower, as well as the exit temperature of gas in downdraft gasifiers being higher than that of updraft gasifiers.

Fluidized-bed gasifiers contain a bed of solid particles into which air is blown. Once the temperature within the fluidized bed is high enough, fuel is introduced into the bed. Fuel reaches the surrounding ambient temperature quickly, resulting in a faster pyrolysis reaction. These gasifiers have the advantages of higher feedstock flexibility, easier control of the process, and lower response time. Due to these designs being more sophisticated, the initial costs are higher. Fluidized-bed gasifiers can come in different designs such as bubbling, circulating, or dual fluidized bed gasification reactors.

Entrained-bed gasifiers operate at higher temperatures. Fuel, gasification agent, and/ or steam are fed into the gasifier co-currently. The gasification agent is mostly oxygen. Air is not as common in these gasifiers as the oxidant. Oxygen and steam surround the pulverized fuel particles and the reaction takes place very quickly resulting in short gasifier residence time, which is one of the strengths of these systems. A disadvantage of these designs is the shorter expected lifetime of certain components within the systems due to high operating temperatures.

7.1.3.8 ALGAE FUEL

Algae can be converted into different kinds of biofuels through thermochemical or biological conversions and different processes including gasification, pyrolysis, anaerobic digestion, transesterification, or hydrothermal liquefaction, which is a thermal depolymerization process converting wet algae into biofuel. If seaweed (macroalgae) is used as the biomass source, the product is called seaweed fuel or seaweed oil. Algae are a potential source for biodiesel production. They contain fat pockets on their surfaces that help them float. Fats from these pockets can be collected and processed to produce biodiesel.

FURTHER LEARNING

Algae Technology Educational Consortium
https://algaefoundationatec.org/

FURTHER LEARNING

Interactive Biofuels Atlas
https://maps.nrel.gov/biomass/

FIGURE 7.8 Biomass gasification facility. Source: Office of Energy Efficiency and Renewable Energy, Department of Energy. www.energy.gov/eere/fuelcells/hydrogen-production-biomass-gasification.

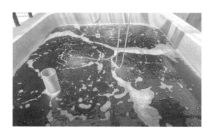

FIGURE 7.9 Green algae being cultured in a tank. *Algaculture* is the farming of various algae species. The cultivated algae can have different uses including algal fuel, fertilizers, pharmaceuticals, and bioplastics. Source: Cherdchanok Treevanchai/Moment via Getty Images.

FIGURE 7.10 Mechanical grabber transferring wastes at a municipal waste management facility. After the waste is sorted and recyclable waste materials are separated, the non-recyclable portion is sent to the plant to generate process heat or electricity. Source: AzmanL/E+ via Getty Images.

7.1.4 Municipal Solid Waste

Municipal solid waste (MSW) is a source of biomass that can be used for producing process heat and electricity. It is also called trash or garbage in the United States, and rubbish in British English. It consists of items such as food, bottles, newspapers, yard trimmings, fabrics, electronics, and packaging. MSW can include residential waste and commercial and institutional wastes, such as wastes from businesses, schools, and hospitals. The EPA's definition of MSW does not refer to industrial, hazardous, and construction waste. The European Statistical Office, Eurostat, defines municipal waste to be mainly produced by households, through similar wastes such as from commercial fields, offices, and public institutions.

Once MSW is produced, it needs to be accumulated and managed properly. The amount of MSW produced consists of waste collected by or on behalf of municipal authorities. After being collected, it goes through a waste management system which can involve different methods such as recycling, composting, combustion with energy recovery, and landfilling (**Figure 7.10**).

7.2 Theory

The stoichiometric equation for combustion of biomass fuels is in a similar form to that for fossil fuels:

$$\text{Fuel} + \text{Oxygen} \rightarrow \text{Carbon dioxide} + \text{Water} + \text{Energy}$$

Wood, as discussed, is one of the most common biomass fuels. It is mostly composed of cellulose which has the formula $C_6H_{10}O_5$. Hence, combustion of wood can be expressed as:

$$C_6H_{10}O_5 + 6O_2 \rightarrow 6CO_2 + 5H_2O \tag{7.1}$$

The two products of combustion of wood are CO_2 and H_2O. Carbon dioxide is released in the form of gas. Due to the temperature of the reaction, water is released in the vapor phase. Therefore, the reaction has a solid reactant (wood) and a gas (oxygen), yielding a gas (carbon dioxide) and water vapor along with some ash, which is formed by minor components and remaining solids with incomplete burning.

Municipal solid waste can consist of food, paper, plastics, yard waste, fabric, glass, metals, and other kinds of components. Each of these components has its own chemical formula, moisture content, and heating value. Therefore, theoretical analysis of input energy for an MSW plant can be intricate depending on the mix in the waste, and moisture content of each component.

Heating values of some selected biomass fuels are listed in **Table 7.1**. Heating values of some materials found in MSW are listed in **Table 7.2**.

Biomass fuels are most efficiently utilized when they provide steam for both useful heat and electricity generation. This is achieved through biomass *cogeneration* systems, also known as *combined heat and power* (CHP) systems. The term cogeneration defines production of more than one useful form of energy (e.g., process heat and electricity) from the same energy source, which can be a fossil fuel such as natural gas, or biomass if this is a waste-to-energy power plant.

Two common configurations of CHP systems exist: a combustion turbine (or reciprocating engine) with heat recovery unit, and a steam boiler with steam turbine. Schematics of both are given in **Figures 7.11** and **7.12**, respectively.

Table 7.1

Average heat content* of selected fuels [5]		
Fuel	Heating value (imperial)	Heating value (SI units)
Biodiesel	5.36 million Btu/barrel	5.65 GJ/barrel
Digester gas	0.62 million Btu/thousand cubic feet	23.1 GJ/thousand cubic meter
Ethanol	3.56 million Btu/barrel	3.76 GJ/barrel
Landfill gas	0.49 million Btu/thousand cubic feet	18.26 GJ/thousand cubic meter
Methane	0.49 million Btu/thousand cubic feet	18.26 GJ/thousand cubic meter
Peat	8.00 million Btu/short ton	9.30 GJ/metric ton
Sludge waste	7.51 million Btu/short ton	8.73 GJ/metric ton
Sludge wood	10.07 million Btu/short ton	11.70 GJ/metric ton
Utility poles	12.50 million Btu/short ton	14.54 GJ/metric ton
Waste alcohol	3.80 million Btu/barrel	4.01 GJ/barrel

* Heating values provided in the report have also been converted to SI units.

Table 7.2

Heat content* of materials in municipal solid waste [6]		
Material	Heating value (million BTU/short ton)	Heating value (GJ/metric ton)
Plastics, PET	20.5	23.8
Plastics, PVC	16.5	19.2
Leather	14.4	16.7
Textiles	13.8	16.0
Wood	10.0	11.6
Food	5.2	6.0
Yard trimmings	6.0	7.0
Newspaper	16.0	18.6
Corrugated cardboard	16.5	19.2
Mixed paper	6.7	7.8

* Heating values provided in the report have also been converted to SI units.

A detailed schematic and the *T–S* diagram of an ideal steam turbine CHP power plant configuration are illustrated in **Figure 7.13**.

In the system depicted in **Figure 7.13**, biomass fuel such as MSW is combusted in the grate furnace to generate steam within the boiler section. Depending on the need for

FIGURE 7.11 Combustion turbine or reciprocating engine CHP system. Source: EPA, www.epa.gov/chp/what-chp

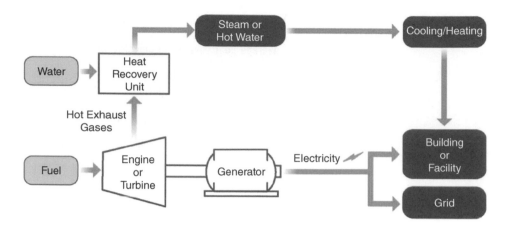

FIGURE 7.12 Steam turbine CHP system. Source: EPA, www.epa.gov/chp/what-chp

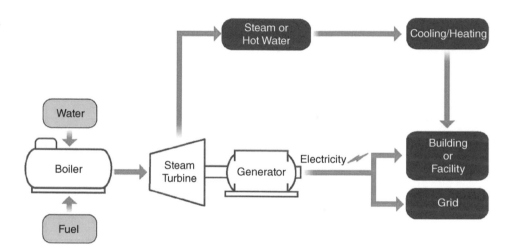

FIGURE 7.13 Schematic and *T–S* diagram for an ideal cogeneration system utilizing biomass fuel.

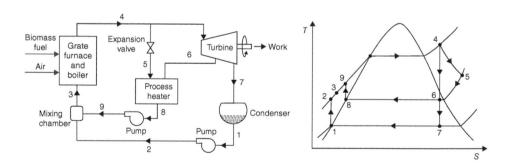

process heat or electricity generation, steam flow through the expansion valve directly into the process heater and or the extraction flow rate from the turbine can be adjusted.

The efficiency of cogeneration plants can be identified by the *utilization factor*, which is the fraction of input energy used for process heating and/or power generation. The utilization factor is given by:

$$\epsilon_U = \frac{\dot{W}_{net} + \dot{Q}_{process}}{\dot{Q}_{in}} = \frac{\left(\dot{W}_{turb} - \dot{W}_{pumps}\right) + \dot{Q}_{process}}{\dot{Q}_{furnace}} \tag{7.2}$$

Control volume analysis on each component yields the energy balance through each unit:

$$\dot{Q}_{in} = \dot{Q}_{furnace} = \dot{m}_3(h_4 - h_3) \tag{7.3}$$

$$\dot{W}_{turb} = (\dot{m}_4 - \dot{m}_5)(h_4 - h_6) + \dot{m}_7(h_6 - h_7) \qquad (7.4)$$

$$\dot{W}_{pumps} = v_{f1}(p_2 - p_1) + v_{f8}(p_9 - p_8) \qquad (7.5)$$

$$\dot{Q}_{process} = \dot{m}_5 h_5 + \dot{m}_6 h_6 - \dot{m}_8 h_8 \qquad (7.6)$$

$$\dot{Q}_{out} = \dot{Q}_{condenser} = \dot{m}_7(h_7 - h_1) \qquad (7.7)$$

Due to the parallel flow of steam at different sections of the cycle, the mass flow rate of steam will not be equal to the total value at each section:

$$\dot{m}_{total} = \dot{m}_3 = \dot{m}_4 = \dot{m}_5 + \dot{m}_6 + \dot{m}_7 = \dot{m}_2 + \dot{m}_9 \qquad (7.8)$$

Example 7.1 Determining Energy Equivalence

One metric ton of bituminous coal has an energy equivalence of approximately 25 GJ. Using **Table 7.2**, determine the amount of each material listed below to have an energy equivalent to 1500 kg of bituminous coal:

a. Polyethylene terephthalate (PET) plastic
b. Textiles
c. Wood
d. Yard trimmings

Solution

First of all, let's get the heating value data for the listed materials from **Table 7.2**:

Material	Heating value (GJ/metric ton)
Plastics, PET	23.8
Textiles	16.0
Wood	11.6
Yard trimmings	7.0

If one ton of bituminous coal has 25 GJ equivalent energy, then 1500 kg (1.5 ton) would have (1.5) × (25 GJ) = 37.5 GJ of energy. Then, the amount of each material listed to get the same value of energy can be calculated:

a. $m_{PET} = \dfrac{37.5 \text{ GJ}}{23.8 \dfrac{\text{GJ}}{\text{ton}}} \cong 1.58 \text{ ton}$

b. $m_{textile} = \dfrac{37.5 \text{ GJ}}{16 \dfrac{\text{GJ}}{\text{ton}}} \cong 2.34 \text{ ton}$

c. $m_{wood} = \dfrac{37.5 \text{ GJ}}{11.6 \dfrac{\text{GJ}}{\text{ton}}} \cong 3.23 \text{ ton}$

d. $m_{yard} = \dfrac{37.5 \text{ GJ}}{7 \dfrac{\text{GJ}}{\text{ton}}} \cong 5.36 \text{ ton}$

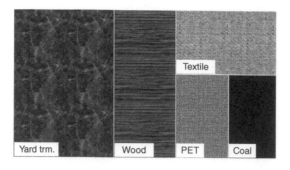

Example 7.2 Carbon Dioxide Production in Combustion of Biomass

Determine the amount of CO_2 produced by combustion of 1 metric ton of wood.

Solution

The combustion process for wood can be written as given in **Equation 7.1**:

$$C_6H_{10}O_5 + 6O_2 \rightarrow 6CO_2 + 5H_2O + \text{Energy}$$

As can be seen in the equation, for every mole of wood (approximated by the chemical formula for cellulose, $C_6H_{10}O_5$) combusted, six moles of CO_2 are released.

Let's look at the molar mass of each compound:

Molar mass of $C_6H_{10}O_5$ is : $(6 \times 12) + (10 \times 1) + (5 \times 16) = 162 \text{ kg/kmol}$

Molar mass of CO_2 is : $(1 \times 12) + (2 \times 16) = 44 \text{ kg/kmol}$

Hence, for every 162 kg of wood burnt, approximately $6 \times 44 = 264$ kg CO_2 is released.

Then, the amount of CO_2 produced during combustion of 1 ton of wood is:

$$m_{CO_2} = \left(\frac{264 \text{ kg } CO_2}{162 \text{ kg wood}} \right) (1000 \text{ kg wood}) = 1629.6 \text{ kg} \cong 1.63 \text{ ton}$$

For every ton of wood burnt, 1.63 tons of CO_2 is released. It should be noted that dry wood is primarily composed of cellulose, lignin, and hemicellulose. The chemical formula of wood was approximated to be the same as the formula for cellulose as it makes up about 50% of the wood by weight.

Cross-sectional view of a tree stump. The main layers of a tree trunk are the outer bark, phloem, cambium, sapwood, growth (annual) ring, and heartwood. Source: Brett Wrightson/ EyeEm via Getty Images.

Example 7.3 Analysis of a Biomass CHP Plant

A biomass CHP plant has steam leaving the boiler at 6 MPa and 480 °C with a mass flow rate of 12 kg/s. Thirty percent of this steam is directed to the process heater through an expansion valve with pressure reducing to 400 kPa. Twenty percent of the total steam is then extracted from the turbine at 400 kPa to contribute to the process heating. The remaining 50% expands to 10 kPa. Steam leaves both the process heater and the condenser as saturated liquid and is pumped to the boiler pressure of 6 MPa through both pumps in the system. The turbine and pumps are assumed to operate isentropically.

a. Sketch the T–S diagram
b. Determine the total rate of process heat (MW)
c. Find the net power output (MW)
d. Calculate the utilization factor (%)

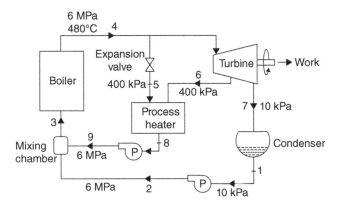

Solution

a. The T–S diagram for the CHP plant is given below along with enthalpy values for the states.

$h_1 = 191.83$ kJ/kg **(Table B.2)**

$h_2 = h_1 + v_1(p_2 - p_1) = 191.83 + 5.99 = 197.8$ kJ/kg

$h_3 = \dfrac{(\dot{m}_2 h_2) + (\dot{m}_9 h_9)}{\dot{m}_3} = 404.1$ kJ/kg

$h_4 = h_5 = 3375.4$ kJ/kg **(Table B.3)**

$h_6 = 2706.6 \text{ kJ/kg (isentropic, } s_4 = s_6, x_6 = 0.99)$

$h_7 = 2160.5 \text{ kJ/kg (isentropic, } s_4 = s_7, x_7 = 0.82)$

$h_8 = 604.74 \text{ kJ/kg (Table B.2)}$

$h_9 = h_8 + v_8 (p_9 - p_8) = 604.74 + 5.6 = 610.3 \text{ kJ/kg}$

b. The rate of process heat can be determined by applying the energy balance equation to the process heater as the control volume:

$$\dot{Q}_{process} = (\dot{m}_5 h_5) + (\dot{m}_6 h_6) - (\dot{m}_8 h_8)$$

$$= \left[\left(3.6 \frac{\text{kg}}{\text{s}} \right) \left(3375.4 \frac{\text{kJ}}{\text{kg}} \right) \right] + \left[\left(2.4 \frac{\text{kg}}{\text{s}} \right) \left(2706.6 \frac{\text{kJ}}{\text{kg}} \right) \right] - \left[\left(6 \frac{\text{kg}}{\text{s}} \right) \left(604.65 \frac{\text{kJ}}{\text{kg}} \right) \right]$$

$$= 15{,}019.4 \frac{\text{kJ}}{\text{s}} \left| \frac{1 \text{ MW}}{10^3 \frac{\text{kJ}}{\text{s}}} \right| \cong 15.02 \text{ MW}$$

c. The net power output of the plant is calculated by:

$$\dot{W}_{net} = \dot{W}_{turb} - \dot{W}_{pumps}$$

$$\dot{W}_{turb} = (\dot{m}_4 - \dot{m}_5)(h_4 - h_6) + \dot{m}_7(h_6 - h_7)$$

$$= \left[(12 - 3.6) \frac{\text{kg}}{\text{s}} (3375.4 - 2706.6) \frac{\text{kJ}}{\text{kg}} \right] + \left[\left(6 \frac{\text{kg}}{\text{s}} \right) (2706.6 - 2160.5) \frac{\text{kJ}}{\text{kg}} \right]$$

$$= 8894.5 \text{ kW}$$

$$\dot{W}_{pumps} = \dot{m}_1(h_2 - h_1) + \dot{m}_8(h_9 - h_8) = \left[\left(6 \frac{\text{kg}}{\text{s}} \right) \left(5.99 \frac{\text{kJ}}{\text{kg}} \right) \right] + \left[\left(6 \frac{\text{kg}}{\text{s}} \right) \left(5.6 \frac{\text{kJ}}{\text{kg}} \right) \right]$$

$$= 69.54 \text{ kW}$$

Then,

$$\dot{W}_{net} = 8894.5 - 69.54 \cong 8825 \text{ kW} = 8.825 \text{ MW}$$

d. The utilization factor can be obtained by using **Equation 7.2**:

$$\in_U = \frac{\dot{W}_{net} + \dot{Q}_{process}}{\dot{Q}_{in}} = \frac{(8.825 + 15.02) \text{ MW}}{\dot{m}_3(h_4 - h_3)} = \frac{23.845 \text{ MW}}{\left[\left(12 \frac{\text{kg}}{\text{s}} \right) (3375.4 - 404.1) \frac{\text{kJ}}{\text{kg}} \right] \left| \frac{1 \text{ MW}}{10^3 \frac{\text{kJ}}{\text{s}}} \right|}$$

$$= \frac{23.845 \text{ MW}}{35.656 \text{ MW}} \cong 0.67 \quad (67\%)$$

Note: *The utilization factor implies that 67% of the input energy was useful in achieving heat and power. Unlike thermal efficiencies, utilization factors can go above 80%, depending on the components used in the system and the percentage distribution of steam for process heat and power generation. The greater the amount of steam used for process heating, the higher the $\in_U$ will be as the amount of heat rejected from the condenser to the outside environment will be smaller.*

Example 7.4 Plotting Biomass Data Using Python

Fuel ethanol data can be obtained from the EIA website. Plot fuel ethanol production and consumption in the United States between 1981 and 2021 based on EIA data.[1]

Solution

The data can be retrieved from: www.eia.gov/totalenergy/data/browser/xls.php?tbl=T10.03&freq=m.

Then, the source of data can be embedded into the code.

```
import matplotlib.pyplot as plt
import pandas as pd
url="https://www.eia.gov/totalenergy/data/browser/xls.php?
tbl=T10.03&freq=m"
df=pd.read_excel(url, 1,skiprows=10)
df=df.drop([0])

plt.title("Fuel Ethanol Production and Consumption (Trillion Btu)")
plt.xlabel("Year")
plt.ylabel("Trillion Btu")

plt.plot(df["Annual Total"],
df["Fuel Ethanol Production"], label="Production")
plt.plot(df["Annual Total"],
df["Fuel Ethanol Consumption"], label="Consumption")
plt.legend()


<matplotlib.legend.Legend at 0x7f285041bfd0>
```

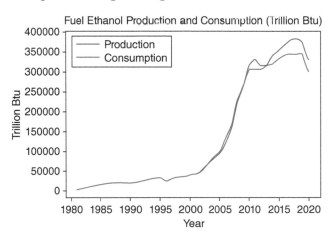

[1] U.S. Energy Information Administration, Monthly Energy Review December 2021.

7.3 Applications and Case Study

In this section, we will observe real-world examples via two applications and a case study. The applications include a cellulosic ethanol plant in Brazil, and a municipal solid waste (MSW) power plant in Ethiopia. The case study is about a combined heat and power (CHP) plant in Naantali, Finland.

Application 1 Cellulosic Ethanol Plant (Brazil)

Bioflex 1 Cellulosic Ethanol Plant
Location: Alagoas, Brazil
Owner: GranBio
Annual ethanol production capacity:
    82 million liters
Annual electricity generation
    capacity[1]: 135,000 MWh
Data: www.granbio.com.br/en/ [7]
www.d-maps.com

Bioflex 1 is the largest second-generation (2G) ethanol plant in the world. It has a production capacity of 82 million liters of ethanol per year. Besides production of ethanol as a biofuel, bioenergy is also generated. Lignin, a byproduct of second-generation ethanol, is burnt together with sugar cane bagasse to produce steam for electricity generation. The cogeneration system, which is a partnership between GranBio and Usina Caeté, is built next to the Bioflex 1 unit and has 200 tons/h steam generation capacity. The boiler of the cogeneration system runs for 11 months of the year and supplies 135,000 MWh/year to the grid [7]. In addition to being an alternative source for the energy demand in Brazil, electricity generation with wastes significantly reduces CO_2 emissions.

[1] In partnership with Usina Caeté

Application 2 Municipal Solid Waste Power Plant (Ethiopia)

Reppie Waste-to-Energy Plant
Location: Addis Ababa, Ethiopia
Owner: Ethiopian Electric Power
    (State-run)
Power Capacity: 20 MW
MSW Capacity: 1400 tons/day
Data: https://ramboll.com/projects [8]
www.d-maps.com

Reppie is the first waste-to-energy plant on the African continent. This power plant uses municipal solid waste (MSW). The waste is burnt in two grate-firing systems with vertical serpentine economizers to produce steam. The steam is then sent to turbines to generate electricity. Each waste combustion unit has a capacity of 700 tons/day. Flue gas treatment involves selective non-catalytic reduction (SNCR), dry flue-gas desulfurization (FGD), and a baghouse filter to lessen nitrogen oxide and sulfur dioxide emissions and to filter the dusty gas.

One of the advantages of waste-to-energy power plants is that, besides providing energy from waste, they open up land area which otherwise would be used for landfill. In the Koshe area in Addis Ababa, a landslide of waste caused more than 100 fatalities. Treatment of waste for biomass energy helps mitigate such risks.

Case Study Combined Heat and Power Plant (Finland)

Naantali CHP Plant!
Location: Naantali, Finland
Owner: Turun Seudun Energia
Power Capacity: 256 MW
District Heating Capacity: 1400 GWh/yr
Data: www.fortum.com [9]
www.d-maps.com

INTRODUCTION Finland has the most forest land in Europe with forests covering almost 75% of its land area. These forests provide rich resources for utilizing biomass in heating and energy generation. Finland's average monthly temperatures can go below $-10\,^{\circ}$C, resulting in high heating loads for buildings for a considerable fraction of the year. District heating therefore is a common practice in Finland. In recent years, district heating from biomass (wood fuels) has been the prior resource, followed by coal and peat.

OBJECTIVE The objective of this case study is to learn about a CHP plant as part of the goal of a Finnish city in becoming carbon neutral.

METHOD This case study was conducted through researching articles and reports.

RESULTS The Finnish Government National Energy and Climate Strategy outlines actions at the national level such as prohibiting coal in energy generation by law in 2029 and increasing the share of transportation biofuels to 30%. At the local level there are efforts aligned with these goals. The city of Turku aims to become a carbon neutral city by 2029 [10]. This is planned to be achieved by three key measures:

— Increasing the share of renewables in both district heating and energy generation
— Improving energy efficiency in all operations
— Promoting sustainable modes of transportation

An example action that addresses the first measure is the Naantali Combined Heat and Power (CHP) Plant with 256 MW capacity in Turku. The plant was commissioned to become a multi-fuel plant where it can use biomass in addition to coal as fuel. The plant produces 1400 GWh district heat, 200 GWh process steam, and 800 GWh electricity annually. Biomass has become the main fuel of the plant with a share of 60–70% of all fuel [10].

DISCUSSION The significance of such an effort by a city is that it can set a good example for other cities in the world to take measures towards becoming carbon neutral. CHP plants have already proven to be more efficient and use of biomass as a local energy resource can help harvest energy with reduced emissions, less reliance on imported fuel, and lower fuel transportation costs.

RECOMMENDATIONS Some of the common problems that slow down the adoption of CHP power plants using biomass are financial limitations, interconnection barriers, and reluctant utility companies. All these challenges need to be addressed at local or national scales for widespread use of multi-fuel CHP plants.

 CLASSROOM DEBATE

Compare the benefits of biomass to the deforestation risks associated with it. Debate the strengths vs weaknesses of biomass energy.

FIGURE 7.14 A biogas facility and a wind turbine next to a corn field. Source: Jan-Otto/E+ via Getty Images.

7.4 Economics

The economics of biomass can be studied in two areas: biofuel production, and power generation from biomass. As for biofuel production, these renewable fuels compete with fossil fuels such as gasoline and diesel on price. As prices differ among various countries based on the reserves and trading arrangements of fossil fuels and the technology for biofuel production, competition of biofuels with traditional fuels varies. Significant decreases in prices of natural gas (since ~2010) and crude oil (since ~2014) have affected investments in biofuels. This is one of the reasons for the modest, rather than inspiring increase in biofuel production globally, as discussed in the Global Overview section of this chapter. A comparative analysis by the IEA in **Table 7.3** portrays ethanol and biodiesel production costs, and the oil prices for biofuel costs to break even in price competition [11].

On the power generation side, biomass has been performing better. The economics of biomass power generation has three components:

Total installed cost: This is the total cost accounting for designing and construction of the plant, machinery for fuel processing, and significant components such as the prime mover and fuel conversion systems, process heating units, and energy generation system. Other costs such as connection to the grid, or infrastructure (e.g., roads) also add up to the total installed costs. While the enhanced systems such as CHP designs have higher initial costs, their higher overall efficiency and the services they provide such as heat/steam production for industrial processes or district heating can increase the feasibility of such investments significantly.

Operating and maintenance cost: Operating costs include utilities, wages and salaries, social security, unemployment insurance, workers' compensation, medical insurance, and taxes. Maintenance costs account for routine maintenance,

Table 7.3

Ethanol and biodiesel production costs and break-even oil prices for Brazil and the United States, 2017 [11]				
Country	Ethanol production cost (USD/Lge)	Biodiesel production cost (USD/Lde)	Ethanol break-even (USD/bbl)	Biodiesel break-even (USD/bbl)
Brazil	0.54–0.62	0.73–0.98	50–60	81–120
United States	0.51–0.58	0.76–0.86	64–76	104–120

Notes: Lge = liter of gasoline equivalent; Lde = liter of diesel equivalent; bbl = barrel.

Biofuel production costs have been adjusted to account for the higher calorific values of gasoline and diesel. In 2017, energy-adjusted production costs for ethanol in Brazil were equal to domestically produced gasoline on average, compared with an ethanol premium of 0.15 USD/L in the United States. In Brazil the biodiesel premium was 0.26 USD/L above fossil-based diesel production on average, smaller than in the United States, where the premium was 0.37 USD/L in 2017. Production costs exclude taxes. Sources: F. O. Lichts (2018), F. O. Lichts Interactive Data (database), www.agra-net.com; ANP (2018), "Producer prices," www.anp.gov.br; U.S. EIA (2018), Petroleum and Other Liquids (database), www.eia.gov.

FIGURE 7.15 Global weighted-average total installed costs, capacity factors, and LCOE values for biomass energy, 2010–2020. Source: IRENA Renewable Cost Database.

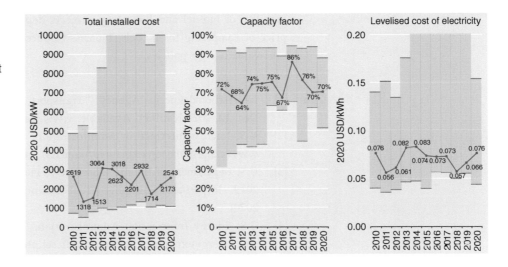

troubleshooting, and replacement of plant components such as boilers, gasifiers, and feedstock handling equipment.

Feedstock cost: This is the fuel cost for biomass plants. This cost depends on several factors such as the geographic location and climate conditions for the plants, their chemical composition and energy density, cost of land used, harvesting, delivery, storage, and disposal costs, and fuel processing technology used.

Global weighted-average total installed costs, along with capacity factors and levelized cost of energy (LCOE) values during 2010–2020 can be seen in **Figure 7.15**. In 2020, the global average for total installed cost was $2543 per kW of capacity. In the same year, weighted-average LCOE ranged from 0.057 $/kWh in India and 0.06 $/kWh in China to 0.087 $/kWh in Europe and 0.097 $/kWh in North America according to IRENA's renewable cost database. These are the nationally weighted averages. As for an individual biomass plant utilizing the industrial process steam or heat onsite, LCOE values can be as low as 0.3 $/kWh [12].

7.5 Summary

Biomass is organic matter that is available on a renewable basis, including agricultural, forestry, animal, and industry residues, municipal solid waste, sewage, and algae. These biomass sources can be utilized for producing biofuels, electricity, process heat, or some bioproducts. In this chapter, biofuels, process heat, and electricity generation were explored including theoretical and practical aspects.

Biofuel analysis and production involves some fundamental knowledge of natural sciences, mainly biology and chemistry, and their subdiscipline, biochemistry. Thermochemical processes such as pyrolysis and gasification require a good understanding of heat transfer and in some applications fluid mechanics, in addition to chemistry. Analysis of power generation from biomass resources relies mostly on thermodynamics principles. Within the chapter, thermodynamic analysis of cogeneration systems executing Rankine cycles was discussed. In biomass power plants, the fuel might be biomass feedstock for direct use or biofuel which would have been produced from biomass. Municipal solid waste is also an important source of bioenergy utilized by municipalities themselves or the private sector all around the world. The economic viability of biofuel and bioenergy projects depends on several factors including the price of competing fossil fuels, biomass resources, available technology, and existing renewable energy policies.

REFERENCES

[1] International Energy Agency, "Transport Biofuels," IEA, Paris, 2020. Accessed: Dec. 12, 2020 [Online]. Available: www.iea.org/reports/transport-biofuels.

[2] U.S. Energy Information Administration, "Biomass Explained," Aug. 28, 2020. www.eia.gov/energyexplained/biomass/ (accessed Dec. 08, 2020).

[3] U.S. Department of Energy, Alternative Fuels Data Center, "Ethanol Fuel Basics," 2019. https://afdc.energy.gov/fuels/ethanol_fuel_basics.html (accessed Dec. 8, 2020).

[4] U.S. Energy Information Administration, "Biodiesel Explained," 2016. www.eia.gov/energyexplained/biofuels/biodiesel.php (accessed Dec. 8, 2020).

[5] U.S. Energy Information Administration, "Renewable Energy Annual," 2009. Accessed: Dec. 9, 2020. Available: www.eia.gov/renewable/annual/trends/pdf/table1_10.pdf.

[6] U.S. Energy Information Administration, "Methodology for Allocating Municipal Solid Waste to Biogenic and Non-Biogenic Energy," EIA, Washington, DC, May 2007. Accessed: Dec. 9, 2020 [Online]Available: www.eia.gov/totalenergy/data/monthly/pdf/historical/msw.pdf.

[7] GranBio, "BioEnergy," GranBio: Bioenergy solutions. www.granbio.com.br/en/conteudos/bioenergy/ (accessed Dec. 9, 2020).

[8] Ramboll, "Reppie in Addis Ababa, Ethiopia: The First Waste-To-Energy Facility on the African Continent," 2018. https://ramboll.com/projects/re/reppie-in-addis-ababa-ethiopia-first-waste-to-energy-facility-african-continent (accessed Dec. 9, 2020).

[9] Fortum, "Naantali CHP Plant." www.fortum.com/about-us/our-company/our-energy-production/our-power-plants/naantali-chp-plant (accessed Dec. 10, 2020).

[10] I. Georgiev and I. Dobre, "Case Study on Biomass Use in CHP in Turku, Finland," European Commission, Brussels, Jun. 2019. Accessed: Dec. 11, 2020.

Available: http://publications.europa.eu/resource/cellar/c74cbafb-c490-11e9-9d01-01aa75ed71a1.0001.01/DOC1.

[11] International Energy Agency (IEA), "How Competitive Is Biofuel Production in Brazil and the United States?" 2019, Accessed: Dec. 12, 2020 [Online]. Available: www.iea.org/articles/how-competitive-is-biofuel-production-in-brazil-and-the-united-states.

[12] IRENA, "Renewable Power Generation Costs in 2020," IRENA, Abu Dhabi, Jun. 2021. Accessed: Aug. 12, 2021 [Online]. Available: www.irena.org/publications/2021/Jun/Renewable-Power-Costs-in-2020.

CHAPTER 7 EXERCISES

7.1 Determine the amount of CO_2 produced by combustion of 60 bushels of corn, assuming the chemical composition of corn is approximated to be $C_{27}H_{48}O_{20}$.

7.2 One barrel of oil equivalent (BOE) converts to 6.12 GJ at 15 °C. Using Table 7.2, determine the amount of each material listed below that would have an energy equivalent to 50 barrels of crude oil:

a. Leather

b. Mixed paper

c. Food

d. Corrugated cardboard

7.3 Sketch a flow diagram for ethanol production with the dry-milling process and briefly explain each step. How does the dry-milling process differ from the wet-milling process?

7.4 Australia is one of the countries promoting use of bioethanol as an alternative fuel for transportation. According to a study in Queensland, the bioethanol yields of agave and sugar cane are as given in the table below. Determine the amount of land required for both crops to provide ethanol fuel (E85) that has equivalent energy to replace gasoline for 3000 flexible fuel vehicles (FFVs) in Queensland, with each car having a fuel tank volume of 40 L. Assume the E85 fuel to contain 83% ethanol ($\rho \approx 790$ kg/m^3) and 17% gasoline ($\rho \approx 770$ kg/m^3) by volume. Energy contents of gasoline and ethanol are also provided in the table.

Crop	Bioethanol yield (L/ha)
Agave	7500
Sugar cane	9900
Fuel	**Energy content (MJ/kg)**
Gasoline	34.2
Ethanol	24

7.5 In a manure-to-energy facility in Texas, approximately 180 m^3 methane (CH_4) is produced per metric ton of dry cattle manure. If the density of methane at 18 °C and standard atmosphere is 0.664 kg/m^3 and its specific energy is 55.5 MJ/kg, determine:

a. the amount of CO_2 released during combustion of all methane produced from 1.25 ton dry manure (kg)

b. the amount of energy released during the combustion process if all methane is burnt (MJ)

Cowboys herding the cattle at a ranch in Texas. Besides meat production, the cattle industry is important for other byproducts, one of which is manure. Cattle manure can be used in soil enrichment and as a source of biofuels. Source: Andy Sacks/The Image Bank via Getty Images.

7.6 Sketch the schematic of a cross-draft gasifier. Explain how it differs from a downdraft and an updraft gasifier.

7.7 The oil yields of various crops are provided in the table below in liter/hectare. *Microalgae* are a potential biodiesel feedstock due to their much higher oil yield per area. If for every 100 kg of oil reacting with alcohol, 100 kg of biodiesel is produced, determine the area required to produce 2 metric tons of biodiesel from each crop. For microalgae, calculate the range of area required based on the oil yield range given. Density of oil can be approximated as 900 kg/m^3.

Crop	Oil yield[1] (L/ha)
Corn	168
Soybean	449
Canola	1188
Jatropha	1890
Coconut	2685
Oil Palm	5949
Microalgae[2]	58,871–136,951

[1] Values converted to SI units from https://farm-energy.extension.org/algae-for-biofuel-production/
[2] The range is due to the oil content (%) of microalgae by mass

7.8 Research the total municipal solid waste generation in the United States in 2022. Explore the components and their shares in the total amount generated. How is the solid waste managed?

7.9 In 2021, the population of Amsterdam in the Netherlands was approximately 1.16 million. Assume that each individual in the city produces municipal solid waste (MSW) at a rate of 1 kg daily with the MSW mix having an average heating value of 11,000 kJ/kg. If this MSW mix is used as biomass fuel at a waste-to-energy (WTE) power plant with an overall thermal efficiency of 38%, determine the rate of power generation at this plant in Amsterdam.

7.10 A biomass CHP plant has steam leaving the boiler at 6 MPa and 480 °C with a mass flow rate of 10 kg/s. Some portion of this steam is directed to the process heater through an expansion valve with pressure reducing to 400 kPa. Some other fraction of the steam is extracted from the turbine at 400 kPa to be directed to the process heater. The remaining steam leaves the turbine at 10 kPa. Steam exits the process heater and the condenser as saturated liquid and is pumped to the boiler pressure of 6 MPa through both pumps in the system. The turbine and pumps are assumed to operate isentropically. Calculate (a) the net power output in MW, (b) the total rate of process heat in MW, and (c) the utilization factor of the cogeneration plant for the following special cases:

a. 0% through the expansion valve, 20% extracted from the turbine, 80% exiting the turbine

b. 80% through the expansion valve, 0% extracted from the turbine, 20% exiting the turbine

Schematics for operating scenarios (a) and (b).

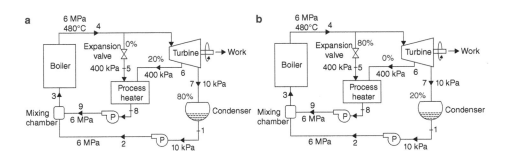

CHAPTER 8
Geothermal Energy

8.1 Introduction

The word geothermal is a combination of two words: *geo* meaning earth, and *thermal* meaning heat related, with both words originating from Greek. Geothermal energy, therefore, is the energy that is extracted from the Earth. Geology is the science that studies the physical structure of the Earth, the substances that the structural features are formed of, and the processes that act on the formations. Besides the Earth, the scope of geology can include study of other planets such as Mars, or natural satellites such as the Moon. To better understand geothermal energy, one would first need to have a good understanding of the geological structure of the Earth. The Earth is made up of three layers, which are the core, the mantle, and the crust, from the center to the surface. Cutaway views of these layers can be seen in **Figure 8.1**. The *core* is at the center of the Earth, approximately 2900 km beneath the surface, and is made up of two layers. These layers are the outer core which is liquid and the inner core which is solid. The inner core has a diameter of about 2400 km and is believed to be composed mostly of iron and nickel. The temperature here is about 5000 °C. The *mantle* lies between the core and the crust and is about 2900 km thick (which is the approximate distance of the core from the surface as the crust's thickness is less than 2% of that thickness). The mantle is made up of hot iron and magnesium-rich solid rocks and makes up about 84% of Earth's volume. The *crust* is the outermost and the thinnest layer. It is of two types, which are the continental crust (30–65 km thick) and the oceanic crust (5–10 km thick). The *lithosphere*, which includes the uppermost part of the mantle and the crust, consists of distinct regions called the *tectonic plates*, which are illustrated in **Figure 8.2**.

Heat transfer in the direction from the core to the crust of the Earth causes convective currents and results in the mobile molten rocks beneath the crust moving in a circular manner. This is similar to free convection of air over a heated vertical wall. As air gets close to the wall, it heats up, gets less dense, and rises. At the top of the wall, as air moves away from the wall, it gets colder, and hence denser, resulting in a descending flow. This repeated cycle causes air to experience a vertical circular flow. Convective currents in the mantle are formed in a similar manner, with the heated molten rocks rising and then flowing in the direction of the core again as they get colder, just like simmering soup where warmer soup rises in the pot and colder soup at the top sinks to the bottom as it is denser. This natural event can cause fracturing at the ocean floor resulting in *mid-ocean ridge* formation when the molten mantle reaches the bottom of the ocean and solidifies. The mantle that melts and spills out of the crust is called *magma*. There is also another natural event that is the inverse of magma making it to the surface. This "recycling" phenomenon is called *subduction*, where the colder lithosphere, which is called the subducting plate, descends back into the mantle. A 3D illustration of mid-oceanic ridges and subducting plates as a result of the movement of tectonic plates can be seen in **Figure 8.3**.

FIGURE 8.1 Cutaway views of the geological structure of the Earth. The smaller image on the left is drawn to scale and the drawing shows that Earth's crust is very thin. The larger image on the right side is not drawn to scale and it outlines the three main layers (crust, mantle, core) in more detail. Source: U.S. Geological Survey.

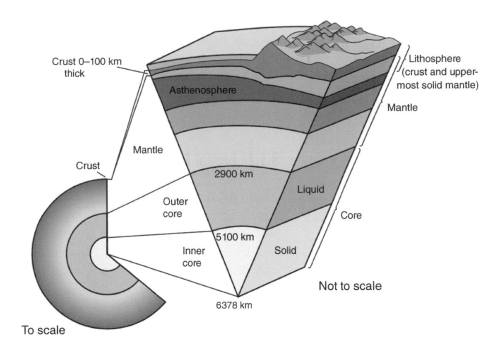

FIGURE 8.2 The Earth's tectonic plates. The plates rip apart at divergent plate boundaries, crash together at convergent plate boundaries, and slide past each other at transform plate boundaries. National Park Services image modified from "Beauty from the Beast: Plate Tectonics and the Landscapes of the Pacific Northwest," by Robert J. Lillie, Wells Creek Publishers, 92 pp., 2015. Source: National Park Services.

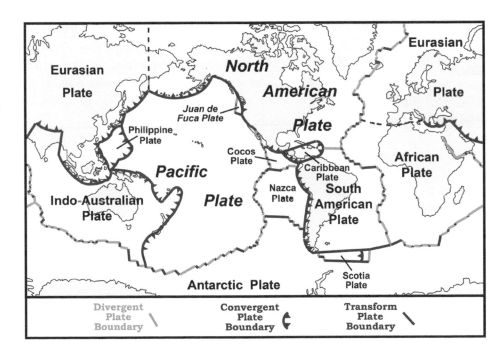

FIGURE 8.3 Three-dimensional illustration of mid-oceanic ridges and subducting plates. Source: Christoph Brugstedt/Science Photo Library via Getty Images.

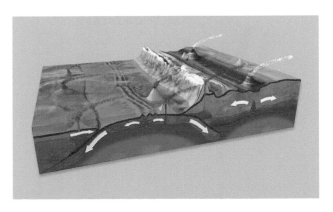

FIGURE 8.4 Fjaðrárgljúfur Canyon in Iceland, dating back to the Ice Age, was formed by erosion caused by glaciers moving through the rocks. Source: Tobias Ackeborn/ Moment via Getty Images.

FURTHER LEARNING

Animation: Tectonic Forces
https://spaceplace.nasa.gov/ tectonics-snap/

FURTHER LEARNING

Geothermal Energy Data
www.irena.org/geothermal

Geothermal energy potential is higher at or near the plate boundaries as the temperature gradients beneath these zones are much larger than in other parts of the crust. The energy can either be used directly for residential, commercial, or industrial needs such as heating, drying, cooling, or hot water; or it can be utilized to generate electricity.

8.1.1 Global Overview

Global geothermal power generation has been increasing steadily from 52 TWh in 2000 to 92 TWh in 2019 [1]. Total installed geothermal global capacity by the end of 2021 was reported to be 15,854 MW [2]. The top 10 countries contributing to this capacity can be seen in **Figure 8.5** with their corresponding capacity values. Countries with installed capacities over 1 GW are the United States, Indonesia, Philippines, Turkey, and New Zealand.

The United States had the largest installed capacity in the world, followed by Indonesia, Philippines, and Turkey as of 2020. All these countries happen to be on or close to plate boundaries where tectonic and volcanic activities are more likely to be observed. Knowing that geothermal energy potential is proportional to the geothermal temperature gradient, it is only logical to guess that the temperature gradients in these regions would be higher than those for other parts of the globe. As an example, a geothermal temperature gradient map of the United States can be seen in **Figure 8.6**. The scale shows the temperature variation in degrees Celsius per kilometer of depth. As can be seen on the map, the highest gradient regions are where most of the fault lines exist. It is a similar case when the geothermal energy potentials of other parts of the world are also examined.

FIGURE 8.5 Top 10 geothermal countries based on installed capacity in MWe by the end of 2021. Total installed capacity was 15,854 MW. Source: ThinkGeoEnergy Research 2022.

Top 10 Geothermal Countries 2021

Installed Capacity in MWe Year-End 2021

Total 15,854 MW

THINK
GEOENERGY

Source: ThinkGeoEnergy Research 2022

FIGURE 8.6 Geothermal temperature gradient for the United States in degrees Celsius per kilometer of depth. Source: U.S. Department of Energy.

Geothermal resources of the United States

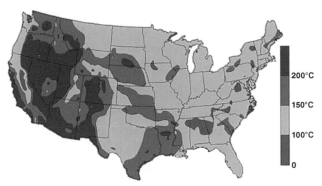

Source: U.S. Department of Energy, Office of Energy Efficiency & Renewable Energy (public domain)

FIGURE 8.7 Turquoise pools in travertine terraces in Pamukkale, a district in the Aegean region in Turkey. Thermal spring water has been coming from underground for over two millennia now. The ancient Romans built the spa city, Hierapolis, here for their citizens for recreational and health purposes. Source: Malcolm P. Chapman/Moment via Getty Images.

FIGURE 8.8 Another travertine formation at the Mammoth Hot Springs in Yellowstone National Park in Wyoming. The travertine pools were formed over thousands of years from the calcium carbonate in the hot springs. Source: Addy Ho/EyeEm via Getty Images.

8.1.2 Geothermal Resources

Geothermal resources can be categorized based on their characteristics as:

- Geothermal gradient
- Hot dry rock
- Hot water reservoirs
- Natural steam reservoirs
- Molten magma
- Geopressured reservoirs

8.1.2.1 GEOTHERMAL GRADIENT

The geothermal gradient is a natural resource of heat to harvest energy from beneath the Earth's surface due to the temperature rise with increasing depth. The normal geothermal gradient can still be useful in utilizing the heat from the depths of the crust in areas that are not necessarily in tectonic regions. Even a comparatively low temperature gradient of 30 °C/km can provide a source temperature of about 180 °C in a well drilled to 6 km depth, which is sufficient for power generation. Much smaller depths can still be functional for heating purposes. A well drilled to a depth of about 2.7 km can provide a source temperature of approximately 80 °C which can be used in direct heating applications. While the energy potential with the normal geothermal gradient is enormous and the technology is available, extracting energy by this method is still not economic due to the challenges in drilling to those depths, where there may exist strata that are not as easy to pass through. Stratigraphic analysis for the area is part of the feasibility studies for such applications, where the order, position, permeability, and porosity of the layers are examined. Having to circulate working fluid in the long piping systems is another economic challenge in extracting energy through the geothermal gradient.

8.1.2.2 HOT DRY ROCK

Hot dry rock is another geothermal resource which is very similar to the normal geothermal gradient. The only difference is that the temperature gradient is higher with these resources, exceeding 40 °C/km. They also possess a great amount of energy potential, and yet they have the same feasibility problems.

8.1.2.3 HOT WATER RESERVOIRS

Underground water trapped between impermeable layers forms reservoirs which can heat up to high temperatures for direct heating, or even power generation, depending on the reservoir temperature.

8.1.2.4 NATURAL STEAM RESERVOIRS

These are the preferred reservoirs as they require less processing to generate electricity. The steam from the reservoir can be sent directly to the turbine for power generation. Some reservoirs can have a small amount of liquid phase which is treated to prevent the liquid reaching the turbine blades. The power plants are dry steam types (different types of power plants are discussed in the next section). An applied example is *The Geysers* in California, which houses multiple power plants extracting natural steam from more than 350 wells.

8.1.2.5 MOLTEN MAGMA

Molten magma or lava can have temperatures above 650 °C, which makes them have great potential. This resource of geothermal energy can be found in magma chambers below volcanoes as molten magma, or as lava which makes it to the surface. Despite the potential that these resources have, the unpredictability of volcanoes is the biggest obstacle in commercializing energy extraction by these means.

8.1.2.6 GEOPRESSURED RESERVOIRS

Geopressured reservoirs are sediment-filled sources of geothermal energy that contain high-pressure hot water. In most of these reservoirs, pressured water also contains methane. Therefore, energy extraction from geopressured reservoirs can be attained by steam generated using the hot water, burning the methane as a fossil fuel, or utilizing the built-up pressure directly to run turbines to generate electricity. Hot water in these reservoirs can also contain brine, which is high-concentration salt dissolved in water and is highly corrosive. This is one of the challenges in harvesting energy from geopressured reservoirs.

8.1.3 Geothermal Energy Systems

Geothermal energy systems can be employed either to harvest energy from geothermal resources directly or to generate electricity through power plants utilizing turbines and generators. **Section 8.1.3.1** covers the applications where geothermal energy is used directly, and **Section 8.1.3.2** discusses different geothermal power plants operating to generate electricity from geothermal resources.

8.1.3.1 DIRECT USE OF GEOTHERMAL ENERGY

Direct use of geothermal energy can be achieved by harvesting the thermal energy from geothermal resources in a variety of applications such as:

- Geothermal heat pump systems (heating and cooling)
- Geothermal residential and commercial heating
- Geothermal district heating

FIGURE 8.9 A horizontal earth heat exchanger with plastic tubing in a trench dug for a residential geothermal heat pump system. This is the traditional horizontal loop where the tubes are laid out in a serpentine heat exchanger form. Source: SimplyCreativePhotography/E+ via Getty Images.

FIGURE 8.10 A 2.5-m-deep trench with looped earth heat exchanger for a residential geothermal heat pump application. Slinky coil geothermal ground loops are more space efficient as they require smaller trenches than traditional horizontal loops. Source: BanksPhotos/E+ via Getty Images.

Geothermal Heat Pump Systems (Heating and Cooling)

Geothermal heat pump systems, also referred to as ground-source or ground-coupled heat pumps, utilize the ground soil as a heat source or a heat sink in winter or summer seasons, respectively. At about 3 m depth, soil temperature remains relatively stable throughout the year. Depending on the local climate, the range of ground temperature is about 7 °C to 21 °C. In the summer season, ground temperature during the day can be significantly cooler than the outside air temperature. Hence, the working fluid (e.g., water) can be circulated through the closed loop of an *earth heat exchanger*, which helps reduce the temperature difference between the air to be air-conditioned and the desired temperature inside the space. Similarly, in winter the heat exchanger beneath the surface enables the air to be pre-heated with water circulating through an ambient at a comparatively higher temperature than the outside air temperature. This results in reducing the temperature difference between the reservoirs, and hence increasing the coefficient of performance (COP). The earth heat exchanger can be in a vertical, horizontal, or looped configuration (**Figures 8.9** and **8.10**). Heat pumps can also exchange heat with a water-based cold reservoir such as a lake or a pond, where the heat exchanger would be placed into the water body.

Geothermal Residential and Commercial Heating

Hot water from geothermal reservoirs can be used directly for heating residential or commercial buildings. Water at temperatures of 60 °C or higher (or sometimes steam) is pumped from the reservoir into either a direct heating system such as a radiator, where it is the only working fluid, or through a heat exchanger over which air can be blown, hence employing two different working fluids. In terms of the heat transfer mode, the first example is a radiation and free convection case while the latter example is mainly a forced convection case. Geothermal heating for buildings can be done on individual buildings, or on multiple buildings, as discussed in the next section.

Geothermal District Heating

Geothermal district heating (GDH) offers residential and commercial heating services collectively over multiple buildings, and can extend to a city at a municipal level. Utilizing geothermal heating at a district scale can offer environmental and technical advantages besides the economic benefits it provides. A well-known GDH example is in Iceland. The largest municipal district heating service is provided in Reykjavik, and throughout the whole country more than 90% of the citizens are served with heating through over 60 district heating systems [3].

8.1.3.2 ELECTRICITY GENERATION

Electricity generation through geothermal energy is achieved in a similar manner to thermal power plants utilizing fossil fuels or nuclear energy. Steam going into the turbine generates mechanical energy, which is then converted into electrical energy. The difference in geothermal power plants is the source of energy and how the working fluid is utilized. Hot water, steam, or a mix of both is extracted from beneath the ground through a production well that is drilled into the geothermal reservoir. A schematic of a generic geothermal power plant is illustrated in **Figure 8.11**. The energy from the working fluid is either directly or indirectly used to send steam into the turbine for power generation. The way the energy is utilized can be different depending on different types of geothermal power plant applications.

FIGURE 8.11 Schematic of a generic geothermal power plant. Source: U.S. Department of Energy, Office of Energy Efficiency and Renewable Energy.

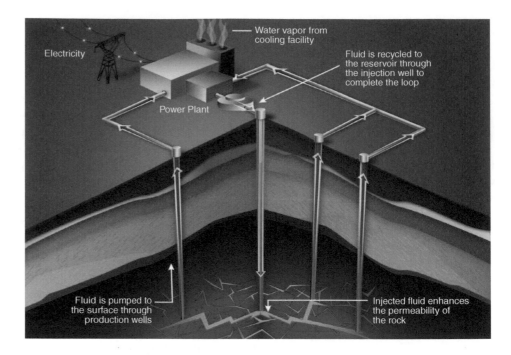

Electricity generation by geothermal energy can be achieved at geothermal power plants in three main types of plants:

- Dry steam power plants
- Flash steam power plants
- Binary cycle power plants

These three types of geothermal power plant based on their operating principles can be seen in **Figure 8.12**. There are also hybrid geothermal power plants which are discussed at the end of this section.

Dry Steam Power Plants

This is the type of geothermal power plant that requires the least amount of processing and is the most desired type. However, there are not many applications due to source limitations. Some geothermal reservoirs have dry steam or steam of high quality (having little liquid in the mix). With minimal to no processing, the steam coming from the production well can be directed to the turbine to generate electricity. After the steam does the work through the turbine, it is condensed and injected back into the reservoir through the injection well. This type can be seen in **Figure 8.12a**.

Flash Steam Power Plants

The majority of geothermal reservoirs have hot water. Therefore, a direct way of producing work within the turbine utilizing the working fluid from the reservoir is not feasible. Hot water coming from the production well needs to be in the vapor phase. Thanks to the natural existence of the hot water underground, it is mostly trapped in high-pressure reservoirs. Hence, as it makes its way to the surface, it vaporizes suddenly in the lower pressure ambient producing *flash steam*. A separator is used to prevent the liquid from entering the turbine, as liquid droplets can cause serious damage to turbine blades and the crank shaft. Liquid water collected in the separator mixes with the condensed steam that has left the turbine and they are injected back into the reservoir through the injection well. This type is a single flash steam power plant and can be seen in **Figure 8.12b**.

FIGURE 8.12 Geothermal power plant types based on their operating principles. (a) Dry steam power plant, (b) single flash steam power plant, (c) binary cycle power plant. Source: U.S. Department of Energy, Office of Energy Efficiency and Renewable Energy.

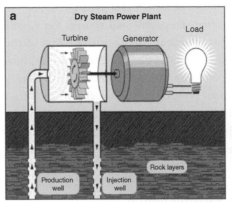

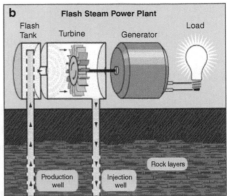

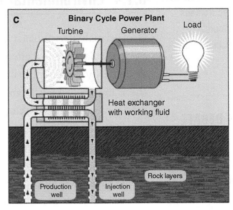

Flash steam power plants can employ double or triple separators as well. A double (or dual) flash steam power plant uses two separators. Water collected in the first separator is sent into a second separator (or flasher) which is at a lower pressure to generate more steam. The steam is then sent into the turbine, entering the turbine at a lower pressure. A triple flash steam power plant has three separators, each one having a lower pressure than the preceding one. A double or triple flash steam power plant has a higher power output as more of the energy from the hot water coming from the reservoir is utilized. The initial costs, as well as the O&M costs for plants with multiple separators, are higher than those for single flash steam power plants due to the use of additional flashers, piping, fittings, and safety components. Energy analyses can be conducted to determine the economic feasibility of double or triple flash steam applications.

Binary Cycle Power Plants

In a binary cycle power plant, hot water coming from the geothermal reservoir flows through a vaporizer where it exchanges heat with a secondary working fluid. The name *binary* is used as there are two working fluids involved. Fluid coming from the geothermal reservoir does not come into contact with the turbine. After releasing its energy to the secondary fluid in the heat exchanger, it is sent back to the reservoir through the injection well. The secondary fluid is typically an organic compound such as butane or pentane hydrocarbon. Once it vaporizes, the vapor phase fluid goes into the turbine to generate electricity. The secondary fluid has its own closed loop. Vapor leaving the turbine is condensed and directed to the heat exchanger to absorb heat from the primary fluid coming from the reservoir, hence completing its cycle. A schematic of a binary cycle power plant is given in **Figure 8.12**c.

Besides these three types, there are also *hybrid geothermal power plants* which operate based on the binary cycle and another method of geothermal power generation such as

FURTHER LEARNING

Global Geothermal Power Plant Map
www.thinkgeoenergy.com/map/

FIGURE 8.13 The world's first commercial geothermal power plant was built in Larderello in the Pisa province of Italy in 1911. This region has been known for its hot springs since ancient times. The Romans used the springs for bathing. Source: Atlantide Phototravel/ Corbis Documentary via Getty Images.

CLASSROOM DEBATE

Form three groups and debate on the advantages and disadvantages of *dry steam*, *flash steam*, and *binary cycle* geothermal power plants against each other.

the dry steam or flash steam principles. The idea in these systems is to get more useful energy from the geothermal source. For example, a hybrid plant operating based on flash steam and the binary cycle will first utilize the flash steam from the hot water and then use the condensate (condensed steam) in a binary cycle to extract more heat from the hot working fluid. This increases the exergetic efficiency of the overall cycle. For geothermal power plants, the term *hybrid* is not necessarily used for a combination of two different geothermal applications. Hybrid geothermal power plant can also refer to a combined version of a geothermal and another fossil or non-fossil plant such as hybrid geothermal–coal-fired, geothermal–solar, or geothermal–biomass power plants.

8.1.4 Environmental Impacts

Geothermal energy has a higher capacity factor than other renewable energy resources such as solar or wind energy, as it is available all the time throughout the year. As with any energy resource, geothermal energy also has some environmental impacts. One of these impacts comes from geothermal heat pumps, which run on electricity and utilize refrigerants as the working fluid. Depending on the type of system, the refrigerant used can have ozone depletion potential (ODP) or global warming potential (GWP) values above safe limits, which can be environmentally detrimental. Another environmental impact of geothermal energy is the potential risk of CO_2 release from the hot water coming from the geothermal reservoir. If the hot fluid coming to the surface contains dissolved CO_2 in the form of bicarbonate ions, these can yield calcite and hence release CO_2. One other concern about geothermal energy use can be the lowering of water levels in aquifers in the long run. There are no CO_2 emissions due to combustion as geothermal plants do not burn fossil fuels. Geothermal power plants emit minimal amounts of NO_x and a nearly zero amount of SO_2 or particulate matter (PM). The H_2S emissions can be kept below the critical threshold with appropriate choices of pollution control equipment [4]. Overall, the environmental impact of geothermal energy can be considered as minimal if the plants have been designed and built to appropriate standards.

8.2 Theory

Theoretical analysis of geothermal energy can involve both direct and indirect use applications. Direct use examples as discussed in **Section 8.1.3.1** include geothermal heat pumps, and residential, commercial, and district heating applications. The theory of these systems is associated with basic heat exchange problems. In geothermal heat pumps, energy is exchanged between the soil (heat sink in summer and heat source in winter) and the working fluid flowing through the ground loop system. In building or district heating applications, energy absorbed from the hot water is:

$$\dot{Q} = \dot{m}c_p(T_i - T_o) \tag{8.1}$$

where T_i and T_o are the inlet and outlet temperatures of hot water flowing through a heat exchanger such as a radiator or an underfloor (*radiant floor*) heating system. Traditionally radiators operate at higher inlet temperatures (60–90 °C), while radiant floor heating systems can operate at much lower inlet temperatures (35–50 °C). Indirect use of geothermal energy as discussed in **Section 8.1.3.2**, on the other hand, pertains to electricity generation at power plants. The theoretical analysis in this chapter focuses mainly on these plants.

It is essential for engineers to have an idea about the theoretical limits for a given plant site. The maximum efficiency that can be achieved in any type of power plant operating between two thermal reservoirs is its Carnot efficiency, which can be determined by:

$$\eta_{Carnot} = 1 - \frac{T_L}{T_H} \tag{8.2}$$

where T_L and T_H are the low (cold reservoir) and high (hot reservoir) temperatures, respectively.

In a real-world application, there will be losses and irreversibilities at almost every section of the geothermal site, starting from the production well beneath the surface all the way to the generator above ground. While some of these losses can be neglected, irreversibilities within the critical elements such as the turbine can result in considerable exergy reduction. Therefore, isentropic turbine efficiency should be accounted for in determining the thermal performance.

Thermal efficiency of a geothermal power plant can be calculated by:

$$\eta_{th} = \frac{\left(\dot{W}_{turb}\right)_{actual}}{\dot{m}_{supply}\left(h_{supply} - h_{f@T_{amb}}\right)} \tag{8.3}$$

In this equation, the numerator is the actual rate of work done by the turbine and the denominator is the amount of energy available from the geothermal fluid. The actual rate of work done by the turbine can be calculated in terms of the ideal rate of work as:

$$\left(\dot{W}_{turb}\right)_{actual} = \left(\eta_t\right)\left(\dot{W}_{turb}\right)_{ideal} \tag{8.4}$$

where the isentropic turbine efficiency and ideal turbine work are:

$$\eta_t = \frac{h_i - h_e}{h_i - h_{e,s}} \tag{8.5}$$

$$\left(\dot{W}_{turb}\right)_{ideal} = \dot{m}\left(h_i - h_{e,s}\right) \tag{8.6}$$

In **Equation 8.6**, subscripts i and e denote the inlet and exit properties and subscript s is associated with the isentropic case.

Schematics and *T–S* diagrams of dry steam, single flash steam, double flash steam, and binary cycle geothermal power plants are illustrated in **Figures 8.14–8.17**. Thermal performance analyses on any of these systems can be conducted with known enthalpy values or properties that can help determine the corresponding enthalpy values. Mass flow rates of the working fluid and isentropic turbine efficiency values would also be needed to be able to obtain the thermal efficiency of the power plant. It should be noted that the working fluid in the dry steam, single flash steam, and double flash steam cycles is water (vapor or liquid). In the binary cycle application, it is the secondary working fluid, which is an organic compound, that does the work through the turbine. Hence, the *T–S* diagram in **Figure 8.17** belongs to the organic fluid.

FIGURE 8.14 Schematic and *T–S* diagram for a dry steam power plant (*T–S* diagram is sketched for the steam as the working fluid).

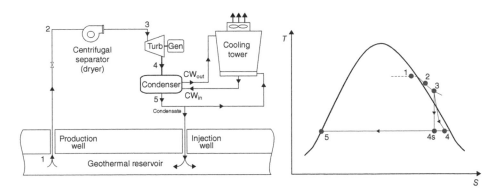

FIGURE 8.15 Schematic and *T–S* diagram for a single flash steam power plant (*T–S* diagram is sketched for the hot water–steam mixture as the working fluid).

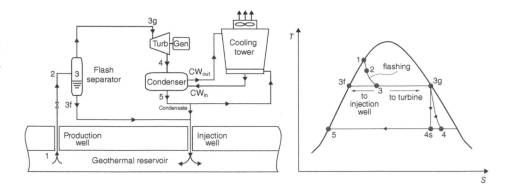

FIGURE 8.16 Schematic and *T–S* diagram for a double flash steam power plant (*T–S* diagram is sketched for the hot water–steam mixture as the working fluid).

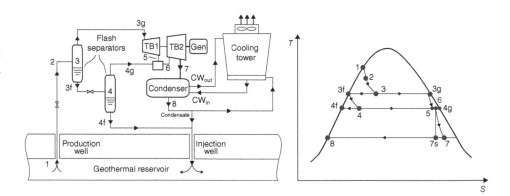

FIGURE 8.17 Schematic and *T–S* diagram for a binary cycle power plant (*T–S* diagram is sketched for the organic compound [e.g., isobutane] as the working fluid).

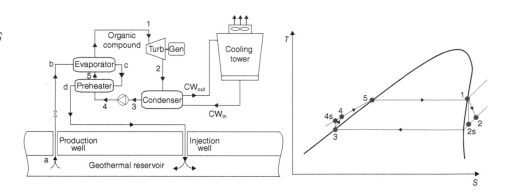

Example 8.1 Carnot Efficiency of a Geothermal Power Plant

Determine the Carnot efficiency of a geothermal power generation plant if the temperature of the fluid coming from the geothermal reservoir is 220 °C and the condensation pressure is 20 kPa.

Solution

The Carnot efficiency of the plant can be determined using **Equation 8.2**:

$$\eta_{\text{Carnot}} = 1 - \frac{T_{\text{L}}}{T_{\text{H}}}$$

where $T_{\text{H}} = 220\,°\text{C} + 273.15 = 493.15\ \text{K}$

T_L can be determined from the given condensation pressure (**Table B.2**):

$$T_L = T_{sat@20\ kPa} = 60.06\,^{\circ}C \equiv 333.21\ K$$

Then,

$$\eta_{Carnot} = 1 - \frac{333.21}{493.15} = 0.324$$

Hence, the Carnot efficiency of the geothermal power plant operating between the given two thermal reservoirs is 32.4%.

Note: *As covered in thermodynamics courses, the Carnot efficiency marks the theoretical limit for the maximum efficiency that can be attained. Hence, for the given thermal reservoirs the maximum fraction of heat that can be used to do work is 32.4% of the heat supplied. In a real-world application, the thermal efficiency of the power plant will be less than this value due to irreversibilities that can result from heat losses, friction, etc.*

Example 8.2 Single Flash Steam Power Plant

A single flash steam power plant receives geothermal fluid from the production well as saturated liquid at 220 °C, at a flow rate of 175 kg/s. During flashing, the fluid isenthalpically expands to 500 kPa. The liquid portion goes back to the injection well while the saturated vapor enters the turbine and exits at 8 kPa. The turbine has an isentropic efficiency of 84%. The capacity of the plant is 15 MW.

a. Sketch the *T–S* diagram.
b. Determine the mass flow rate of steam through the turbine (kg/s).
c. Find the thermal efficiency of the power plant at standard ambient conditions (%).

Solution

a.

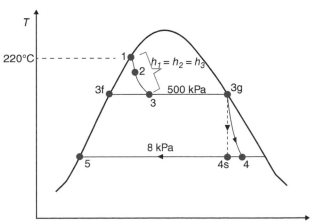

b. Mass flow rate of the steam is determined from the capacity of the power plant:

$$P_{cap} = (\dot{W}_{turb})_{actual} = 15 \text{ MW}$$

$$(\dot{W}_{turb})_{actual} = \eta_t (\dot{W}_{turb})_{ideal}$$

Isentropic efficiency of the turbine is given. Ideal work that could be extracted from the turbine per unit mass flow rate of steam is:

$$\frac{(\dot{W}_{turb})_{ideal}}{\dot{m}_{steam}} = h_{3g} - h_{4s}$$

where 3g is saturated vapor at 500 kPa, and $s_{4s} = s_{3g} = 6.82$ kJ/kg K,

which yields the quality at state 4s (at 8 kPa) as $x_{4s} = 0.815$ (81.5%). Then,

$$h_{3g} = 2748.7 \text{ kJ/kg} \quad \text{and} \quad h_{4s} = 2132.4 \text{ kJ/kg}.$$

Plugging these into the actual turbine output equation, we get:

$$\frac{(\dot{W}_{turb})_{actual}}{\dot{m}_{steam}} = \eta_t \frac{(\dot{W}_{turb})_{ideal}}{\dot{m}_{steam}} = (0.84)(2748.7 - 2132.4) \frac{\text{kJ}}{\text{kg}} = 517.7 \frac{\text{kJ}}{\text{kg}}$$

Solving for mass flow rate in the power capacity equation gives:

$$\dot{m} = \frac{P_{cap}}{(\Delta h_{turb})_{actual}} = \frac{15 \text{ MW}}{517.7 \frac{\text{kJ}}{\text{kg}}} \left| \frac{1000 \frac{\text{kJ}}{\text{s}}}{1 \text{ MW}} \right| = 28.97 \frac{\text{kg}}{\text{s}}$$

c. Thermal efficiency of the geothermal power plant can be calculated using **Equation 8.3**:

$$\eta_{th} = \frac{(\dot{W}_{turb})_{actual}}{\dot{m}_{supply}(h_{supply} - h_{f@T_{amb}})} = \frac{\left(28.97 \frac{\text{kg}}{\text{s}}\right)\left(517.7 \frac{\text{kJ}}{\text{kg}}\right)}{\left(175 \frac{\text{kg}}{\text{s}}\right)(943.62 - 104.89) \frac{\text{kJ}}{\text{kg}}} = 0.102 (10.2\%)$$

Example 8.3 Binary Cycle Power Plant

Consider the same geothermal reservoir as in **Example 8.2**. In this case, water coming from the production well is used as a heat source in a binary cycle where isobutane (C_4H_{10}) is the organic compound. Geothermal fluid enters the evaporator at 180 °C and leaves at 100 °C. Isobutane goes through a constant pressure process within the evaporator, entering at 3 MPa and 40 °C, and leaving as saturated vapor. It then expands through the turbine and leaves the turbine at 300 kPa and 28 °C. Determine:

a. the mass flow rate of isobutane (kg/s)
b. the rate of heat addition to isobutane through the evaporator (kW)
c. the turbine power output (kW)
d. the thermal efficiency of the binary cycle power plant (%)

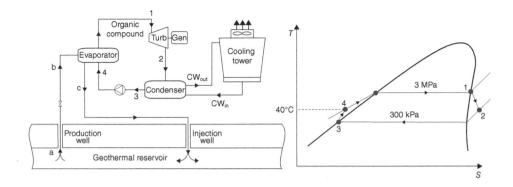

Solution

a. The evaporator is assumed to be a control volume having no stray heat transfer with the surroundings.

Then, the rate of energy given by the geothermal fluid will be equal to that absorbed by the isobutane:

$$\dot{Q}_{\text{water}} = \dot{Q}_{\text{isobutane}}$$
$$\dot{m}_{\text{water}}(h_b - h_c) = \dot{m}_{\text{isobutane}}(h_1 - h_4)$$

From **Table B.1**:

$$h_b = h_{f@180\,°C} = 763.2 \text{ kJ/kg}$$
$$h_c = h_{f@100\,°C} = 419 \text{ kJ/kg}$$

Properties for isobutane can be retrieved from the NIST Chemistry Webbook (https://webbook.nist.gov/chemistry/fluid/).

$$h_1 = h_{g@3\text{ MPa}} = 685.7 \text{ kJ/kg } (P_1 = 3 \text{ MPa and } x_1 = 1)$$
$$h_4 = 297.3 \text{ kJ/kg } (P_4 = 3 \text{ MPa and } T_4 = 40\,°C)$$

Then:

$$\left(175\ \frac{\text{kg}}{\text{s}}\right)(763.2 - 419)\frac{\text{kJ}}{\text{kg}} = \dot{m}_{\text{isobutane}}(685.7 - 297.3)\frac{\text{kJ}}{\text{kg}}$$

Solving for mass flow rate of isobutane, we get:

$$\dot{m}_{\text{isobutane}} = 155.1\ \frac{\text{kg}}{\text{s}}$$

b. Rate of heat addition to isobutane within the evaporator can be calculated by:

$$\dot{Q}_{\text{isobutane}} = \dot{m}_{\text{isobutane}}(h_1 - h_4) = \left(155.1\ \frac{\text{kg}}{\text{s}}\right)(685.7 - 297.3)\frac{\text{kJ}}{\text{kg}} = 60{,}241 \text{ kW}$$

c. Turbine power output is:

$$\dot{W}_{\text{turb}} = \dot{m}_{\text{isobutane}}(h_1 - h_2)$$

h_2 can be determined using the NIST Chemistry Webbook:

$$h_2 = 595.5 \text{ kJ/kg } (P_2 = 300 \text{ kPa and } T_2 = 28\,°C)$$

Then:

$$\dot{W}_{\text{turb}} = \left(155.1\ \frac{\text{kg}}{\text{s}}\right)(685.7 - 595.5)\frac{\text{kJ}}{\text{kg}} = 13{,}990 \text{ kW}$$

d. Thermal efficiency of the binary cycle power plant can be calculated using **Equation 8.3**:

$$\eta_{th} = \frac{\dot{W}_{turb}}{\dot{m}_{supply}\left(h_{supply} - h_{f@T_{amb}}\right)} = \frac{13{,}990 \text{ kW}}{\left(175\ \dfrac{kg}{s}\right)(943.62 - 104.89)\ \dfrac{kJ}{kg}} = 0.095(9.5\%)$$

Note: *Standard ambient conditions were used in calculating the thermal efficiency of the geothermal power plant as in Example 8.2. The efficiency of this power plant could be increased if it were a hybrid geothermal power plant such as single flash steam and binary cycle employed together. Other hybrid applications involving non-geothermal resources such as fossil fuel-based sources can also be explored to see the thermal efficiency and levelized cost of electricity (LCOE) of the power plant with such modifications.*

Example 8.4 Retrieving and Plotting Geothermal Energy Data Using Python

Using EIA data[1]:

a. Tabulate the US renewable energy production and consumption by source for July–August 2021.
b. Plot the monthly geothermal energy consumption between 1973 and 2021.

Solution

Energy data is obtained from the EIA website in spreadsheet form:

www.eia.gov/totalenergy/data/browser/xls.php?tbl=T10.01

a. We perform the following steps in building the code:
   1. Obtain the spreadsheet file from the EIA website.
   2. Read the file, take the first sheet in the file (0), and skip 10 rows from the start (skiprows=10).
   3. Drop the "units" line. (Python takes the header as one line; the second line is truncated to the column values. In the original Excel file, header is the resource but then the next line is the unit for that resource. We will omit the units in this exercise for simplicity.)
   4. Set the index for DataFrame as Month column, which shows the year-month of that data.

```
import matplotlib.pyplot as plt
import pandas as pd

#HTTPS note: Use the flowing link for data in case needed:
https://www.eia.gov/totalenergy/data/browser/xls.php?tbl=T10.01
url="http://www.eia.gov/totalenergy/data/browser/xls.php?tbl=T10.01"
df=pd.read_excel(url, 0,skiprows=10)
df=df.drop([0])
df=df.set_index("Month")
```

Next, let's obtain the data for the last 3-month time period in the spreadsheet:

```
df.tail(3)
```

[1] U.S. Energy Information Administration, Monthly Energy Review December 2021.

Month	Wood Energy Production	Biofuels Production	Total Biomass Energy Production	Total Renewable Energy Production	Hydroelectric Power Consumption	Geothermal Energy Consumption	Solar Energy Consumption	Wind Energy Consumption	Wood Energy Consumption	Waste Energy Consumption	Biofuels Consumption	Total Biomass Energy Consumption	Total Renewable Energy Consumption
2021-07-01	190.908	202.84	429.335	986.802	194.009	17.633	156.893	188.932	183.026	35.587	198.1	416.713	974.18
2021-08-01	189.282	189.777	414.181	1003.26	183.85	17.345	152.963	234.922	178.405	35.122	194.428	407.955	997.035
2021-09-01	182.256	179.835	396.794	964.228	157.616	17.141	141.19	251.488	172.539	34.703	181.626	388.868	956.303

b. Now, let's plot the monthly geothermal consumption between 1973 and 2021:

```
plt.title("Geothermal Energy Consumption (Trillion Btu)")
plt.xlabel("Year")
plt.ylabel("Trillion Btu")

plt.plot(df["Geothermal Energy Consumption"], label="Geothermal")
plt.legend()

<matplotlib.legend.Legend at 0x7f6fac181b50>
```

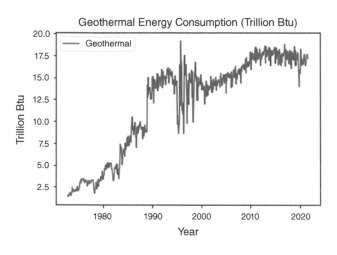

8.3 Applications and Case Study

In this section, we will observe real-world examples via two applications and a case study. The applications include a single flash steam power plant in Indonesia, and a binary cycle combined heat and power (CHP) plant in Iceland. The case study is about a geothermal district heating project in Izmir, an Aegean city in Turkey.

Application 1 Single Flash Steam Power Plant (Indonesia)

Ulubelu Geothermal Plant
Location: Lampung, Indonesia
Operated by: PLN (state-owned) and PGE
Units: 4 × 55 MW
Capacity: 220 MW
Data: https://pge.pertamina.com [5]
www.d-maps.com

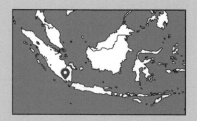

Ulubelu Geothermal Power Plant is located at WKP Gunung Way Panas in Lampung province in Indonesia, which is northwest of Jakarta. The plant has four units, with each unit having a power generation capacity of 55 MW, totaling 220 MW capacity. The units operate based on single flash steam. The first unit started operating in 2012, and the fourth one has been running since 2017 [5].

Application 2 Binary Cycle CHP Plant (Iceland)

Svartsengi Geothermal Power Plant
Location: Keflavik, Iceland
Owner: HS Orka
Heating capacity: 190 MW
Electricity generation capacity: 75 MW
Data: www.hsorka.is/en/power-plants/ [6]
www.d-maps.com

Svartsengi is a combined heat and power (CHP) plant which operates based on a binary cycle. It houses a cogeneration unit enabling it to generate both useful heat and electricity. The plant has an installed capacity of 190 MW for heat and 75 MW for electricity generation. It is the first of its kind in Iceland as a CHP plant. It has six units, with the first unit starting to generate electricity in 1976, and the sixth one joining the fleet in 2007. The plant uses a SCADA (Supervisory Control and Data Acquisition) system for controlling and managing its operation [6].

Case Study Geothermal District Heating (Turkey)

Balcova–Narlidere District Heating System
Location: Izmir, Turkey
Owners: Izmir Jeotermal Enerji and Izmir
Metropolitan Municipality
Hot water capacity: 2020 m³/h
Temperature: 90–144 °C
Data: www.izmirjeotermal.com.tr/ [7]
www.d-maps.com

INTRODUCTION Balcova and Narlidere are two neighboring districts in Izmir, which is a western city of Turkey on the Aegean coast. The Balcova–Narlidere geothermal district heating system is the largest of many district heating systems in Turkey based on its potential of 71.3 MW [8]. The geothermal field is situated at the Izmir Fault Zone surrounded by the *Agamemnon* fault lines (they take their name from the Mycenaean king who led the Greek forces in the *Trojan War*, which took place about 350 km north of Izmir).

OBJECTIVE The objective of this case study is to learn about a local district heating system utilizing geothermal energy.

METHOD This case study was conducted through researching articles and reports.

RESULTS The Balcova–Narlidere geothermal district heating system, with its potential of 71.3 MW, is the largest of its kind in Turkey, followed by Afyon, an inner Aegean city famous for its thermal resorts. The district heating system in Afyon has a capacity of 33.9 MW. The Balcova–Narlidere system is sourced by the geothermal reservoirs in the Balcova district with an average inlet water temperature of 118 °C and an outlet temperature of 60 °C. Production and injection wells reach out to the Izmir flysch sequence consisting of sandstone, clay, and siltstone layers.

The thickness of the sequence is estimated to be over 2 km, in some places extending to 4 km. The Balcova part of the project became functional in 1996, and the Narlidere part started operating in 1998. Over one million square meters of residential housing is being heated by the Balcova–Narlidere district heating project [8].

DISCUSSION The geothermal district heating application discussed in this case study is a good example of local energy savings through a renewable source. Such systems can have higher initial costs; however, the O&M costs are much lower compared to other technologies. The distribution of items making up the O&M costs may vary depending on the application site, utility costs, human resources, and available technology. France, Germany, Hungary, and Iceland are some other countries in Europe with remarkable geothermal district heating applications.

RECOMMENDATIONS As the initial investment costs are higher for geothermal district heating applications, these systems should be designed with long-term commitments. The locations selected should provide high water temperatures and should host denser populations to achieve a better return on investments. Hence, residential areas with multi-story buildings housing individual apartments would yield more effective results than detached, single-family houses as the cost of system connections to each building can be notable.

8.4 Economics

The total cost of a geothermal power plant is inclusive of three main line items, which are similar in most of the other types of power generation plants:

FURTHER LEARNING

Student Design Competition: Geothermal Design Challenge
www.energy.gov/eere/geothermal/geothermal-design-challenge

FIGURE 8.18 Mammoth Geothermal Complex in California. Unlike fossil fuel- or nuclear fuel-based plants, the cost of geothermal electricity does not fluctuate with changing fuel supply and costs as the source of energy is always free and available in locations where geothermal reservoirs are accessible. Source: VisionsofAmerica/Joe Sohm/DigitalVision via Getty Images.

- *Capital cost*: This is the total cost of construction of the entire power plant. It has two parts, which are the overnight and interest costs. Overnight cost in the geothermal power generation sector refers to the cost of the land, construction, drilling of the wells, equipment, and the labor. The other part of the capital cost is the price of interest for the loan.
- *Operating and maintenance cost*: Operating costs include utilities, wages and salaries, social security, unemployment insurance, workers' compensation, medical insurance, and taxes. Maintenance costs include routine maintenance, troubleshooting, and repair costs.
- *Fuel cost*: Unlike fossil fuel- or nuclear fuel-based plants, the cost of geothermal electricity does not fluctuate with changing fuel supply and costs as the source of energy is free and available at all times in locations where geothermal reservoirs are accessible.

Depending on the location of the site, possible drilling challenges, major equipment such as the flasher, turbine or generator, staff on site, taxes, and other line items, the share of costs within the total cost can vary in different geothermal applications.

A levelized cost of energy (LCOE) analysis for geothermal power plants becomes helpful in feasibility analyses. The LCOE value allows comparison of geothermal energy applications with other renewable resources, as well as fossil-based power generation, which all have different operating lifetimes, project costs, and application capacities. Estimated cost values for dispatchable (fossil fuel, nuclear, biomass, geothermal, and battery storage) and non-dispatchable (wind, solar, and hydroelectric) technologies for resources entering service in 2026 are tabulated in **Table 8.1** [9].

Table 8.1

Estimated unweighted levelized cost of electricity (LCOE) and levelized cost of storage (LCOS) for new resources entering service in 2026 (2020 dollars per MWh) [9]								
Plant type	**Capacity factor**	**Levelized capital cost**	**Levelized fixed O&M[1]**	**Levelized variable O&M**	**Levelized transmission cost**	**Total system LCOE or LCOS**	**Levelized tax credit[2]**	**Total LCOE or LCOS including tax credit**
Dispatchable technologies								
Ultra-supercritical coal	85%	$43.80	$5.48	$22.48	$1.03	$72.78	NA	$72.78
Combined cycle	87%	$7.78	$1.61	$26.68	$1.04	$37.11	NA	$37.11
Combustion turbine	10%	$45.41	$8.03	$132.38	$9.05	$194.9	NA	$194.9
Advanced nuclear	90%	$50.51	$15.51	$2.38	$0.99	$69.39	-$6.29	$63.10
Geothermal	90%	$19.03	$14.92	$1.17	$1.28	$36.40	-$1.90	$34.49
Biomass	83%	$34.96	$17.38	$35.78	$1.09	$89.21	NA	$89.21
Battery storage	10%	$57.98	$28.48	$23.85	$9.53	$119.8	NA	$119.8
Non-dispatchable technologies								
Wind, onshore	41%	$27.01	$7.47	$0.00	$2.44	$36.93	NA	$36.93
Wind, offshore	44%	$89.20	$28.96	$0.00	$2.35	$120.5	NA	$120.5
Solar, standalone[3]	29%	$23.52	$6.07	$0.00	$3.19	$32.78	-$2.35	$30.43
Solar, hybrid[3, 4]	28%	$31.13	$13.25	$0.00	$3.29	$47.67	-$3.11	$44.56
Hydroelectric[4]	55%	$38.62	$11.23	$3.58	$1.84	$55.26	NA	$55.26

Source: U.S. Energy Information Administration, Annual Energy Outlook 2021

[1] O&M: operations and maintenance

[2] The tax credit component is based on targeted federal tax credits such as the production tax credit (PTC) or investment tax credit (ITC) available for some technologies. It reflects tax credits available only for plants entering service in 2026 and the substantial phaseout of both the PTC and ITC as scheduled under current law. Technologies not eligible for PTC or ITC are indicated as NA, or not available. The results are based on a regional model, and state or local incentives are not included in LCOE and LCOS calculations.

[3] Technology is assumed to be photovoltaic (PV) with single-axis tracking. The solar hybrid system is a single-axis PV system coupled with a four-hour battery storage system. Costs are expressed in terms of net AC (alternating current) power available to the grid for the installed capacity.

[4] As modeled, EIA assumes that hydroelectric and hybrid solar PV generating assets have seasonal and diurnal storage, respectively, so that they can be dispatched within a season or a day, but overall operation is limited by resource availability by site and season for hydroelectric and by daytime for hybrid solar PV.

8.5 Summary

Geothermal energy is the energy that can be harvested from the Earth by means of direct or indirect applications. Some of the direct applications include geothermal heat pumps and geothermal heating, which can be applied to residential and commercial buildings on an individual basis, or collectively to multiple buildings in the form of district (or municipal) heating. Indirect applications are associated with power generation, which is covered in more depth in this chapter. Geothermal power generation can be conducted in different types of plants such as dry steam, flash steam, or binary cycle plants. Hybrid power plant applications also exist, which utilize either different geothermal plant types or geothermal energy along with other resources such as fossil fuels or biomass. As can be seen in **Table 8.1**, geothermal energy has remarkable economic advantages over some of the other energy resources in terms of levelized cost of energy. It should, however, be noted that regions having sufficient thermal gradients for geothermal power generation are limited in the world. These regions are close to plate boundaries where tectonic and volcanic activities are recorded more than in other parts of the world. Hence the gift of having higher geothermal energy capacity and richer reservoirs also comes with a higher risk of earthquakes, not due to the plants or energy extraction activities necessarily, but because they happen to be located in or near earthquake zones.

REFERENCES

[1] International Energy Agency, "Geothermal Power Generation in the Sustainable Development Scenario, 2000–2030," IEA, Paris, Jul. 2, 2020. www.iea.org/data-and-statistics/charts/geothermal-power-generation-in-the-sustainable-development-scenario-2000-2030 (accessed Aug. 16, 2020).

[2] A. Richter, "The Top 10 Geothermal Countries 2021 – based on installed generation capacity (MWe)," Think GeoEnergy - Geothermal Energy News, Jan. 10, 2022. www.thinkgeoenergy.com/the-top-10-geothermal-countries-2021-based-on-installed-generation-capacity-mwe (accessed Jan. 28, 2022).

[3] Samorka, Icelandic Energy & Utilities, "District Energy in Iceland," Euroheat & Power, Nov. 15, 2019. www.euroheat.org/knowledge-hub/district-energy-iceland (accessed Aug. 16, 2020).

[4] Geothermal Communities, European Commission, "Environmental Impacts of Geothermal Energy." Accessed: Aug. 18, 2020 [Online]. Available: https://geothermalcommunities.eu/assets/presentation/ENVIRONMENTAL_IMPACTS.pdf.

[5] Pertamina Geothermal Energy, "Ulubelu Geothermal Plant." https://pge.pertamina.com (accessed Aug. 18, 2020).

[6] HS Orka, "Power Plants." www.hsorka.is/en/power-plants (accessed Aug. 18, 2020).

[7] İzmir Jeotermal Enerji, "About Geothermal." www.izmirjeotermal.com.tr (accessed Aug. 19, 2020).

[8] B. Erdogmus, M. Toksoy, B. Ozerdem, and N. Aksoy, "Economic assessment of geothermal district heating systems: A case study of Balcova–Narlidere, Turkey," *Energy and Buildings*, vol. 38, no. 9, pp. 1053–1059, 2006, doi: 10.1016/j.enbuild.2006.01.001.

[9] U.S. Energy Information Administration, "Levelized Costs of New Generation Resources in the Annual Energy Outlook 2021," Feb. 2021. Accessed: Feb. 27, 2021 [Online]Available: www.eia.gov/outlooks/aeo/pdf/electricity_generation.pdf.

CHAPTER 8 EXERCISES

8.1 In a geothermal district heating (GDH) application in Boise, Idaho, hot water enters a residential unit at 79 °C and leaves at 51 °C. If the flow rate of water is 2.8 kg/s, determine the rate of heat transfer from the geothermal fluid to the residential building. Obtain the specific heat value of water at the average temperature.

8.2 Calculate the Carnot efficiency of a geothermal power plant if the temperature of the geothermal fluid is 230 °C and the temperature of the cold reservoir is 90 °C.

8.3 Draw the schematic and *T–S* diagram for a triple flash steam geothermal power plant.

8.4 Draw the schematic and *T–S* diagram for a hybrid single flash steam/binary cycle geothermal power plant.

8.5 Saturated steam at 2.8 MPa is extracted from a geothermal reservoir at a power plant. Steam pressures at the inlet and exit of the turbine are 400 kPa and 10 kPa, respectively. Isentropic efficiency of the turbine is 86%. Determine:

    a. the thermal efficiency of the geothermal power plant if the ambient temperature is 24 °C (%)

    b. the mass flow rate of steam if the power generation capacity of the plant is 8 MW and the efficiency of the electric generator is 92% (kg/s)

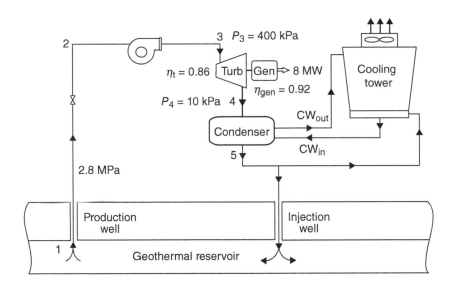

8.6 The schematic of a single flash steam geothermal power plant is given below. Geothermal fluid at 210 °C is extracted as saturated liquid at a flow rate of 190 kg/s. It isenthalpically flashes down to a pressure of 600 kPa. The liquid phase is directed to the injection well. The steam portion enters the turbine as saturated vapor and leaves at 15 kPa pressure. Isentropic efficiency of the turbine is 85%. Determine:

    a. the mass flow rate of steam flowing through the turbine (kg/s)

    b. the power output from the turbine (MW)

    c. the thermal efficiency of the geothermal power plant if the ambient temperature is 20 °C (%)

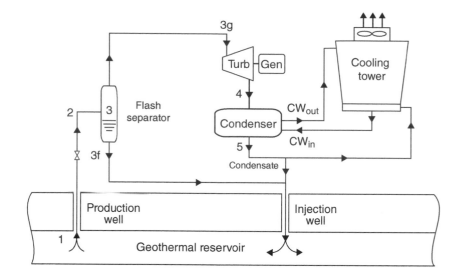

8.7 Consider the single flash steam geothermal power plant in Exercise 8.6. The plant is to be redesigned to have a second flash separator (in other words, the single flash steam system is being converted into a double flash steam plant). Geothermal fluid leaving the first flash separator in liquid phase enters the second flash separator, which is at 200 kPa, to generate more steam. The steam enters the second stage of the turbine at 200 kPa and expands to the same pressure of 15 kPa as the steam coming from the first flash separator. Isentropic efficiency is same for the overall expansion process. Determine:

a. the mass flow rate of steam flowing through the first stage turbine (kg/s)
b. the mass flow rate of steam flowing through the second stage turbine (kg/s)
c. the total power output (MW)
d. the thermal efficiency of the geothermal power plant if the ambient temperature is 20 °C (%)

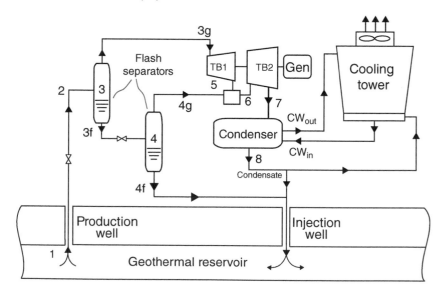

8.8 Consider the single flash steam geothermal power plant in Exercise 8.6. The plant is to be redesigned to be a hybrid single flash steam/binary cycle geothermal power plant. Geothermal fluid coming from the flash separator enters the evaporator at 155 °C and exits at 95 °C while exchanging heat with isobutane (C_4H_{10}) which is the organic compound for the binary cycle. Isobutane enters the turbine as saturated vapor at 3 MPa and leaves at 350 kPa and 30 °C. The organic compound is condensed to saturated liquid and then pumped to evaporator pressure where the pump isentropic efficiency is 88%. Determine:

a. the mass flow rate of steam flowing through the steam turbine (kg/s)
b. the mass flow rate of isobutane flowing through the isobutane turbine (kg/s)
c. the power output from the steam turbine (MW)
d. the net power output from the binary cycle (MW)
e. the thermal efficiency of the hybrid power plant if the ambient temperature is 20 °C (%)

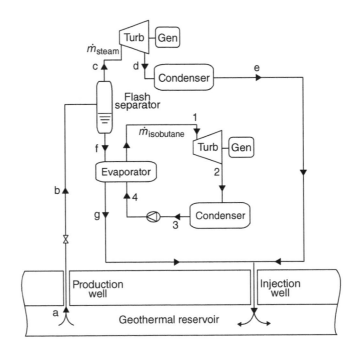

CHAPTER 9
Hydropower

9.1 Introduction

Hydropower is one of the renewable technologies that has been utilized for centuries. In the early stages of the use of hydropower, energy from water was harnessed to do mechanical work in agriculture and forestry. In ancient Greece, water wheels were used for grinding wheat. The Egyptians used the *Archimedes screw* for irrigation purposes. The foundation for transitioning to electrical energy from mechanical energy dates back to the 1750s when the French engineer *Bernard Forest de Bélidor* published *L'architecture hydraulique* (*Hydraulic architecture*) [1]. It then evolved into use of turbines for generating electricity by the conversion of potential energy into kinetic, mechanical, and electrical energy sequentially. The amount of potential energy that is accumulated is the key source of hydropower. This source relies on the water cycle (or the *hydrologic cycle*) which is the continuous movement of water on the Earth, including underground water, ice sheets, and glaciers as well (**Figure 9.1**). It is indeed the solar energy that carries water to higher elevations providing potential energy to it. This energy is first gathered for controlling water and obtaining an elevation difference to generate electricity. This is done by dams, which are barrier structures built between a reservoir and the afterbay. Dams can differ by means of construction, such as arch dams, gravity dams, or a hybrid structure called the arch–gravity dam. Some other types include embankment dams, fixed-crest dams, and barrages. The uses of dams are not limited to hydroelectric

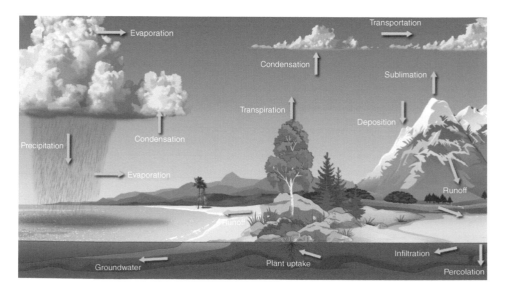

FIGURE 9.1 Water cycle. Source: NOAA.

power. In fact, hydropower has a fairly small share amongst the other purposes of dams including recreation, flood control, fire and farm ponds, irrigation, and tailings.

9.1.1 Global Overview

Hydroelectric power is the largest global renewable energy resource, making up almost half of the total installed renewable capacity. Between 2000 and 2020, the majority of the increase in hydroelectric energy consumption was dominated by the Asia–Pacific region, mainly due to the additions in China [2]. **Figure 9.2** illustrates the history of hydroelectric power use in different regions of the world. Installed capacity has also increased significantly since the new millennium, and yet the capacity of the planned and ongoing projects that have not yet been completed is estimated to be more than the existing global capacity. Total installed hydropower capacity for top ranking countries is graphed in **Figure 9.3**. China has the largest total installed capacity with a number of globally top-ranking hydroelectric power stations based on capacity. The world's largest hydroelectric dams based on their installed power capacity are listed in **Table 9.1**, including their capacity, height, and reservoir volume.

The first industrial use of hydropower in the United States was for powering brush-arc lamps at the Wolverine Chair Factory in Grand Rapids, Michigan, in 1880. The first time a hydroelectric facility started selling electricity in the United States was in 1882, on the Fox River in Wisconsin. There are currently about 1450 conventional and 40 pumped hydro energy storage (PHES) plants in the United States [4]. In terms of power generation, Grand Coulee Dam is the largest dam in the United States with a capacity of 6.809 GW. The dam was built in 1942 and is a concrete gravity type. Hoover Dam, which was completed in 1936, houses another of the major hydroelectric power

FIGURE 9.2 Global hydroelectric energy consumption history between 1995 and 2020, in EJ. Source: BP Statistical Review of World Energy 2021.

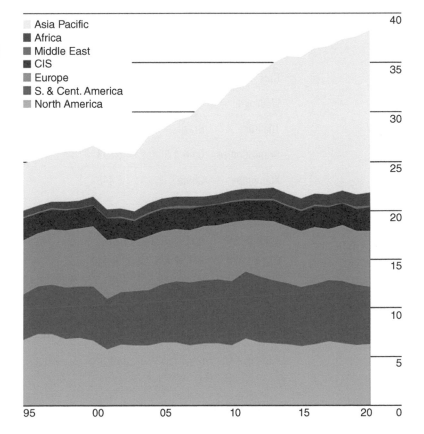

FIGURE 9.3 Total installed hydropower capacity by country. Source: Graph generated based on data from Hydropower Status Report 2021, International Hydropower Association.

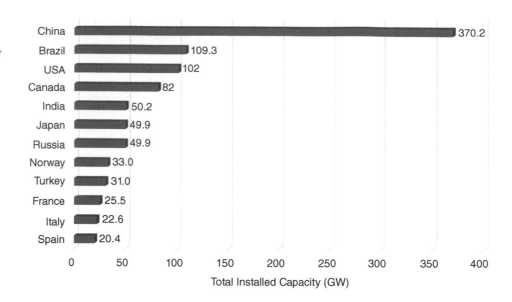

Table 9.1

World's top hydroelectric plants by power capacity [3]				
Dam	Country	Installed capacity (MW)	Height (m)	Reservoir volume (km^3)
Three Gorges	China	22,500	181	39.3
Baihetan	China	16,000	277	17.9
Itaipu	Brazil–Paraguay	14,000	196	29.0
Xilodu	China	13,860	286	12.7
Belo Monte	Brazil	11,233	90	3.9
Guri	Venezuela	10,235	162	135
Wudongde	China	10,200	240	7.4
Tucurui	Brazil	8,370	78	45
Grand Coulee	United States	6,809	168	12
Xiangijaba	China	6,448	161	5.2

stations in the United States. It is an influential structure in terms of its history, the time it was built, and the changes it has introduced to the region. It is an arch–gravity type dam with a height of 221 m, which makes it the second tallest dam in the United States, following the 235 m tall Oroville Dam in California.

9.1.2 Hydropower System Categories

Dams can be categorized based on different criteria. In terms of structure, types of dams are listed below:

- Arch dam
- Buttress dam

FIGURE 9.4 Hydroelectric power plant types based on different criteria.

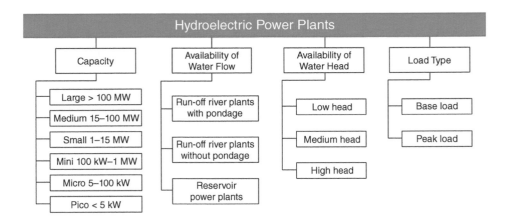

- Coffer dam
- Diversion dam
- Embankment dam
- Gravity dam
- Hydropower dam
- Industrial waste dam
- Masonry dam
- Overflow dam
- Saddle dam

There may be overlaps between these types depending on the application. Amongst the listed types, hydropower dams are of interest in this chapter due to their nature of generating electrical energy. These hydropower systems are also categorized within themselves based on various parameters, as illustrated in **Figure 9.4**.

In respect of capacity, hydroelectric power systems can range from fairly small values of less than 5 kW to high-capacity systems exceeding 100 MW. Most of the small-scale DIY hydroelectric applications fall into the pico hydro category. Micro hydro systems are the next step with a power range of 5 100 kW. These systems are typically dammed pools at a high elevation. If the head is smaller, a waterwheel or an Archimedes hydrodynamic screw is employed. Mini, small, and medium hydroelectric sites combined are in the range of 100 kW to 100 MW power capacity. These can be stations powering small communities to larger cities. Large-scale hydropower plants have capacities over 100 MW. Three Gorges Dam in China has a capacity of 22,500 MW, which is one example at the high end of this type of station. Ruskin Dam in British Columbia, Canada, houses a smaller power station with a capacity of 105 MW, which puts it only just into the large hydro category.

Availability of water flow is another criterion for classifying hydroelectric systems. Some of the hydropower plants utilize water directly from a river without storing it in pondage. In such plants, water can be wasted following heavy rains, or there might be little power generation when the water level of the river is low in dry seasons. Hence, electricity production can be undependable. Run-off river plants with pondage on the other hand have the advantage of storing water and providing more consistent power output. Most hydropower plants store water by using a dam which builds up a reservoir. Reservoir power plants have the advantage of having enough water supply even during dry seasons. However, depending on drought duration and the depth of water available, the amount of water can fall below critical levels for power generation.

Loss of water through evaporation to the atmosphere in such dry seasons can be another concern. Still, these types of dams can provide power generation throughout the year if a long drought is not observed.

In terms of the available head, hydroelectric power plants can be classified as *low-*, *medium-*, or *high-head* systems. While the threshold values for these three types can be indistinct, one of the commonly assumed boundary sets considers low-head to be below 30 m, medium-head to be in the range of 30–250 m, and high-head systems to be above 250 m. Low-head dams require high flow rates as the head is not high. Because of the limited length, pipes with larger diameters are preferred to be able to handle the high flow rate. For low-head systems, propellers, Francis turbines, or Kaplan turbines are better options due to the operating conditions. In medium-head dams, water is collected at a forebay from which it flows into the powerhouse through the penstock. Having lower flow rates compared to low-head systems, crossflow or Turgo turbines can also be used in addition to the Francis turbine option. High-head dams have the reservoir at a much higher elevation than the lower reservoir. This type takes its name from the head being high due to this elevation difference, not necessarily due to the dam structure being tall. As the height is large, compact turbine designs can be implemented in these kinds of power stations. Pelton turbines are more suitable for high-head systems.

According to the load, a power plant can be *base load* or *peak load* type. Base load power plants operate as a source for the base load demand. These types can provide constant load and they have a high load factor. Typically, run-off river plants without pondage and reservoir power plants are used to supply base load. Peak load power plants are used to complement supply when the load demand peaks at certain hours of the day. They enable the grid to be able to compensate for the increase in demand. Run-off river power plants with pondage can be used as such stations to account for some or all of the peak curve depending on the amount of water available in the pondage. In **Chapter 13**, there is additional discussion of energy demand management methods in addressing peak loads.

FURTHER LEARNING

International Hydropower Association
www.hydropower.org/

FURTHER LEARNING

Hydropower Glossary
www.energy.gov/eere/water/glossary-hydropower-terms

9.1.3 Turbine Classification and Designs

Turbines are key components as turbomachines in energy systems for generating electricity. In hydropower systems, turbines can be classified based on the flow of water through them. In *axial flow* types, water flows through the turbine parallel to the axis of rotation of the shaft. In *radial flow* designs, water flows perpendicular to the axis of rotation. *Mixed flow* types are a hybrid version of the previous two designs. Water experiences both axial and radial flow through these turbines. Another parameter used in classifying turbines is the pressure change across the turbomachine. According to this criterion, turbines are classified into two types which are *reaction turbines* and *impulse turbines*. Both of these types and some common designs associated with both types are discussed in this section.

9.1.3.1 REACTION TURBINES

Reaction turbines use the pressure of the water to rotate the blades, and hence the turbine shaft to generate electricity. The pressure of the water changes during this process. As the pressure of the water is utilized for harnessing energy from the water, the turbine needs to be immersed in water. These designs require high durability as they operate under high pressures. Therefore, material selection and manufacturing

FIGURE 9.5 A Kaplan turbine design being assembled. Source: Voith Hydro.

FIGURE 9.6 3D render of a Francis turbine design. Source: Voith Hydro.

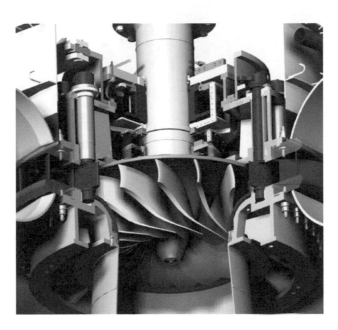

techniques are of importance for reliability, as well as for efficiency. Two common types of reaction turbines are *Kaplan turbines* and *Francis turbines*. Kaplan turbines are axial flow type systems. This design, invented by Viktor Kaplan, is a propeller type. It is placed in a cylindrical volume so that the water is forced to flow through the turbine blades, hence improving the efficiency of the turbine. A Kaplan turbine design is illustrated in **Figure 9.5**. Francis turbines are another type of reaction turbines which differ from the Kaplan turbines in design such that water enters these radially and leaves axially. Hence, Francis turbines are considered as mixed flow type systems. The design takes its name from James Francis, a civil engineer, who improved existing turbines and invented the Francis turbine. The major components of this design are the volute (spiral) casing, guide vanes, runner, and the main shaft. A 3D drawing of a Francis turbine can be seen in **Figure 9.6** (see also **Figure 9.10**).

9.1.3.2 IMPULSE TURBINES

Unlike in reaction turbines, there is no noticeable change in pressure in impulse turbines. Energy from the water is extracted by use of nozzles which generate a jet

FIGURE 9.7 A Pelton turbine runner design. Source: apomares/E+ via Getty Images.

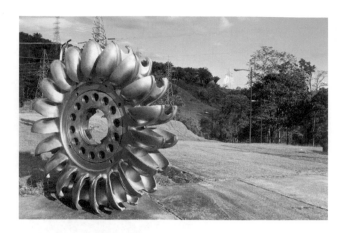

FIGURE 9.8 A mini Turgo turbine runner design. Source: Hartvigsen-Hydro.

flow directed into bucket- (or spoon-) shaped pockets, forcing the runner to rotate. As the water hits the runner blades, its velocity is reduced, and the momentum resulting from the energy transfer is utilized by the turbine generator to produce electricity. The free jet is achieved open to the atmosphere; hence these designs are not immersed in water like the reaction turbines. Common designs of this class include *Pelton turbines*, *Turgo turbines*, and *crossflow turbines*. Pelton turbines (or Pelton wheels), invented by Lester Pelton, are designed to receive the water jet parallel to the direction of wheel rotation. The buckets are divided by a splitter making them look like two side-by-side spoons. The function of the splitter is to cut through the water jet, splitting it to exert force onto both sides. The sharpness of the splitter is important as the amount of water used to do work depends on it. The duller it gets, the more water will be wasted to the surroundings without doing useful work in the buckets. A Pelton turbine runner is depicted in **Figure 9.7**. In Turgo turbines, the incoming jet makes an angle with the horizontal. The spoons do not have a splitter. Instead, they have arc-shaped notches on one side to let the incoming jet in. In small hydro applications where head is high and flow rate is low, these designs can be favored over Pelton turbines due to their lower costs. A mini Turgo turbine runner can be seen in **Figure 9.8**. Crossflow turbines receive water in the form of a slot rather than a point strike as in Pelton and Turgo turbines. These designs can be analogous to crossflow tidal turbines used in ocean

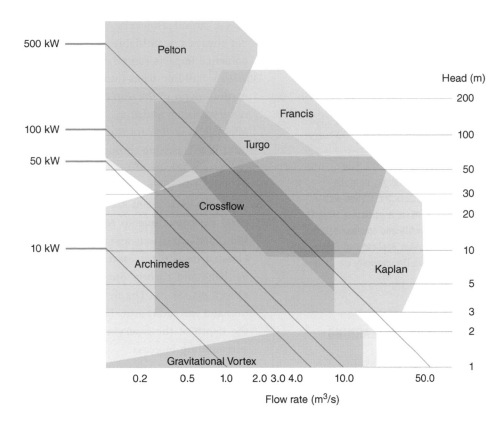

FIGURE 9.9 Turbine selection chart. Source: Adopted from www.currenthydro.com.

FIGURE 9.10 Francis turbines in the generator room of Hoover Dam, which can be viewed by visitors touring the facility. Source: YinYang/E+ via Getty Images.

energy systems, or Savonius rotors utilized in wind energy applications, only being oriented horizontally rather than vertically.

As with any turbomachine, turbine selection is also critical and can be project specific. Based on the available head and flow rate, a turbine design may be more advantageous than another design. A selection chart mapping the appropriate ranges of head and flow rates for different turbine designs is given in **Figure 9.9**. As can be seen from the chart, there are indistinct regions on the chart where more than one design option can exist. In such cases, the budget of the project, structural specifications of the dam, accessibility, and durability requirements become instrumental in making the optimum selection.

9.1.4 Environmental Impacts

Hydropower is a renewable energy resource. It yields an enormous reduction in fossil fuel use for the amount of electricity generated at hydroelectric power plants. This results in reduced carbon emissions. If hydropower was replaced with thermal power plants burning coal as fuel, up to four billion tonnes of additional greenhouse gases (GHGs) would be emitted to the atmosphere each year and global emissions from fossil fuels and industry would be at least 10% higher [5]. Besides these advantages of hydropower, there are some negative impacts of it on the environment. These impacts can be listed as:

Land use: Hydroelectric power plants can have very large reservoirs depending on the construction of the dam and the upstream landscape condition. These reservoirs cover large areas of land, which can include residential areas, agricultural land, historic sites, or wildlife habitats. A quantitative approach for land use effectiveness is to determine the hectares of flooded land area per MW of

FIGURE 9.11 Three Gorges Dam, the world's largest hydroelectric station with a capacity of 22.5 GW, is a particular example for discussing benefits and environmental impacts of hydropower applications. Source: Xiaoyang Liu/Corbis Documentary via Getty Images.

FURTHER LEARNING

Virtual Reality Visit to a Hydroelectric Facility
www.hydroquebec.com/visit/video-360-lg2.html

FURTHER LEARNING

Fishway (Fish Ladder) Designs
www.dpi.nsw.gov.au/fishing/habitat/rehabilitating/fishways

power output. However, the impact of land use on the aforementioned areas is not always a quantifiable measure. Therefore, a holistic assessment of the environmental impacts can be challenging.

Wildlife: Building dams on rivers most of the time interferes noticeably with the natural course of the water. This can block fish migration paths, which results in reduced fish population. This poses a risk for humans as the food supply is affected, as well as for the ecosystem. Another risk is the water turbines, which can kill fish passing through the blades. A common practice in addressing these concerns is the use of *fish ladders*, which provide a detour path for migrating fish past the dam to the upstream water.

GHG emissions: Even though hydropower yields a significant reduction in GHG emissions due to reduced fossil fuel use to generate power, there are some GHG emissions associated with hydropower. The reservoirs emit GHGs. Flooded dead plants and some other organic materials decompose and emit methane and carbon dioxide. Reservoirs can also act as a carbon sink and absorb emissions. The net amount of GHG emissions depends on the absorption and emission rates. These can differ based on different parameters such as the depth of the reservoir, sunlight penetration through the water, and the surrounding biomass.

9.2 Theory

Theoretical analysis of hydropower plants involves two important topics which we discuss in this section:

- Hydropower capacity
- Turbine specific speed

9.2.1 Hydropower Capacity

The power output capacity calculation of a hydropower plant starts with the energy analysis of the flow of water through the penstock and the turbine into the downstream river. For steady, incompressible flow, the energy expression that involves the Bernoulli equation can be written as:

$$\frac{p_1}{\gamma} + \frac{V_1^2}{2g} + z_1 = \frac{p_2}{\gamma} + \frac{V_2^2}{2g} + z_2 + h_L + h_t \tag{9.1}$$

where h_L is the total head loss and h_t is the change in head across the turbine. Total head loss is the sum of major and minor losses:

$$h_{L_{major}} = f\frac{L}{D}\frac{V^2}{2g} \tag{9.2}$$

$$h_{L_{minor}} = K_L\frac{V^2}{2g} \tag{9.3}$$

FIGURE 9.12 Schematic of the dam for hydropower calculation. Source: Adapted from U.S. Department of Energy, www.energy.gov/eere/water/types-hydropower-plants.

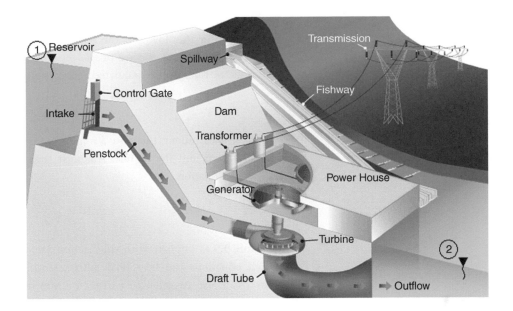

$$h_{\mathrm{L}} = \frac{V^2}{2g}\left(f\frac{L}{D} + \sum_{i=1}^{n} K_{\mathrm{L}_i}\right) \tag{9.4}$$

where f is the friction factor and K_{L} is the loss coefficient. Substituting the loss term in **Equation 9.1** with the expression in **Equation 9.4** gives:

$$\frac{p_1}{\gamma} + \frac{V_1^2}{2g} + z_1 = \frac{p_2}{\gamma} + \frac{V_2^2}{2g} + z_2 + \frac{V^2}{2g}\left(f\frac{L}{D} + \sum_{i=1}^{n} K_{\mathrm{L}_i}\right) + h_t \tag{9.5}$$

A convenient way of selecting the nodes for the energy analysis is shown in **Figure 9.12**. Selecting the inlet and exit nodes at the free surfaces on both the upper and lower reservoirs comes with simplifications. According to the given schematic, we can list the known values of the variables.

At the reservoir side (node 1):

$p_1 = 0$ (open to atmosphere)
$V_1 = 0$ (very large volume, negligible velocity)
z_1 = height of the reservoir surface with reference to a datum

At the afterbay side (node 2):

$p_2 = 0$ (open to atmosphere)
$V_2 = 0$ (very large volume, negligible velocity)
z_2 = height of the afterbay surface with reference to a datum

Then, **Equation 9.5** can be rearranged as:

$$h_t = (z_1 - z_2) - \frac{V^2}{2g}\left(f\frac{L}{D} + \sum_{i=1}^{n} K_{\mathrm{L}_i}\right) \tag{9.6}$$

The velocity of the water through the penstock and the turbine into the afterbay can be determined for a known value of volume flow rate and the cross-sectional area of the flow by:

$$V = \frac{Q}{A} = \frac{4\,Q}{(\pi D^2)} \tag{9.7}$$

The friction factor, f, can be determined using the *Moody chart* based on the Reynolds number (Re) for the flow and the relative roughness, ε/D, where ε is the equivalent roughness of the inner surface of the pipe or channel the water flows through. The friction factor can also be calculated theoretically. If the flow is laminar:

$$f = \frac{64}{Re} \tag{9.8}$$

If the flow is turbulent, there are a number of equations that can be used based on the flow conditions and the regime. Amongst these, the Colebrook equation is the most widely used relation:

$$\frac{1}{\sqrt{f}} = -2 \, \log \left(\frac{\varepsilon/D}{3.7} + \frac{2.51}{Re \, \sqrt{f}} \right) \tag{9.9}$$

Equation 9.9 is implicit with the friction factor on both sides of the relation. Such implicit equations can be solved with scientific calculators or using a spreadsheet. Factor f can also be calculated using Python or MATLAB. With known values of f and K_{L}, the turbine head (h_{t}) can be calculated using **Equation 9.6**.

Power extracted from the fluid flowing through the turbine is:

$$P_{\mathrm{ext}} = \rho \, g \, Q \, h_{\mathrm{t}} \tag{9.10}$$

Obviously, not all of the energy extracted from the fluid is converted into electrical energy. Power generated depends on the turbine-generator efficiency, η_{t}:

$$P_{\mathrm{gen}} = \eta_{\mathrm{t}} P_{\mathrm{ext}} = \eta_{\mathrm{t}} \rho \, g \, Q \, h_{\mathrm{t}} \tag{9.11}$$

Hence, the power capacity for a hydroelectric plant can be estimated utilizing **Equation 9.11**. It can be interpreted from this equation that the capacity is proportional to the turbine head. The same argument, although it looks like it, is not valid for the flow rate. This is due to the relation between the turbine head and the flow rate as given in **Equations 9.6** and **9.7**. An increase in flow rate translates into a decrease in the h_{t} value, which suggests that there is an optimum flow rate for maximum power output.

Example 9.1 Determining Power Capacity of a Hydroelectric Facility

Feasibility and preliminary design studies for a proposed hydroelectric facility yielded the following input values. If the turbine efficiency is 90%, determine the power output capacity for this facility in kW.

- $z_1 = 124$ m
- $z_2 = 78$ m
- $Q = 0.25$ m^3/s
- $L = 300$ m
- $D = 360$ mm
- $\varepsilon = 0.050$ mm
- $K_{\mathrm{inlet}} = 0.8$
- $K_{\mathrm{exit}} = 1.2$
- $K_{\mathrm{gate}} = 0.3$
- $K_{\mathrm{elbow}} = 0.35$ (two elbows)

Solution

Starting with **Equation 9.5**, we have:

$$\frac{p_1}{\gamma} + \frac{V_1^2}{2g} + z_1 = \frac{p_2}{\gamma} + \frac{V_2^2}{2g} + z_2 + \frac{V^2}{2g}\left(f\frac{L}{D} + \sum_{i=1}^{n}K_{L_i}\right) + h_t$$

If we select the inlet and exit nodes at the reservoir and the afterbay as given in **Figure 9.12**, and replace the velocity term in **Equation 9.6** with the expression given in **Equation 9.7**, we get:

$$h_t = (z_1 - z_2) - \frac{8Q^2}{g\pi^2 D^4}\left(f\frac{L}{D} + \sum_{i=1}^{n}K_{L_i}\right)$$

The total loss coefficient is the sum of coefficients from each component:

$$\sum K_L = 0.8 + 1.2 + 0.3 + (2 \times 0.35) = 3.0$$

The friction factor, f, can be determined by using either the *Moody chart* or the *Colebrook equation*. For either method, we need to find Re first:

$$Re = \frac{VD}{\nu}$$

The velocity is:

$$V = \frac{Q}{A} = \frac{4Q}{\pi D^2} = \frac{4(0.25\ \text{m}^3/\text{s})}{\pi(0.36\ \text{m})^2} = 2.46\ \text{m/s}$$

Taking the kinematic viscosity of water, $\nu = 0.89 \times 10^{-6}\ \text{m}^2/\text{s}$, the Reynolds number is:

$$Re = \frac{(2.46\ \text{m/s})(0.36\ \text{m})}{0.89 \times 10^{-6}\ \text{m}^2/\text{s}} = 9.95 \times 10^5$$

Hence, the flow is turbulent. Using the Colebrook relation as expressed in **Equation 9.9**:

$$\frac{1}{\sqrt{f}} = -2\ \log\left(\frac{\varepsilon/D}{3.7} + \frac{2.51}{Re\sqrt{f}}\right) = 0.014$$

Then, the change in head across the turbine, h_t, is:

$$h_t = (124 - 78)\ \text{m} - \frac{8\left(0.25\ \frac{\text{m}^3}{\text{s}}\right)^2}{\left(9.81\ \frac{\text{m}}{\text{s}^2}\right)\pi^2(0.36\ \text{m})^4}\left(0.014\frac{300\ \text{m}}{0.36\ \text{m}} + 3.0\right) = 41.52\ \text{m}$$

Now, the power generation capacity can be calculated employing **Equation 9.11**:

$$\begin{aligned}
P_{\text{gen}} &= \eta_t\, \rho\, g\, Q\, h_t \\
&= (0.90)(999\ \text{kg/m}^3)(9.81\ \text{m/s}^2)(0.25\ \text{m}^3/\text{s})(41.52\ \text{m}) \\
&= 91{,}553.4\ \text{kg m}^2/\text{s}^3 \left|\frac{1\ \text{kW}}{10^3\ \text{kg m}^2/\text{s}^3}\right| = 91.55\ \text{kW}
\end{aligned}$$

9.2.2 Turbine Specific Speed

In turbomachinery, appropriate pump or turbine selection is a critical step. This selection can be done based on a non-dimensional group that is obtained through dimensional analysis. Assuming similar mechanical efficiency, the performance of a pump or a turbine can be evaluated based on seven parameters which are listed in **Table 9.2**.

According to the Buckingham Π theorem with seven variables and three independent dimensions (mass, length, and time) associated with these variables, four independent dimensionless groups exist that relate all the variables to each other. These groups are:

$$\Pi_1 = \frac{Q}{\omega D^3} \tag{9.12}$$

$$\Pi_2 = \frac{gH}{\omega^2 D^2} \tag{9.13}$$

$$\Pi_3 = \frac{P}{\rho \omega^3 D^5} \tag{9.14}$$

$$\Pi_4 = \frac{\rho \omega D^2}{\mu} \tag{9.15}$$

The dimensionless groups listed in **Equations 9.12–9.15** are known as the flow coefficient, head coefficient, power coefficient, and Reynolds number, respectively. For a turbine, head and power coefficients Π_2 and Π_3 can be used to develop another dimensionless group by eliminating rotor diameter, D. This yields the *specific speed* of the turbine which is:

$$N_s = \frac{\omega \sqrt{P}}{\rho^{1/2} (gH)^{5/4}} \tag{9.16}$$

where ω is the rotational speed in radians per second, P is the power generated in watts, and H is the head in meters. In the United States, customary units used are revolutions per minute (rpm) for rotational speed, horsepower (hp) for power, and feet (ft) for height.

Table 9.2

Variables used in dimensionless pump or turbine analysis		
Variable	Symbol	Dimensions
Volume flow rate	Q	$L^3 T^{-1}$
Pressure change	Δp	$ML^{-1}T^{-2}$
Power[*]	P	$ML^2 T^{-3}$
Rotor diameter	D	L
Rotational speed	ω	T^{-1}
Fluid density	ρ	ML^{-3}
Fluid velocity	μ	$ML^{-1}T^{-1}$

[*] Power input for pumps (energy consumption), power output for turbines (energy generation)

FIGURE 9.13 Application ranges of different turbine types based on turbine specific speed.

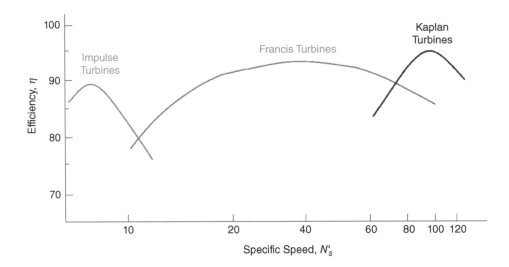

Introducing these into **Equation 9.16** along with known values of ρ and g gives the specific speed in US customary units:

$$N'_s = \frac{\omega\sqrt{P}}{H^{5/4}} \qquad (9.17)$$

In this expression, ω is in rpm, generated power is in hp, and available head is in ft. The specific speed of the turbine can be regarded as the rotational speed needed to generate one unit of power through one unit of head available. Specific speed ranges for different turbine designs can be seen in **Figure 9.13**. These illustrations enable the engineer or the analyst to determine the right turbine selection for the project of interest. Note that the plot is to give approximate guidance on different designs. In an actual project, the efficiency of the turbine will depend on multiple parameters that can be specific to each design. As can be seen from the figure, impulse turbines are better fitting at low specific speeds, meaning higher heads and lower flow rates. If the head is low and flow rate is high, then this yields higher specific speeds where reaction turbines provide better performance.

Example 9.2 Identifying Appropriate Turbine Design by Calculated Specific Speed

Consider the hydroelectric facility in **Example 9.1**. If the rotational speeds for the turbines are to be 1200 rpm, determine a turbine design that would best fit this project.

Solution

Turbine selection can be done based on the specific speed value. Use **Equation 9.17** to get the specific speed with US customary units:

$$N'_s = \frac{\omega\sqrt{P}}{H^{5/4}}$$

where

$\omega = 1200$ rpm

$P_{\text{gen}} = 91.55$ kW $\equiv 122.8$ hp

$H = 46$ m $\equiv 150.9$ ft

Then,

$$N'_s = \frac{1200\sqrt{122.8}}{(150.9)^{5/4}} = 25.1$$

From **Figure 9.13**, the *Francis turbine* seems to be the design that best fits with the calculated specific speed value of 25.1. Corresponding efficiency can also be approximated to be ~92% from the same plot.

Example 9.3 Calculating Specific Speed, Turbine Type, and Flow Rate

A group of turbines are to be installed at a new hydroelectric power station which is designed to generate 28,000 hp with an available head of 25 m. Rotational speed will be maintained at 70 rpm.

a. Determine the specific speed using customary units
b. Suggest a turbine design for this application
c. If 88% of the available head (H) does work across the turbine, calculate the flow rate in m^3/s

Solution

a. Specific speed with customary units is obtained by employing **Equation 9.17**. Head is given in SI units and needs to be converted into imperial units.

$$H = 25\,\text{m} \equiv 82\,\text{ft}$$

Then,

$$N'_s = \frac{\omega\sqrt{P}}{H^{5/4}} = \frac{70\sqrt{28,000}}{(82)^{5/4}} = 47.5$$

b. Based on the calculated specific speed, the *Francis turbine* is a better fit for this project according to the plot in **Figure 9.13**.
c. Flow rate can be calculated from **Equation 9.11**.

$$P_{\text{gen}} = \eta_t\,\rho\,g\,Q\,h_t$$

Then,

$$Q = \frac{P_{\text{gen}}}{\eta_t\,\rho\,g\,h_t}$$

Turbine efficiency can be approximated to be ~92% from **Figure 9.13** based on the calculated specific speed. Change of head across the turbine is given as 88% of the available head.

$$h_t = 0.88\,H = (0.88)(25\,\text{m}) = 22\,\text{m}$$

Then,

$$Q = \frac{28,000\,\text{hp}}{(0.92)(999\,\text{kg/m}^3)(9.81\,\text{m/s}^2)(22\,\text{m})}\left|\frac{1\,\text{kW}}{1.34\,\text{hp}}\right|\left|\frac{10^3\,\text{kg m}^2/\text{s}^3}{1\,\text{kW}}\right| = 105.3\,\text{m}^3/\text{s}$$

Example 9.4 Determining Power Output Based on Turbine Efficiency and Water Temperature Using Python

The upper reservoir of a hydropower dam has an elevation of 445 m while the afterbay has an elevation of 373 m. At a flow rate of 2100 m³/h, total head losses add up to 7 m.

a. Plot the generated power for this hydroelectric station for a turbine efficiency range of 78–92%.

b. Density is one of the variables determining the power generation capacity. As the density of water changes with temperature, obtain a plot showing the relation between power generation and water temperature for a range of 0–25 °C and a fixed turbine efficiency of 80%.

Solution

From **Equation 9.6**, change of head across the turbine, h_t, is:

$$h_t = (z_1 - z_2) - \frac{V^2}{2g}\left(f\frac{L}{D} + \sum_{i=1}^{n} K_{L_i}\right)$$

Hence, with the given information:

$$h_t = (445 - 373)\,\text{m} - 7\,\text{m} = 65\,\text{m}$$

Power generation is calculated employing **Equation 9.11**:

$$P_{\text{gen}} = \eta_t \rho\, g\, Q\, h_t$$

where

$\eta_t = 0.78\text{–}0.92$
$\rho = 999\ \text{kg/m}^3$
$g = 9.81\ \text{m/s}^2$
$Q = 2100\ \text{m}^3/\text{h} = 0.583\ \text{m}^3/\text{s}$
$h_t = 65\ \text{m}$

Then,

$$P_{\text{gen}} = \eta_t\left(999\ \frac{\text{kg}}{\text{m}^3}\right)\left(9.81\ \frac{\text{m}}{\text{s}^2}\right)\left(0.583\ \frac{\text{m}^3}{\text{s}}\right)(65\ \text{m})\left|\frac{1\ \text{kW}}{\dfrac{10^3\ \text{kg m}^2}{\text{s}^3}}\right| = \eta_t(371.4\ \text{kW})$$

a. We can build the code to calculate and plot the power output for varying turbine-generator efficiency values based on the relation we obtained:

```
import matplotlib.pyplot as plt


# Net change of head across the turbine
h_t=(445-373)-7
```

```
# Power Generation = efficiency * density * gravity * flow rate * turbine
head
ro=999
eff= 78/100
g=9.81
Q = 2100/(60*60) # convert m3/hour to m3/second by dividing
by 60*60
P_gen = eff * ro * g * Q * h_t

print("Power generation is = ", P_gen, "Watts")
Power generation is = 289840.61925000005 Watts
```

Now, we can apply the efficiency change over the given range and plot the graph:

```
eff_array=[]
P_array=[]
for eff in range(78, 93,1):
  P_gen = (eff/100) * ro * g * Q * h_t
  eff_array.append(eff/100)
  P_array.append(P_gen)

plt.title("Hydropower generation vs efficiency")
plt.plot(eff_array,P_array, linestyle="dashdot")
plt.xlabel("Efficiency")
plt.ylabel("Power Generation (Watts)")

plt.title("Hydropower generation vs efficiency")
plt.plot(eff_array,P_array, linestyle="dashdot")
plt.xlabel("Efficiency")
plt.ylabel("Power Generation (Watts)")
```

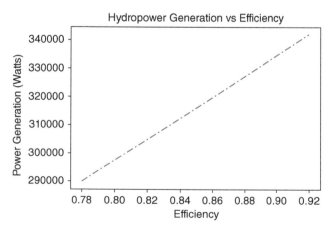

b. Density of water as a function of temperature can be expressed as [6]:

$$\rho(T) = a_5 \left[1 - \frac{(T + a_1)^2 (T + a_2)}{a_3 (T + a_4)} \right]$$

where the five coefficients are determined by the chi-square method. Then,

```
a1 = -3.983035
a2 = 301.797
a3 = 522528.9
a4 = 69.34881
a5 = 999.974950

T_array=[]
P_array=[]
eff=80

for T in range(0, 25,1):
  ro= 1/(1+(0.0002*(T-20)))
  ro=a5*(1- ( (((T+a1)**2)*(T+a2))  / ( a3*(T+a4))))
  P_gen = (eff/100) * ro * g * Q * h_t
  T_array.append(T)
  P_array.append(P_gen)

plt.title("Temperature vs Power Generation")
plt.plot(T_array, P_array)
plt.xlabel("Temperature (°C)")
plt.ylabel("Power Generation (Watts)")
Text(0, 0.5, 'Power Generation (Watts)')
```

As can be seen, power generation is highest when water is at 4 °C, where the density of water has its peak value.

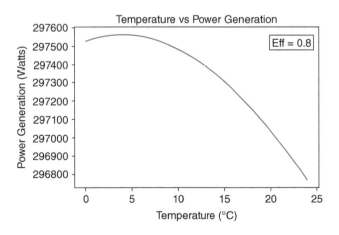

9.3 Applications and Case Study

In this section, two applications and a case study are discussed as real-life examples. The applications include Itaipu Dam on the Parana River between Brazil and Paraguay, and Hoover Dam on the Colorado River between Nevada and Arizona in the United States. The case study is about world's largest embankment dam, Aswan Dam, which is on the Nile River in Egypt.

Application 1 Itaipu Dam (Brazil–Paraguay)

Itaipu Binacional
Between: Foz do Iguacu and
    Hernandarias
Source: Paraná River
Structure: Hollow gravity
Height: 196 m
Turbines: Francis type
Capacity: 14 GW
Capacity factor: 73%
www.d-maps.com

Itaipu Binacional Dam (**Figure 9.14**) houses a hydroelectric power station on the Paraná River which runs between Brazil and Paraguay, hence the term *binational*. In 1973, Brazil and Paraguay signed the Itaipu Treaty. As the river not only runs between the two countries, but also flows along the Argentinian–Paraguayan border, an agreement was also carried out in tripartite form in 1979. Itaipu started generating electricity in 1984. The dam is a hollow gravity structure and has a height of 196 m. The hydroelectric station had 18 Francis type turbines, each having a capacity of 700 MW. With the capacity expansion in 2006 and 2007, two more turbines were added to the fleet, yielding 14 GW of total capacity. The dam has produced more than 2.8 million GWh since its launch. It has had the world record in both cumulative electricity production and annual production categories. It provides 10.8% of the energy consumption in Brazil, and 88.5% of the energy consumption in Paraguay [7]. The reservoir of Itaipu Dam covers an area of 1350 km^2, with a volume of 29 km^3. Approximately 12.3 million cubic meters of concrete were used to build the dam. An interesting fact about the dam is that, of the 20 turbine-generators, 10 of them generate electricity at 60 Hz for Brazil, and the other 10 units generate electricity at 50 Hz for Paraguay, as the two countries use different frequencies.

FIGURE 9.14 Itaipu Dam from the lens of skydivers. Source: Graiki/Moment via Getty Images.

Application 2 Hoover Dam (United States)

Hoover Dam
Between: Arizona and Nevada
Source: Colorado River
Structure: Concrete arch–gravity
Height: 221.4 m
Turbines: Francis type
Capacity: 2.08 GW
Capacity factor: 23%
www.d-maps.com

Hoover Dam (**Figure 9.15**) is probably one of the most intriguing hydroelectric power stations in the world with its history and the challenges that were experienced during its construction. It was built between 1931 and 1936, during the Great Depression. The project was led by *Frank Crowe*, a civil engineer, who used different construction techniques to make the project move faster, such as introducing pneumatic systems to transport concrete within the construction site. Another brilliant engineering technique used during the construction was the cooling of cement by steel pipes through which cold water was flowing. Convective heat transfer into the pipes acting as heat sinks helped dissipate the heat created by the cement. The foundation of Hoover Dam is volcanic rock which is very hard and durable. The dam is the concrete arch–gravity type. The load of the water is distributed by both the arched surface and the gravity effect. The wall is curved against the reservoir, distributing the hydrostatic forces to the sides towards the *abutments*. Abutments are the naturally existing or artificially made substructures at both ends of the dam, supporting the structure. The distribution of abutment forces depends on the geometry of the arch dam. The canyon walls from both sides exhibit reaction forces which compress the arched dam, hence *squeezing* it and making the concrete structure more rigid. The dam has 17 Francis type turbines and an installed capacity of 2080 MW. Some interesting facts about the dam help quantify certain properties of it. The dam is 221.4 m tall, which is 52 m (171 ft) taller than the Washington Monument in Washington DC. At its base, the dam is 201 m thick which is a little over two football field lengths end-to-end. There is 3.44 million cubic meters of concrete in Hoover Dam, which is enough to build a two-lane road from Seattle to Miami. During peak demand hours, the amount of water passing through the turbine-generators is enough to fill 15 average sized swimming pools every second [8].

FIGURE 9.15 Part of Lake Mead, the reservoir formed by Hoover Dam, and the arch bridge over the downstream river. Source: Jennifer_Sharp/E+ via Getty Images.

Case Study Aswan Dam and Nubian Monuments (Egypt)

Aswan Dam
Location: Aswan, Egypt
Source: River Nile
Structure: Embankment
Turbines: Francis type
Capacity: 2.10 GW
www.d-maps.com

INTRODUCTION Aswan Dam is the largest embankment dam in the world. The dam impounds the River Nile, forming Lake Nasser which is the reservoir. The dam has a wide base with a width of 980 m, and a height of 111 m. The reservoir volume is 132 km3, which makes it one of the top ten largest dams by volume. It covers an area larger than the annual average size of the Great Salt Lake in Utah. The hydroelectric power station houses 12 Francis type turbines with each of them having a capacity of 175 MW, yielding a total installed capacity of 2.10 GW. The Aswan Dam project is a comprehensive case study example involving engineering and agriculture, as well as history, archeology, and anthropology due to the historic sites in the region that are considered as world heritage.

OBJECTIVE The objective of this case study is to learn about a hydropower application and measures that can be taken to minimize its impacts on historic sites.

METHOD This case study was conducted through researching articles and reports.

RESULTS As in any energy project, Aswan Dam has its own advantages and disadvantages. The dam has benefited the region in terms of reducing flooding and drought risks, providing electricity and employment, and enhancing agriculture. However, the reservoir has caused erosion and change in salinity levels over the years. It also flooded many historic sites, including the monuments of Nubia, which is one of the oldest civilizations in Africa. UNESCO's campaign in relocating the Abu Simbel Temples to higher ground is one of the biggest archeological projects in history. Today, the sites are open for tourists. In addition to the sites, the Nubia Museum established as part of the safeguarding campaign also houses many monuments of this ancient civilization.

DISCUSSION Hydroelectric dams benefit the environment significantly due to the reduction in GHG emissions had the energy been generated using fossil fuels. One of the major concerns about hydropower is the impact of the flooding on the environment. Loss of farming lands, people being relocated, changing ecosystem, and soil erosion are some of these concerns. In areas where there are historic sites, it is of great importance that the heritage is preserved. UNESCO's effort for Abu Simbel is a good example of such an approach.

RECOMMENDATIONS The environmental impact assessment of a hydropower project should involve experts from fields other than engineering, such as biologists, botanists, zoologists, historians, archeologists, architects, as well as members of the nearby communities including farmers, teachers, business owners, and environmental organization representatives. Such a collective approach will yield a diverse pool of potential risks and solutions and can help reduce the impacts of the project on various grounds.

FURTHER LEARNING

Safeguarding Heritage: UNESCO and Abu Simbel
https://en.unesco.org/70years/abu_simbel_safeguarding_heritage

CLASSROOM DEBATE

Split into two groups and debate on the advantages and disadvantages of hydropower from an environmental impact point of view.

9.4 Economics

The economics of hydroelectric power generation has two major components which are total installed costs and O&M costs. Total installed costs include feasibility, planning, design, permitting, dam and reservoir construction, tunnel and canal construction, powerhouse construction, infrastructure, grid connection and equipment. The equipment mainly includes turbines, generators, transformers, and control units. Operational costs include utilities, wages and salaries, workers' compensation, medical insurance, and taxes. Maintenance costs account for routine maintenance, troubleshooting, and replacement of equipment or parts. According to the IEA, capital costs make up about 80–90% of hydropower LCOE. This is similar to other renewable energy technologies that do not require a fuel supply. Hence, O&M costs are comparatively very low, averaging about 2% of the initial investment cost annually [9].

Global weighted-average total installed costs, along with capacity factors and LCOE values during 2010–2020 are given in **Figure 9.16**. In 2020, the global average for total installed cost was $1870 per kW of capacity. In the same year, the global weighted-average capacity factor was 46%, slightly below the 10-year average of 47.3%. The LCOE for 2020 was 0.044 $/kWh, which was above the preceding decade average of 0.040 $/kWh [10].

The LCOE of a hydropower project is calculated by

$$\text{LCOE} = \frac{I_A + \sum_{t=1}^{n} \dfrac{\text{OM}_t}{(1+r)^t}}{\sum_{t=1}^{n} \dfrac{E_t}{(1+r)^t}(1-\delta)^t} \qquad (9.18)$$

FIGURE 9.16 Global weighted-average total installed costs, capacity factors, and LCOE values for hydropower projects between 2010 and 2020. Source: IRENA Power Generation Costs in 2020.

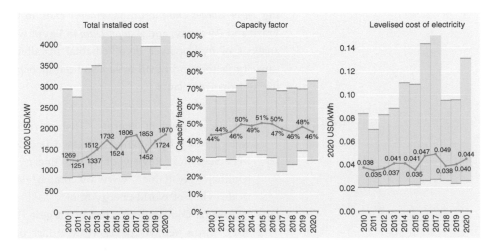

where

I_A: annualized investment costs

OM_t: operating and maintenance costs in year t

E_t: electrical energy generated in year t

r: annual effective discount rate (amount of interest paid or earned as a percentage of the balance at the end of the annual period)

n: estimated lifespan of the hydropower system

δ: system degradation rate

FURTHER LEARNING

HydroSource: Water-Energy Digital Platform
https://hydrosource.ornl.gov/

In this expression, the numerator is the total costs at the time being and in the future, as its *net present value*, NPV. The denominator is the amount of energy produced each year. As the power plant will be aging over time, a *degradation factor*, δ, is introduced into the equation. Annualized investment cost can be calculated using the total investment cost (I_T), annual discount rate (r), and anticipated lifespan (n) of the hydropower system, as given in **Equation 9.19**:

$$I_A = I_T \frac{r\,(1+r)^n}{(1+r)^n - 1} \tag{9.19}$$

9.5 Summary

Hydropower plants depend on the hydrologic cycle. Energy from the water that is elevated due to the water cycle driven by solar energy can be harnessed to generate electricity using hydroelectric systems. These systems can vary in size from very large hydropower plants to mini, micro, or even pico hydro applications. In this chapter, different types of dams, hydropower applications, and turbine designs were presented. A hydropower project can differ based on the construction of the dam, available head and flow rate, scale of the power output, and type of turbine(s) employed. Reaction and impulse turbines and turbine designs pertaining to both categories were discussed. Environmental impacts of hydropower applications in terms of land use, wildlife, and GHG emissions were also covered, addressing both the advantages and potential disadvantages of these projects. The theoretical analysis in the chapter focuses on calculation of the power output capacity of a hydropower plant and turbine selection based on specific speed considerations. Some of the monumental power plants including *Three Gorges Dam* in China, *Hoover Dam* in the United States, *Itaipu Dam* between Brazil and Paraguay, and *Aswan Dam* in Egypt were discussed in the chapter. Economic analysis of hydropower involved total installed costs, and O&M costs. In addition, capacity factors and LCOE have also been incorporated along with calculation of LCOE for hydropower plants.

REFERENCES

[1] U.S. Department of Energy, "History of Hydropower." www.energy.gov/eere/water/history-hydropower (accessed Dec. 2021).

[2] BP, "Statistical Review of World Energy," Jul. 2021. Accessed: Dec. 2, 2021 [Online]. Available: www.bp.com/en/global/corporate/energy-economics/statistical-review-of-world-energy.html.

[3] Wikipedia Contributors, "List of Largest Hydroelectric Power Stations," Dec. 27, 2019. https://en.wikipedia.org/wiki/List_of_largest_hydroelectric_power_stations (accessed Dec. 2, 2021).

[4]   U.S. Energy Information Administration, "Hydropower Explained," EIA, Mar. 30, 2020. https://www.eia.gov/energyexplained/hydropower/ (accessed Dec. 3, 2021).

[5]   International Hydropower Association, "2020 Hydropower Status Report," IHA, London, 2020. Accessed: Dec. 5, 2021 [Online]. Available: www.hydropower.org/publications/2020-hydropower-status-report.

[6]   M. Tanaka, G. Girard, R. Davis, A. Peuto, and N. Bignell, "Recommended table for the density of water between 0 °C and 40 °C based on recent experimental reports," *Metrologia*, vol. 38, no. 4, pp. 301–309, Aug. 2001. doi: 10.1088/0026-1394/38/4/3.

[7]   Itaipu Binacional, www.itaipu.gov.br/en (accessed Dec. 17, 2021).

[8]   U.S. Bureau of Reclamation, "Hoover Dam," 2019. www.usbr.gov/lc/hooverdam (accessed Dec. 17, 2021).

[9]   International Energy Agency, "Hydropower Special Market Report: Analysis and Forecast to 2030," IEA, Paris, Jun. 2021. Accessed: Dec. 18, 2021 [Online]. Available: www.iea.org/reports/hydropower-special-market-report.

[10] International Renewable Energy Agency, "Renewable Power Generation Costs in 2020," IRENA, Abu Dhabi, 2021. Accessed: Dec. 18, 2021 [Online]. Available: www.irena.org/publications/2021/Jun/Renewable-Power-Costs-in-2020.

CHAPTER 9 EXERCISES

9.1 Explain each water cycle process below in one sentence:

    a. Evaporation

    b. Transpiration

    c. Percolation

    d. Precipitation

9.2 Compare a buttress dam, an embankment dam, and a gravity dam. Provide a real-world example for each.

9.3 Calculate the total head loss in meters for a hydropower dam with the following specifications. Assume kinematic viscosity of water as $v = 0.89 \times 10^{-6} \text{ m}^2/\text{s}$.

- Flow rate = 0.55 m^3/s
- Penstock length = 35.3 m
- D = 750 mm
- ε = 0.060 mm
- K_{inlet} = 1.0
- K_{exit} = 0.5
- $K_{\text{elbow,1}}$ = 0.40
- $K_{\text{elbow,2}}$ = 0.45

9.4 Determine the power output (kW) for a hydropower plant with the following values. Turbine efficiency is 91%.

- z_1 = 315 m
- z_2 = 258 m
- Q = 0.3 m^3/s
- L = 180 m
- D = 400 mm
- ε = 0.045 mm
- K_{inlet} = 0.9
- K_{exit} = 1.2
- K_{gate} = 0.4
- K_{elbow} = 0.30 (two elbows)

9.5 The elevation difference between the reservoir and afterbay is 85 m for a hydropower dam. The total head losses are calculated to be 9 m for a flow rate of 1800 m^3/h. Plot the generated power for this hydroelectric station for a turbine efficiency range of 80–90% at 2% increments.

9.6 A hydroelectric dam has a power generation capacity of 20,000 kW with an available head of 38 m. Turbine angular velocity is 6.3 rad/s. What type of turbine would be a better choice for this dam?

9.7 Turbines are to be selected for a hydropower plant which is designed to generate 30,000 hp with an available head of 20 m. Rotational speed will be maintained at 60 rpm.

    a. Calculate the specific speed using US customary units.

    b. Recommend a turbine design for this application.

    c. If 92% of the available head does work across the turbine, find the flow rate in m^3/s.

9.8 A hydropower project has a target power output of 33 MW with an available head of 15 m. Rotational speed of the turbines is 90 rpm. With a specific speed of 40, how many turbines would be needed to provide the targeted power output?

9.9 Research the "Five Cs" of the UNESCO World Heritage Convention. Discuss the safeguarding campaign on the *Abu Simbel Temples* case, relating it to the convention.

9.10 Give three global examples of hydropower dam projects, other than the Aswan Dam project, where a historic town or site has been flooded by the reservoir. Provide the name of the dam, river, and the country of the project.

9.11 List four biological and four physical impacts of hydroelectric facilities on the surrounding environment.

9.12 Explain how hydropower and pumped hydro energy storage (PHES) are related. What makes a PHES system different from a conventional hydroelectric power plant?

CHAPTER 10
Ocean Energy

LEARNING OBJECTIVES

After reading this chapter, you should have acquired knowledge on:

- Global ocean energy capacities
- Small Island Developing States and the *blue economy*
- Thermal energy distribution in oceans and OTEC systems
- Osmotic power and salinity gradient energy systems
- Formation of tides and tidal energy applications
- World wave energy availability and applications
- Eco-friendliness of marine energy technologies
- Economics of ocean energy

10.1 Introduction

The oceans possess a vast amount of energy in different ways. All these forms are studied under the generic term of ocean energy, which can also be referred to as *marine energy* or *marine and hydrokinetic energy*. Energy from the ocean can be harnessed through kinetic, potential, thermal, and osmotic energies with each form of energy encompassing a variety of technologies. Energy in kinetic and potential forms can come from tides or waves. Tidal energy can be categorized within itself into tidal barrages and tidal streams. Barrages work on the idea of converting potential energy to electricity, while streams allow making use of kinetic energy. Wave energy technologies also can operate based on converting kinetic or potential energy into mechanical and then electrical energy. The thermal energy potential that the oceans have is another significant resource. Solar radiation on the oceans causes the upper segments of the water to heat up, while the deeper segments where sunlight penetration gradually reaches minimal values are much cooler. This naturally forming temperature gradient allows us to build thermal energy conversion plants to generate electricity or to desalinate water. This form of energy conversion can also be achieved in different configurations of power plants such as open-cycle, closed-cycle, and hybrid applications. Osmotic power is another resource that can be utilized to generate electricity from the salinity gradient energy, which can be harnessed from the difference in salt concentrations between two water bodies, such as rivers and oceans.

The criteria in selecting any ocean energy form and technologies associated with those forms of energy can be listed as efficiency, durability, accessibility, and environmental impact. The energy conversion efficiency of the technology being used is a significant parameter, especially in terms of determining the payback duration and annual capacity. Durability of ocean energy technologies is another important criterion, considering the large forces the water exerts on the systems and the corrosive nature of saltwater. Accessibility is another factor in evaluating an ocean energy project. The system or the project site should be easily accessible for regular maintenance, troubleshooting, and repairs. An accessible system comes with lower maintenance costs, increased safety for the technical crew, and reduced response time during any system failure. Finally, environmental impact is another guideline that is used in evaluating a marine energy project. Any potential adverse effects of an ocean energy application on the fauna, water quality, and sediment transport need to be considered during the preliminary surveying and site exploration phases, to minimize the detrimental impacts of the project on the environment. A global overview of ocean energy and different forms of ocean energy technologies are discussed in the following sections, followed by theoretical analysis, applications, and a case study.

10.1.1 Global Overview

According to the UN, about 40% of the global population lives within 100 kilometers of the coast, as cited in IRENA's *Innovation Outlook: Ocean Energy Technologies* [1]. This percentage is significant and manifests the importance of utilizing ocean energy in combating climate change. Different coastlines of the globe have different energy potential in one or more of the forms of marine energy briefly discussed in the previous section. **Figure 10.1** illustrates global ocean energy use capacities based on different technologies. The four main categories are ocean thermal energy conversion (OTEC), salinity gradient, tidal, and wave energy. Tidal barrages and other tidal technologies are differentiated based on how water possesses the energy. A tidal barrage is considered to be a *tidal range* application where the actual potential difference between the high and low tides provides the energy. Tidal stream technologies are those that utilize the energy from the *tidal currents*. Global total ocean energy capacity, and capacities for countries ranking top can be seen in **Figure 10.2**. South Korea and France make up most of the global sum with the *Sihwa Lake* and *La Rance* plants, the two largest tidal power stations in the world (see **Figure 10.13**).

A benefit of ocean energy technologies is that they can address the energy and drinking water shortages of islands that are at risk due to their geographic location,

FIGURE 10.1 Global ocean energy deployment capacities based on the technologies used. Source: IRENA, Innovation Outlook: Ocean Energy Technologies, 2020.

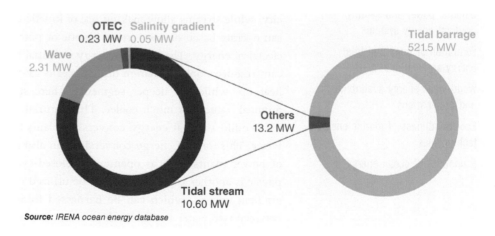

FIGURE 10.2 Ocean energy capacities for countries. The letters next to the numbers indicate how the data was obtained (o: official, u: unofficial, and e: estimated). Source: IRENA Renewable Capacity Statistics 2021.

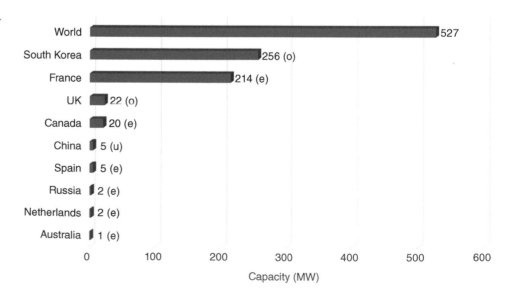

Table 10.1

List of Small Island Developing States (SIDS) [3]		
UN Members		
1. Antigua and Barbuda	14. Guyana	27. Singapore
2. Bahamas	15. Haiti	28. St. Kitts and Nevis
3. Bahrain	16. Jamaica	29. St. Lucia
4. Barbados	17. Kiribati	30. St. Vincent and the Grenadines
5. Belize	18. Maldives	31. Seychelles
6. Cabo Verde	19. Marshall Islands	32. Solomon Islands
7. Comoros	20. Fed. States of Micronesia	33. Suriname
8. Cuba	21. Mauritius	34. Timor-Leste
9. Dominica	22. Nauru	35. Tonga
10. Dominican Republic	23. Palau	36. Trinidad and Tobago
11. Fiji	24. Papua New Guinea	37. Tuvalu
12. Grenada	25. Samoa	38. Vanuatu
13. Guinea-Bissau	26. São Tomé and Príncipe	
Non-UN Members/Associate Members of the Regional Commissions		
1. American Samoa	8. Cook Islands	15. New Caledonia
2. Anguilla	9. Curacao	16. Niue
3. Aruba	10. French Polynesia	17. Puerto Rico
4. Bermuda	11. Guadeloupe	18. Sint Maarten
5. British Virgin Islands	12. Guam	19. Turks and Caicos Islands
6. Cayman Islands	13. Martinique	20. US Virgin Islands
7. C.W. of Northern Marianas	14. Montserrat	

FURTHER LEARNING

NREL Marine Energy Atlas
https://maps.nrel.gov/marine-energy-atlas

FURTHER LEARNING

Ocean Energy Europe
www.oceanenergy-europe.eu/

distance to the continental shelf, and infrastructure. Some islands can be closer to land, situated on the continental shelf, as opposed to some that rise from the ocean bottom due to volcanic or seismic activity. These formations are considered as *oceanic islands*. At the *United Nations Conference on Environment and Development* (UNCED) in Brazil in 1992, a group of island countries was defined and recognized as *Small Island Developing States* (SIDS) [2]. This group includes 38 UN member states and 20 non-UN members/associate members of UN regional commissions that have social, economic, and environmental challenges. These islands are listed in **Table 10.1**. Ocean energy is an important alternative for these countries in terms of providing reliable electricity and desalinated water. It plays an important role in contributing to the *blue economy* of these countries.

FIGURE 10.3 Sectors and opportunities of blue economy that can be powered by marine energy technologies. Source: NREL.

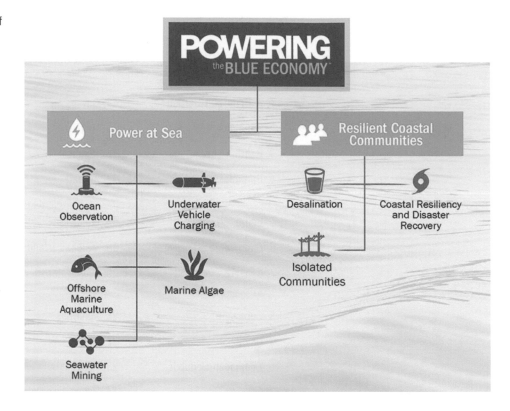

Blue economy is a concept that aims at supporting sustainable economic growth of regions or states through fields and activities related to oceans, including renewable energy, clean water, fisheries, marine transportation, aquaculture, and tourism (**Figure 10.3**).

The marine energy potential estimated for the United States can be seen in **Figure 10.4**. Alaska has the highest potential as an individual state, followed by Hawaii. The capacity of the inland states comes from rivers. In the overall sum, wave energy makes up about 61% of the total potential with 1400 TWh/year annual estimated capacity. This amount corresponds to 34% of the total annual electricity generation in the United States. The overall sum of 2300 TWh/year is equivalent to 57% of the total US electricity production. These estimates are done based on the capacity factor assumptions listed in **Table 10.2**. Capacity factor for OTEC is assumed to be 100%, as this technology is anticipated to be highly consistent [4].

FIGURE 10.4 Estimated energy potential of US marine energy resources (TWh/year). Source: NREL (Image adapted from L. Kilcher, M. Fogarty, and M. Lawson. 2021. Marine Energy in the United States: An Overview of Opportunities. Golden, CO: National Renewable Energy Laboratory. NREL/TP-5700-78773).

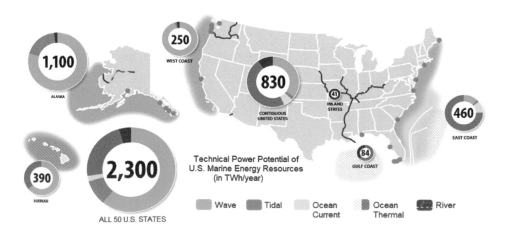

Table 10.2

Capacity factors for each marine energy resource [4]	
Resource type	**Capacity factor**
Wave	30%
Tidal	30%
Ocean-current	70%
OTEC	100%
River	30%

The U.S. Department of Energy's *Marine and Hydrokinetic Technology Database*, which is developed by National Renewable Energy Laboratory (NREL), offers detailed statistics on marine energy projects in the United States and around the world. This *OpenEI* platform is a wiki which allows its own audience to build and edit the content, providing up-to-date and global information.

10.1.2 Ocean Thermal Energy Conversion (OTEC)

Solar radiation on the ocean surface results in a temperature gradient from the free surface to the deeper layers of the ocean. This temperature profile allows utilizing a heat engine to generate power. The temperature profile of the ocean water with depth is illustrated in **Figure 10.5**. The upper layer of the ocean is called the *epipelagic zone* or the mixed layer, which reaches down to approximately 200 m from the free surface of the ocean. This layer is followed by the *thermocline*, which is a transitional layer between the surface water and colder deep ocean water. The thermocline exists to a depth of approximately 1000 m. Scuba divers can observe the thermocline due to both the temperature gradient and the change in visibility. Water becomes turbid, yielding a wrinkled-glass appearance.

Ocean thermal energy conversion (OTEC) can be used in different configurations of power plants such as open-cycle, closed-cycle, and hybrid applications to generate electricity and to desalinate water.

Open-cycle OTEC power plants generate electricity from steam that is produced from surface ocean water which is pumped into an evaporator by a pump. Evaporation is achieved by a controlled flash evaporation process during which

FIGURE 10.5 Temperature gradient of ocean with depth. Source: Adapted from NOAA.

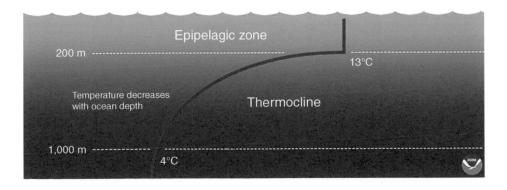

the saturated liquid from the ocean boils under reduced pressure. A vacuum pump maintains the required low pressure. The steam flows through the turbine doing work, after which it is directed into the condenser. The portion of the ocean water that did not evaporate inside the flash evaporator is sent back to the ocean. This is a much larger portion as the steam has low quality at the end of the flashing process. Condensation can take place either in a *direct-contact condenser* or in a *surface condenser*. In direct-contact condensers, the steam is brought in direct contact with the deep ocean water. This type of condenser has a simpler design, and the initial cost is lower. Also, there are no fouling or corrosion concerns with these systems as there is no need for tubes to separate the fluids. Another advantage of this types is that they provide enhanced turbine efficiency. One of the disadvantages of these condensers, however, is that drinking water cannot be produced with them as the desalinated water mixes with the ocean water. Surface condensers are *shell and tube heat exchangers* in which the steam and the coolant (ocean water) do not mix. They exchange heat through the external walls of the tubes within the outer shell. Due to the added thermal resistance introduced, these systems result in reduced overall efficiency. Fouling formation and the corrosive effects of saltwater are other concerns. The advantage of these systems is that the condensate can be used as freshwater as it is not mixing with the ocean water. Similar to single- and double-flash steam geothermal power plants, open-cycle OTEC cycles can also be two-stage, where the warm water exiting the first stage flash evaporator is directed into a secondary evaporator for utilizing more of the usable energy from the surface water. A schematic and *T–S* diagram for a single-stage open-cycle OTEC system utilizing a direct-contact condenser can be seen in **Figure 10.6**.

Closed-cycle OTEC power plants operate based on the principle of the Rankine cycle. A working fluid with a low boiling point is selected. This working fluid can be an organic compound such as propane (C_3H_8), or an inorganic compound such as ammonia (NH_3). Surface and deep ocean water are used as the heat source and heat sink, without mixing with the working fluid. Warm water from the ocean surface flows through an evaporator and boils the working fluid of the closed cycle. The fluid goes through the turbine at high pressure, generating electricity. It is then sent into the condenser, which it leaves as liquid to be pumped back into the evaporator, hence completing the cycle. The condenser in these systems is a surface type as the ocean water and the working fluid cannot mix with each other. A schematic and *T–S* diagram for a closed-cycle OTEC system are illustrated in **Figure 10.7**.

Hybrid OTEC power plants are a mix of open- and closed-cycle configurations. They offer the drinking water producing capability of open-cycle systems along with the higher power generation capacity of closed-cycle systems. Warm surface ocean water

FIGURE 10.6 Schematic and *T–S* diagram for an open-cycle OTEC power plant.

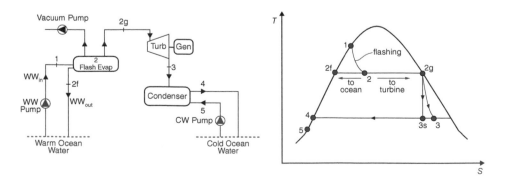

FIGURE 10.7 Schematic and *T–S* diagram for a closed-cycle OTEC power plant.

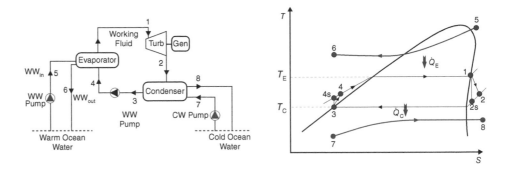

FIGURE 10.8 Schematic of a hybrid OTEC power plant.

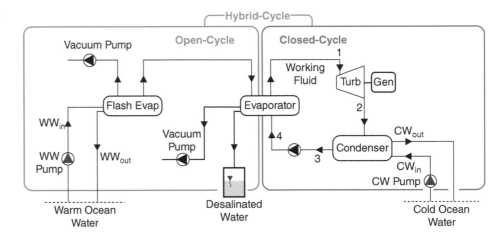

first flows into a flash evaporator where a portion of it is converted into steam. This steam is then directed into a secondary evaporator in which it exchanges heat with the working fluid of the closed cycle without mixing with it. It then condenses into desalinated water. The boiling working fluid completes its own closed cycle, hence generating electricity. A schematic of a hybrid OTEC configuration is given in **Figure 10.8**.

10.1.3 Ocean Salinity Gradient Energy

Salinity gradient energy forms when two fluids with different salt concentrations meet. The most common example of this in nature is a freshwater river flowing into the ocean. Salinity gradient can also be observed within the same water body. This forms a layer similar to the thermocline, with salt concentration being the determinant instead of temperature. This layer is called a *halocline*, which is a separation zone in a body of water where a drastic vertical salinity gradient exists. As the salinity changes the density of water, the separation occurs through this transitional layer. Just as the thermocline, this layer can also be visible to scuba divers. Depending on the position of it with respect to the observer, it may look like a hazy cloud layer above, or a river flowing beneath. Both are remarkable visual experiences for observers (**Figure 10.9**).

There are two different techniques for generating electricity from ocean salinity gradient. These are the pressure-retarded osmosis and reverse electrodialysis methods. *Pressure-retarded osmosis (PRO)* is an osmotically driven process in which freshwater penetrates through a semipermeable membrane and migrates into the saltwater side. The semipermeable membrane allows water molecules to pass while retarding the Na^+

FIGURE 10.9 A scuba diver in a cave, between the bottom of the cave and a *halocline* above. Source: Rodrigo Friscione/ Image Source via Getty Images.

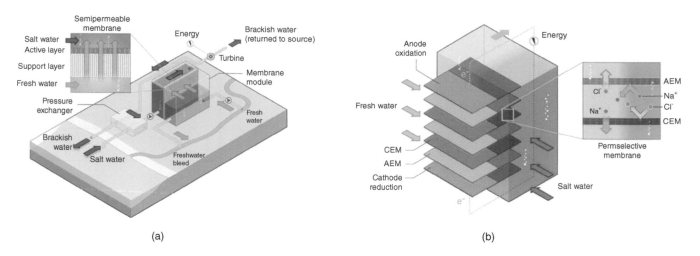

FIGURE 10.10 Salinity gradient energy technologies. (a) Pressure-retarded osmosis (PRO), and (b) reverse electrodialysis (RED). Source: Pacific Northwest National Laboratory (PNNL).

and Cl^- ions. Within a defined volume, this process causes an increase in the column height of the concentrated side, which is potential energy generated by natural means. This energy can then be converted into electrical energy through a turbine-generator system. The capacity of power generation depends on the difference in salt concentrations of two mixing bodies of water as the salinity gradient is the driving force. Membrane type and back pressure are other parameters that affect the energy production. Membrane material can be a limiting factor as it needs to withstand the pressure difference between both the solvent (seawater) and the diluent (freshwater). The first osmotic power plant employing the PRO technique was built in Tofte, Norway, where freshwater flows into the Oslofjord. The plant was a pilot project with a capacity of 10 kW.

Reverse electrodialysis (RED) is a more direct technique of generating electricity from the salinity gradient. Instead of first obtaining mechanical energy as in the PRO method, electricity is generated directly from the chemical potential difference. In RED, concentration difference is the driving force as opposed to electrodialysis (ED) which is derived from electric potential difference. In ED, electricity is used to transport salt ions from one solution to another solution through an ion-exchange membrane. A well-known application of ED is desalination. In RED, on the other hand, the purpose is to generate electricity utilizing the concentration difference. A number of cation-exchange membranes (CEMs) and anion-exchange membranes (AEMs) are stacked together to form high and low salinity volumes between two bodies of water, one being freshwater, and the other one saltwater. As these two bodies with a salt concentration difference are introduced between the ion exchange membranes, positively charged ions (Na^+) and negatively charged ions (Cl^-) are separated. The transport of these ions through the respective membranes yields an ionic current which is captured by the electrodes on both ends of the stack, converting this chemical energy into electrical energy. The operating principles of both PRO and RED techniques are illustrated in **Figure 10.10**.

10.1.4 Tidal Energy

Harvesting energy from the tides dates back to the seventh century. In Northern Europe, energy from the rising water was used in tide mills. Unlike regular water mills, which utilize energy from the flowing fluid directly, tidal mills used the tidal rise of seawater to fill up a basin first and then let it flow back when the tides were low. These basins could be considered as "water batteries" that were charged during the high tides. Back then the energy was used for simpler purposes such as grinding grains. Today, tidal energy can be used for many purposes ranging from small-scale (e.g., farming or agricultural) to large-scale (e.g., power generation) applications.

The main advantages of tidal energy are that it is a clean renewable energy technology that has no GHG emissions, is reliable, and predictable. Now let's discuss these attributes via the physical phenomenon of tides and some terminology associated with tidal energy. Tides are formed by the gravitational fields of the Moon and Sun acting on the Earth which encompasses oceans. The Moon's gravitational force exerted on the oceans causes two bulges on the oceans, with one bulge on the side of the Earth facing the Moon, and the other bulge on the opposite side. As the Earth is rotating, these bulges keep being observed at different parts of the ocean, resulting in rising and falling heights at different times during the day. If the Earth was a perfect sphere and there were no continents to interfere with currents, there would be two high and two low tides each lunar day, with the tides having equal amplitudes. These are called *semidiurnal* (semi-daily) tides which complete a cycle in half a lunar day, hence having two cycles in one day. There are however parts of the world where one high tide and one low tide are observed during a lunar day. These are called *diurnal* (daily) tides. There is also another type called mixed semidiurnal tides, which also experience two high and two low tides during the day, but they have different tidal heights between the two high tides, as well as between the two low tides.

Besides the cycles of the tides during a day, the range or amplitude also plays an important role in the total capacity of energy that can be utilized. *Spring tide* is when the maximum tidal range is observed. The term does not relate to the Spring season. It takes its name from the springing of the tide forward, as in leaping. Spring tides occur during new moon and full moon. *Neap tide* is observed when the Sun and Moon make a right angle with respect to the Earth. Neap tides also occur twice a month, but during the first and third quarter phases of the Moon. A lunar month is equivalent to 29.5 days, yielding a lunar day of 24 hours and 50 minutes, which is about 24.8 hours. As the Earth is rotating, the tidal bulge around it moves with a period of half of the lunar day, which is 12 hours and 25 minutes (~12.4 hours). Since a solar day is 24 hours and we use this as our reference time, each day the times of high and low tide are shifted by 50 minutes. Tidal ranges also differ based on geographic location. National Oceanic and Atmospheric Administration (NOAA) provides an online tool for determining the ranges for a selected location and time. In **Figure 10.11**, tidal heights for St. Petersburg in Florida are projected for April 2026. Corresponding lunar phases are also superimposed on the plot for a better interpretation of the ranges with the lunar time.

In addition to the time shift, the magnitude of the tidal height also changes each day from one new moon to another new moon during the lunar month. The currents caused by the tides also differ based on the rising and falling phases. *Flood current* is the rising tide moving towards the shore. *Ebb current* is the outgoing tide which moves away from the shore. *Slack tide* is the weakest form of the tidal current between flood and ebb. That transitional duration takes about 20 to 30 minutes. In deep waters, tidal currents are not as strong. They become faster in shallow or narrow regions such as estuary entrances or straits. Water accelerates in these regions as it is incompressible, and the mass flow rate remains constant. Hence, it compensates for the reduction in cross-sectional area of the flow by increasing its velocity. This is why when observing a

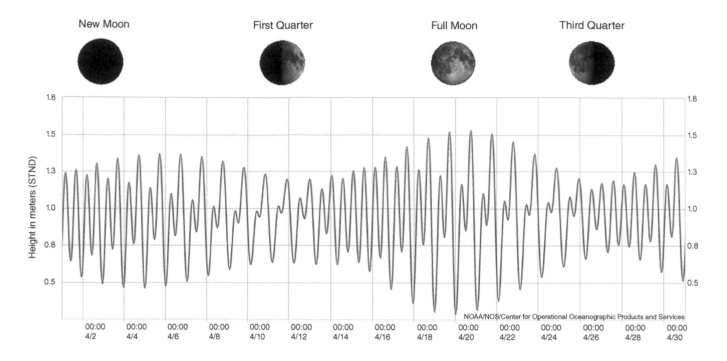

NOAA/NOS/CO-OPS
Tide Predictions at 8726724, Clearwater Beach FL
From 2030/04/01 00:00 GMT to 2030/04/30 23:59 GMT

FIGURE 10.11 Tidal heights for St. Petersburg, Florida, projected for April 2026 with lunar phases superimposed over the plot. Source: National Oceanic and Atmospheric Administration (NOAA).

tidal bore from the shore, the rate of the water getting closer to the observer increases. This can be misleading to the observer, who may think they can get away from the shore in time to secure themselves.

In fluid mechanics, there is another formation called a *vortex*, which is the flow of a fluid in a revolving manner about an axis that is normal to the direction of rotation. Different examples of vortex formations can be observed in marine or large water bodies. A *whirlpool* is an example of such a fluid body that is rotating. Whirlpools can be formed by various physical disturbances such as opposing currents, flow around an object (or obstacle), tides, and at the downstream of dams or weirs. Larger scale whirlpools that are observed in the oceans or seas are named *maelstroms*. Whirlpools can be seen at tidal barrages when the gates are opened to generate electricity.

The majority of the total ocean energy production comes from tidal barrages. Tidal lagoons are a variation of barrages. Tidal turbines are another group of key players in tidal energy deployment. These technologies are further discussed in this section. There are several other tidal energy technologies that are available. Some of these applications are mature while there is ongoing research in improving others and developing more alternatives. Tidal kites, oscillating hydrofoils, Archimedes screw designs, and Venturi-shaped ducts are some examples of these technologies.

FURTHER LEARNING

NOAA Tide and Current Predictions
https://tidesandcurrents.noaa.gov/

Tidal Barrages

In civil engineering or construction, the term *barrage* refers to a structure that is built across a river to hold water. This is similar to the idea of a dam. In fact, the word *dam* translates into a word that has a sound similar to the word barrage in different languages, such as *le barrage* in French and *baraj* in Romanian and in Turkish.

A tidal barrage is a structure that is built to harness energy from rising tides. A schematic of a tidal barrage during the inflow and outflow of tidal water can be seen in **Figure 10.12**. Tidal power with barrages can be generated by three different methods. One of these methods is *flood generation*. This application utilizes sluice gates which control the tidal flow to drive a turbine which is placed in a channel that connects the ocean water to the tidal basin. The gates are opened as the tide reaches its maximum on the ocean side and water flows into the estuary, while generating electricity. The other method is *ebb generation*, which is in the opposite direction to flood generation. The gates close when the tide is high, hence storing energy. Once it is low tide in the ocean, the gates are opened, hence the potential energy stored is converted into mechanical, then to electrical energy. This is similar to a pumped hydro energy storage (PHES) system, which can be considered as a "water battery." In ebb generation, this battery is being charged as the tides are rising and discharged when the sluice gates are open to release the higher elevation water in the basin into the ocean. The third method is *two-way generation*. These systems require a more precise control of sluice gates to allow water to build up sufficient potential energy. Electricity is produced during both flood and ebb tides. As the tide rises, water flows into the basin, and it flows back as the tides come down. It is a bidirectional flow where the turbine operates in both directions. The power capacity of these systems is smaller than flood or ebb generation systems as a larger potential is not accumulated before harnessing it. Another drawback of these systems is that the bidirectional turbine generators cost more than unidirectional units. However, the advantage of two-way systems is that they offer energy generation over a longer period of time during the day.

Another tidal energy technology is *tidal lagoons*. Tidal lagoons operate in a similar manner as tidal barrages. The main difference is that instead of an already existing basin that is utilized in barrage applications, seawalls need to be constructed in tidal lagoon applications to provide an artificial reservoir. The seawall can be in the form of an arc that is attached to the land, or in a ring-shape if operated offshore. During the flood tide, the turbine wicket gates are kept closed to hold the water outside the lagoon so that a potential energy difference builds up. Then, the wicket gates open and water flows into the lagoon through the turbine, generating electricity. Eventually, the water level inside the lagoon reaches an equilibrium with the outside ocean water. This is when the head is approximately zero and there is no power generation capacity until the next elevation change. Similar to a flood tide, during the ebb tide, the wicket gates of the turbine are closed until there is a potential difference between the lagoon and the

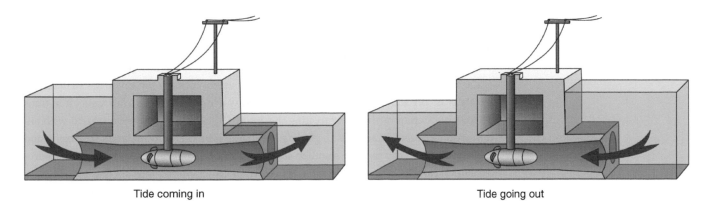

Tide coming in   Tide going out

FIGURE 10.12 Operation of a tidal barrage with the tide coming into the basin and going out. Source: NASA.

FIGURE 10.13 La Rance Tidal Barrage in northwestern France where the Rance River meets the Atlantic Ocean at the exit of the English Channel. The tidal power station has an installed capacity of 240 MW. Whirlpool formation caused by tidal water flowing through the turbines can be observed at the site. Source: Vincent Jary/Moment Open via Getty Images.

ocean water free surfaces. The gates are then opened and the process is repeated in the opposite direction, with electricity being generated during the flow of water through the turbine-generator again. Lagoon wall construction is also as critical as tidal barrage construction. It needs to be durable to withstand the drag acting on it. In addition, as a lagoon typically encompasses a much smaller volume of water than tidal barrages, it is important that its walls do not leak water. Leakage of water from the higher elevation side to the lower level side would reduce the amount of available energy. Therefore, the wall should have a very low permeability value.

Tidal Turbines

Tidal barrages use the potential energy of ocean water gained during the tidal rise. Tidal lagoons work in a similar fashion as discussed earlier. Besides the potential energy that can be utilized, tidal streams also possess a significant amount of kinetic energy which can be harnessed. A *tidal turbine* operates based on the conversion of this flow energy into electrical energy, similar to wind turbines. Although tidal currents are much slower than wind, tidal turbines are still capable of generating competitive amounts of energy due to the density of ocean water being much higher than that of air. Tidal turbines do not have the damage risks that wind turbines have due to high flow speeds; however, the drag exerted on the turbine and the blades under water is still much higher due to the density difference between saltwater and air. The main components include the rotor blades, gearbox, generator, and the tower (or *pile*). The turbines can be attached to the ocean floor with different techniques, some of which are sunken pile, gravity, and tripod. The sunken pile method is performed by immersing the pile into the ocean floor like a nail into wood or a stick into soil. The gravity method utilizes a large heavy block (e.g., concrete) into which the tower is stationed. This method provides a stronger hold than the sunken pile due to the heavy block being more rigid than the ocean floor. The tripod technique distributes the forces on each circumferential leg of the design, hence keeping the tower and the turbine in place. The tubular frame that sits at the ocean bed covers a large surface area, providing a higher truss. Although tidal turbines are similar to wind turbines from a configuration point of view, tidal turbines require a more detailed structural analysis compared to wind turbines as the water drag is much higher than the drag caused by air. Therefore, in tidal turbine projects, a multidisciplinary approach involving civil, construction, or structural engineers is important in alleviating the drifting, tilting, or breaking risks. There are two different types of tidal turbines by configuration. These are horizontal-axis and vertical-axis tidal turbines.

Horizontal-axis tidal turbines (HATTs) have their axis of rotation parallel to the tidal stream. This is similar to horizontal-axis wind turbines interacting with air. HATTs can have blade diameters in the range of 5–30 m and they are designed to rotate 180° or reverse their rotation so that they can generate electricity during both flood and ebb currents. Some of the designs may not need blade adjustment. A typical concern for any metal structure in saltwater is corrosion. In addition, the moving seawater can wash the lubricants off in time. There are different methods for treating this issue. High-viscosity lubricants used with oil-tight rubber seals can address the leakage problem, minimizing or delaying corrosion and rusting of the components. Another solution can be using seawater-lubricated hydro-dynamic bearings. This approach has its own tribological limitations such as the seawater being much less viscous than conventional liquid (i.e., synthetic oil) or semi-solid (i.e., grease) lubricants. Tribology is the field of science that studies friction, wear, and lubrication. A HATT configuration is illustrated in **Figure 10.14**.

FIGURE 10.14 Horizontal-axis tidal turbine (HATT) configuration. Source: https://aquaret.com/.

FIGURE 10.15 Vertical-axis tidal turbine (VATT) configuration. Source: https://aquaret.com/.

Vertical-axis tidal turbines (VATTs), on the other hand, have their axis of rotation perpendicular to the tidal current. These are similar to vertical-axis wind turbines. They can also be classified as *cross-flow turbines*. Different types of VATTs exist. Some of these are *Savonius, Darrieus, squirrel cage Darrieus, H-Darrieus*, and *Gorlov helical* designs. A *tidal fence* is a type of tidal turbine design that utilizes vertical-axis turbines which are lined up in a row on the ocean floor. It takes the name from its resemblance to a fence structure. They can be of *Savonius* or *Darrieus* designs. A VATT configuration is depicted in **Figure 10.15**. Other tidal energy technologies that harness energy from tidal streams can be listed as *reciprocating* (or *oscillating*) *hydrofoils, venturi-shaped ducts, tidal kites*, and *Archimedes screws*.

Eco-friendliness of these marine energy technologies, whether they are tidal turbines or any other systems, is of high importance. The influence of such systems on the marine habitat and ocean bed should be evaluated with care. Blade strike and entanglement are some of the possible issues. The noise these systems generate attenuates at a much higher speed than sound travels in air. The sound waves may interfere with the echoes marine animals use to communicate with each other. This method of communication is called *echolocation* or *bio sonar*. Any possible mechanical or electrical failure may also cause a risk for the habitat near these ocean energy technologies. Therefore, a thorough analysis that involves a multidisciplinary group of professionals is crucial for these applications, just as for any energy system, onshore or offshore.

10.1.5 Wave Energy

Ocean energy can be harnessed from thermal energy (OTEC), salt concentration difference of fresh and saltwater (salinity gradient), and kinetic and potential energy changes that the ocean water can experience (tidal and wave). Wave energy, in principle, utilizes kinetic and/or potential energy of water like tidal energy. However, the physics and concepts are different from tidal energy. Waves are mainly formed due to wind flow over the oceans. Therefore, it is reasonable to state that wave energy is also a derivative of energy of the Sun. Waves can sometimes go in the same direction as tidal currents, resulting in enhanced energy potential. They may equally move in the opposite direction to tidal currents. This can result in an interesting surface view of water. *Seymour Narrows*, which flows between Vancouver Island and Quadra Island in Canada, is known for its strong tidal currents and exhibits compelling surface patterns when the waves and tides are moving in opposing directions. Waves form on all water bodies exposed to outside air, including oceans, inner seas, lakes, and rivers. Oceans have much higher wave energy potential than other water bodies, and the magnitude of the available power varies across the world. Global wave energy potential is shown in **Figure 10.16**. The color contours represent the available wave power per unit crest length. As can be seen from the figure, the Bering Sea, the Gulf of Alaska, northern part of the Atlantic arching towards Greenland, Iceland, the United Kingdom, and Ireland, southern coasts of Chile, South Africa, and most of southern Australia have higher wave power capacities than other parts of the world. For a wave energy application, the desired scenario is to have a minimum wave power of 50 kW per unit crest length for the project to be feasible. Therefore, the analysis of wave energy maps is important before initiating a project. The NREL has an online tool called the *Marine Energy Atlas*, which displays US wave potential (**Figure 10.17**).

Energy analysis of waves requires an understanding of wave characteristics. Depending on the movement of the oscillation profile with reference to the ocean water, a wave can be *progressive* or *standing*. A *progressive wave* is a wave that travels in a horizontal direction continuously without experiencing a change in its amplitude. Progressive waves consist of *crests* and *troughs*. A crest is the highest point to which the wave rises, while a trough is the lowest point the wave reaches while traveling. *Amplitude* is the difference in height between the highest value and the reference (or equilibrium) value. Wavelength is the length of one full wave cycle, which can be measured from one crest to the next one, or from one trough to the following trough. A *standing wave*, on the other hand, does not travel in the horizontal direction. The movement in standing waves is vertical, where the maxima and minima do not move in the *x*-direction. Standing waves consist of *nodes* and *antinodes*. A node is the point

FIGURE 10.16 World wave energy potential map. Source: Ingvald Straume/ Wikimedia Commons/CC-0.

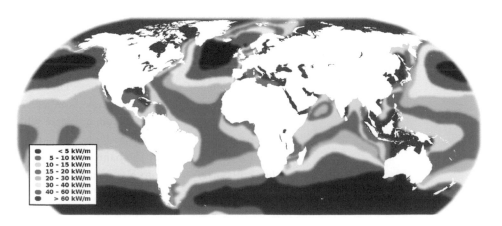

FIGURE 10.17 US coastlines wave and ocean thermal energy potential. Source: NREL Marine Energy Atlas.

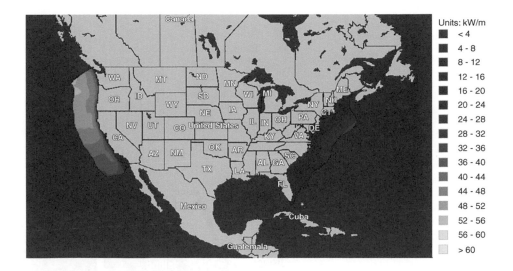

Units: kW/m
- < 4
- 4 - 8
- 8 - 12
- 12 - 16
- 16 - 20
- 20 - 24
- 24 - 28
- 28 - 32
- 32 - 36
- 36 - 40
- 40 - 44
- 44 - 48
- 48 - 52
- 52 - 56
- 56 - 60
- > 60

FIGURE 10.18 Progressive and standing waves.

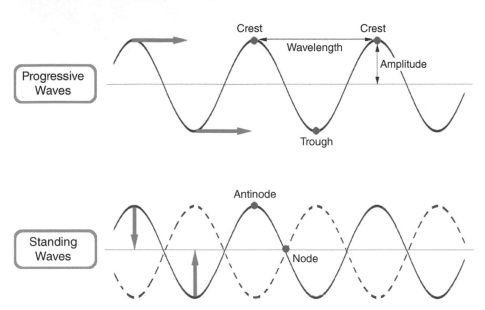

FURTHER LEARNING

Aqua-RET e-Learning Tool
https://aquaret.com/

FIGURE 10.19 Surfer on a plunging breaker where the circular orbit of the wave becomes elliptical, followed by the crest curling over before it collapses. Source: RF/ Justin Lewis via Getty Images.

where the amplitude of the wave is minimum. An antinode of a wave is the point on the standing wave where the amplitude reaches its maximum. Progressive and standing waves and physical terms associated with them can be seen in **Figure 10.18**.

Energy generation with waves can be performed using different technologies. Oscillating water columns, point absorbers, and surface attenuators are some of the well-known and comparatively more mature marine technologies in generating power from waves. Some other alternatives include overtopping devices, oscillating water surge energy converters, rotating masses, and submerged pressure differential systems.

Oscillating Water Columns

An *oscillating water column* (OWC) operates based on the idea of compressing air and forcing it through a turbine in a sealed chamber. This hollow structure can be fixed onshore, or it can be a floating design offshore. As the waves come in and go out, the air is alternately compressed and decompressed, being forced to flow through a turbine bidirectionally at each wave period. A *Wells turbine* or pressure generating valves can be used to harness the energy from the bidirectional flow of air during the rising (air compression) and falling (air suction) phases of the water column. Hence the direction

FIGURE 10.20 Schematic of an oscillating water column. Source: NREL.

of the airflow does not matter as the turbine blades rotate in the same direction. A schematic of an OWC application is illustrated in **Figure 10.20**. As opposed to some other ocean energy technologies, OWC systems do not have any moving parts within the water which reduces the risk to the habitat. A drawback of these systems can be the flow-induced noise, which may be a concern if the site is close to a residential area. Building these systems away from such areas onshore or anchoring them at further points offshore can eliminate the noise problem. Since the moving parts are not in water, maintenance and repair for these systems are easier.

Point Absorbers

Another wave energy application is a *point absorber* which in principle is an oscillating body technology. Energy is available due to the relative motion between a moving body and a fixed or anchored body. The moving body can be floating on the ocean surface, or it could be partially or fully submerged in water. The other body can be fixed on the ocean bed, or it can also be freely moving as an anchored system where its movement would be much less than the moving body so that the relative motion is sufficient to make the project viable. Electricity can be generated directly using generator coils and magnets or indirectly by pumping a fluid to rotate a turbine. In direct generation, linear or rotary generators produce electricity by electromagnetic induction, which results from the relative motion between coils and magnets. The indirect method involves conversion of wave energy first into mechanical energy by compressing a gas or pumping a fluid, and then generating electricity by passing that fluid through a turbine generator. These systems have lower efficiencies since more moving parts are involved, with each part contributing to conversion losses. A point absorber with the moving body floating and the other side fixed on the ocean bed can be seen in **Figure 10.21**. While point absorbers typically have lower sound pressure levels than OWCs, these systems can be more harmful for the marine habitat due to the existence of moving parts in the water. Another environmental impact of point absorbers is the electromagnetic field (EMF)

FIGURE 10.21 Schematic of a point absorber. Source: NREL.

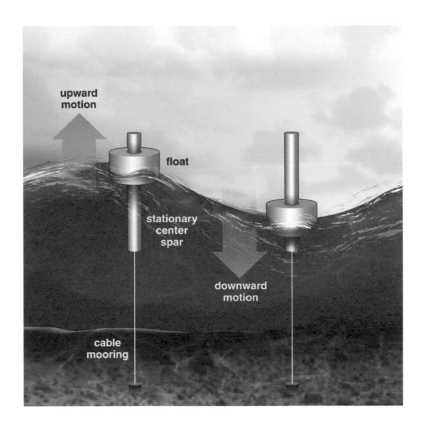

generation under water. A natural EMF provides certain conveniences for some ocean habitat species such as communication, navigation, hunting as a predator, and protecting themselves as prey. An artificially generated EMF can interfere with the naturally existing field, hence confusing or misleading the habitat species.

Surface Attenuators

A *surface attenuator* absorbs energy from the waves over a line or a surface as opposed to a point absorber harvesting energy at a point. Therefore, a surface attenuator can also be described as a linear attenuator. These systems involve multiple pieces that are linked to each other. Energy from the waves can be harnessed by obtaining a rotational motion or pumping a hydraulic system during the flexing motion at the hinges. The attenuators can have multiple or single degrees of freedom. A design with multiple degrees of freedom can absorb the energy from both the up-and-down motion in the vertical direction, and side-to-side motion in the horizontal direction. On the 3D Cartesian coordinate system, both of these movements are perpendicular to the direction of the wave, hence they are transverse to the wave and to each other. A surface attenuator can also have only one degree of freedom. A nodding motion is observed in such systems where energy is absorbed from the up-and-down motion of the links that are attached to each other. Schematics of surface attenuators with both multiple and single degrees of freedom are shown in **Figure 10.22**. Some of the potential environmental impacts of surface attenuators include entanglement and collision concerns. Marine animals can also become trapped at the hinges. A well-known example of this technology is the *Pelamis wave energy converter* (**Figure 10.23**). Pelamis was the first offshore wave energy converter in the world to provide energy to the grid. It had a capacity of 750 W. It was utilized in Orkney, Scotland, as well as in Agucadoura, Portugal. It took its name from the yellow-bellied sea snake as it resembled a snake.

FIGURE 10.22 Schematic of attenuator configuration with **(a)** multiple degree of freedom, and **(b)** single degree of freedom. Source: NREL.

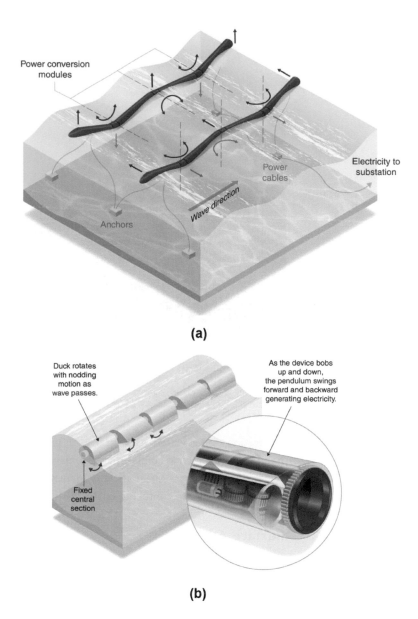

(a)

(b)

10.2 Theory

Theoretical analysis of ocean energy involves four topics that we discussed in the previous section. These topics are:

- Ocean thermal energy conversion (OTEC)
- Ocean salinity gradient energy
- Tidal energy
- Wave energy

FIGURE 10.23 Pelamis wave power device off the coast of Agucadoura, Portugal. It was the world's first commercial wave energy system. Source: Scottish Government, Wikimedia Commons author, https://commons.wikimedia.org/wiki/File:Pelamis_P2_wave_energy_device_%287020981211%29.jpg

10.2.1 Ocean Thermal Energy Conversion (OTEC)

In OTEC applications, knowing the performance limit is as important as it is in any other power generation technology. Maximum possible efficiency for an OTEC power plant is determined by the Carnot efficiency:

$$\eta_{\text{Carnot}} = 1 - \frac{T_{\text{C}}}{T_{\text{H}}} \tag{10.1}$$

where T_{C} is the cold reservoir (deep water) temperature, and T_{H} is the hot reservoir (surface water) temperature. Hence the desired case is that the temperature difference between the cold and hot reservoirs is as high as possible. The temperature difference can be higher if the cold water comes from a greater depth; however, pumping cold water from such depths to the surface also has an energy cost. Therefore, the economic feasibility study of an OTEC power station should involve all factors including the depths water needs to be transported from and any irreversibilities the components of the system such as the turbine and the pumps have.

Example 10.1 Determining Maximum Possible Efficiency for an OTEC Power Plant

Calculate the maximum overall efficiency that can be achieved for an OTEC power plant if the surface water temperature is 26 °C and the deep water temperature is 4 °C.

Solution

Maximum efficiency is defined by the Carnot expression for power cycles. Using **Equation 10.1**:

$$\eta_{\text{Carnot}} - 1 - \frac{T_{\text{C}}}{T_{\text{H}}} = 1 - \frac{(4 + 273.15)\,\text{K}}{(26 + 273.15)\,\text{K}} = 0.0735$$

Maximum efficiency is 7.35% for the given cold and hot reservoir temperatures. An actual power plant executing the cycle between these two reservoirs would have lower efficiency due to the irreversibilities within each component employed.

Thermal efficiency and pertinent relations for an open-cycle OTEC system according to the schematic and T–S diagram in **Figure 10.6** can be listed as follows:

$$\eta_{\text{th}} = \frac{\dot{W}_{\text{net}}}{\dot{Q}_{\text{in}}} = \frac{\dot{W}_{\text{turb}} - \sum \dot{W}_{\text{pumps}}}{\dot{Q}_{\text{evap}}} \tag{10.2}$$

$$\dot{W}_{\text{turb}} = \dot{m}_{\text{steam}} \left(h_{2\text{g}} - h_3 \right) \tag{10.3}$$

$$\dot{W}_{\text{WW pump}} = \dot{m}_{\text{WW}} (h_1 - h_{\text{WW}}) \tag{10.4}$$

$$\dot{W}_{\text{CW pump}} = \dot{m}_{\text{CW}} (h_5 - h_{\text{CW}}) \tag{10.5}$$

$$\dot{Q}_{\text{evap}} = \dot{m}_{\text{WW}} (h_{\text{WW,in}} - h_{\text{WW,out}}) \tag{10.6}$$

For closed-cycle OTEC systems, the thermal efficiency equation is the same as that of open-cycle plants, as given in **Equation 10.2**. It should be remembered that as a binary-cycle design, closed-cycle systems utilize two different fluids, one being the ocean water and the other one being a working fluid with a low boiling point such as ammonia or propane. Hence, the heat transfer rate at the evaporator can be expressed in terms of either fluid. According to the schematic and T–S diagram in **Figure 10.7**:

$$\dot{Q}_{evap} = \dot{m}_{wf}(h_1 - h_4) = \dot{m}_{ww}\, c_p(T_5 - T_6) \tag{10.7}$$

where $\dot{m}_{wf}$ is the mass flow rate of the working fluid. **Equation 10.7** assumes that the ocean water is incompressible and there is no heat loss to the surroundings, and that all heat given by the warm ocean water is absorbed by the working fluid. The heat exchange rate within the evaporator can also be expressed in terms of the UA value:

$$\dot{Q}_{evap} = (UA)_{evap}\, \text{LMTD}_{evap} \tag{10.8}$$

In **Equation 10.8**, U is the overall heat transfer coefficient, A is the heat transfer area, and LMTD is the logarithmic mean temperature difference across the evaporator. The LMTD for the evaporator can be obtained by:

$$\text{LMTD}_{evap} = \frac{(T_5 - T_E) - (T_6 - T_E)}{\ln \dfrac{(T_5 - T_E)}{(T_6 - T_E)}} \tag{10.9}$$

Similarly, for the condenser:

$$\dot{Q}_{cond} = \dot{m}_{wf}(h_2 - h_3) = \dot{m}_{CW}\, c_p(T_8 - T_7) \tag{10.10}$$

$$\dot{Q}_{cond} = (UA)_{cond}\, \text{LMTD}_{cond} \tag{10.11}$$

$$\text{LMTD}_{cond} = \frac{(T_C - T_7) - (T_C - T_8)}{\ln \dfrac{(T_C - T_7)}{(T_C - T_8)}} \tag{10.12}$$

Example 10.2 Theoretical Analysis of a Closed-Cycle OTEC Power Plant

A closed-cycle OTEC plant operates employing ammonia (NH_3) as the working fluid. Warm surface water enters the evaporator at 31 °C and leaves at 27 °C. Cold water, on the other hand, enters the condenser at 5 °C and leaves at 9 °C. The evaporating temperature of ammonia in the evaporator is 20 °C, and the condensation temperature is 10 °C. The capacity of this thermal power plant is 100 kW. Isentropic efficiency of the turbine is given as 86%. The overall heat transfer coefficient for both heat exchangers (evaporator and condenser) is 1100 W/m²K. Determine:

a. the mass flow rate of ammonia (kg/s)
b. the evaporator surface area (m²)
c. the condenser surface area (m²)
d. the warm (surface) water mass flow rate (kg/s)
e. the cold (deep) water mass flow rate (kg/s)
f. the thermal efficiency of the power station (%)

Solution

First, the properties of ammonia are determined from *NIST Chemistry WebBook*.
At 20 °C, $h_1 = 1623.3$ kJ/kg and $s_1 = 5.848$ kJ/kg K

For an isentropic expansion, at $10\,°C$, $s_{2s} = s_1$ yielding quality of $x_{2s} = 0.973$ (or 97.3%)

Then, $h_{2s} = h_f + 0.973\,(h_g - h_f) = (389.7) + 0.973\,(1615.3 - 389.7) = 1582.2$ kJ/kg

Actual turbine exit enthalpy, h_2, can be calculated from the isentropic turbine efficiency:

$$\eta_t = \frac{h_1 - h_2}{h_1 - h_{2s}} = \frac{1623.3 - h_2}{1623.3 - 1582.2} = 0.86$$

Hence, $h_2 = 1588$ kJ/kg

a. Mass flow rate of ammonia can be calculated with known values of enthalpy change through the turbine and the power output:

$$\dot{m}_{NH_3} = \frac{\dot{W}}{(h_1 - h_2)} = \frac{100 \text{ kW}}{(1623.3 - 1588)\dfrac{\text{kJ}}{\text{kg}}}\left|\frac{1\,\dfrac{\text{kJ}}{\text{s}}}{1 \text{ kW}}\right| = 2.83\,\frac{\text{kg}}{\text{s}}$$

b. Evaporator surface area is calculated using **Equation 10.8**. First, the left-hand side of this equation is obtained from **Equation 10.7**, and then LMTD is determined from **Equation 10.9**:

$$\dot{Q}_{evap} = \dot{m}_{NH_3}(h_1 - h_4)$$

Assuming ammonia enters the evaporator at a midpoint temperature of $15\,°C$ (see **Figure 10.7**), $h_4 = 413.3$ kJ/kg

Then,

$$\dot{Q}_{evap} = \left(2.83\,\frac{\text{kg}}{\text{s}}\right)(1623.3 - 413.3)\frac{\text{kJ}}{\text{kg}} = 3424.3 \text{ kW}$$

$$\text{LMTD}_{evap} = \frac{(31 - 20) - (27 - 20)}{\ln\dfrac{(31 - 20)}{(27 - 20)}} = 8.85\,°C$$

Then,

$$\dot{Q}_{evap} = \left(1100\,\frac{\text{W}}{\text{m}^2\text{K}}\right)(A_{evap})(8.85\,°C) = 3424.3 \text{ kW}$$

Hence, giving $A_{evap} = 351.8$ m^2

c. Condenser surface area is calculated by using **Equation 10.11**, with a similar approach as practiced in the previous part of the problem:

$$\dot{Q}_{cond} = \dot{m}_{NH_3}\,(h_2 - h_3)$$

Assuming ammonia exits the condenser as saturated liquid, $h_3 = h_{f@10°C} = 389.7$ kJ/kg.

Then,

$$\dot{Q}_{cond} = \left(2.83\,\frac{\text{kg}}{\text{s}}\right)(1588 - 389.7)\frac{\text{kJ}}{\text{kg}} = 3391.2 \text{ kW}$$

$$\text{LMTD}_{cond} = \frac{(10 - 5) - (10 - 9)}{\ln\dfrac{(10 - 5)}{(10 - 9)}} = 2.48\,°C$$

Plugging these into **Equation 10.11**, we get:

$$\dot{Q}_{cond} = \left(1100 \ \frac{W}{m^2 K}\right)(A_{cond})(2.48\,°C) = 3391.2 \text{ kW}$$

Hence, yielding $A_{cond} = 1243.1 \text{ m}^2$

d. Warm water mass flow rate is calculated by **Equation 10.7**:

$$\dot{Q}_{evap} = \dot{m}_{WW} c_p (T_5 - T_6) = 3424.3 \text{ kW}$$

c_p of seawater at its average temperature through the evaporator is 4.01 kJ/kg K. Then,

$$\dot{m}_{WW} = \frac{3424.3 \text{ kW}}{\left(4.01 \ \frac{kJ}{kg \ K}\right)(31 - 27)\,°C} = 213.5 \ \frac{kg}{s}$$

e. Cold water mass flow rate is calculated by **Equation 10.10**:

$$\dot{Q}_{cond} = \dot{m}_{CW} c_p (T_8 - T_7) = 3391.2 \text{ kW}$$

c_p of seawater at its average temperature through the condenser is 4.00 kJ/kg K. Then,

$$\dot{m}_{CW} = \frac{3391.2 \text{ kW}}{\left(4.00 \ \frac{kJ}{kg \ K}\right)(9 - 5)\,°C} = 212 \ \frac{kg}{s}$$

f. Thermal efficiency of the power station is:

$$\eta_{th} = \frac{\dot{W}_{net}}{\dot{Q}_{evap}} = \frac{100 \text{ kW}}{3424.3 \text{ kW}} = 0.029 \ (2.9\%)$$

10.2.2 Ocean Salinity Gradient Energy

As discussed in the introduction section, the osmotic pressure caused by the salinity gradient in areas where freshwater meets with saltwater can be used to generate power. Osmotic pressure, π, can be calculated using the *Morse equation* which is expressed as:

$$\pi = i\bar{R}MT \tag{10.13}$$

where i is the van't Hoff factor which is an empirical constant associated with the solute, $\bar{R}$ is the universal gas constant, M is the molar concentration of the solution, and T is the absolute temperature. For sea salt (NaCl), i is 2.0, and the molarity of the seawater is approximately 0.6 mol/L. In **Figure 10.24**, the height difference h is caused by the osmotic pressure. When the system is in equilibrium, the pressure, which is the weight of the saltwater above the freshwater level per unit area, is equal to the osmotic pressure:

$$p = \frac{W}{A} = \frac{mg}{A} = \frac{(\rho V)g}{A} = \frac{\rho(A\,h)g}{A} = \rho g h = \pi \tag{10.14}$$

Combining **Equations 10.13** and **10.14**, we get:

$$i\bar{R}MT = \rho g h \tag{10.15}$$

FIGURE 10.24 Potential difference between freshwater and ocean water caused by the osmotic pressure.

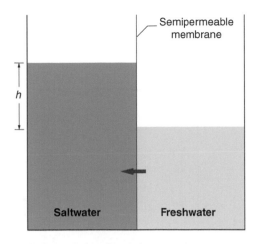

Hence, the height difference can be approximated as:

$$h = \frac{i\bar{R}MT}{\gamma}$$

(10.16)

where γ is the specific weight of saltwater ($\gamma = \rho g$).

Example 10.3 Calculating Height Difference Caused by Osmotic Pressure

Determine the height difference between seawater and freshwater that are separated by a semipermeable membrane at a salinity gradient energy facility. The mean temperature of water is recorded to be 15 °C and the specific weight of seawater, γ, is approximately 10.05 kN/m³.

Solution

The height difference can be calculated by employing **Equation 10.16**:

$$h = \frac{i\bar{R}MT}{\gamma} = \frac{(2.0)\left(8.314\ \dfrac{\text{kJ}}{\text{kmol K}}\right)\left(0.6\ \dfrac{\text{kmol}}{\text{m}^3}\right)(15 + 273.15)\text{K}}{10{,}050\ \dfrac{\text{N}}{\text{m}^3}} = 286\ \text{m}$$

The column height difference will be 286 m. If the temperature of water increases, this will benefit the height not only due to the numerator value increasing, but also because of the denominator on the right-hand side getting smaller as a result of specific weight decreasing with temperature.

10.2.3 Tidal Energy

A schematic of a tidal barrage is illustrated in **Figure 10.25** for the theoretical analysis. The amount of potential energy stored in the basin is given by:

$$E_{\text{cycle}} = \frac{1}{2}A\rho g R^2$$

(10.17)

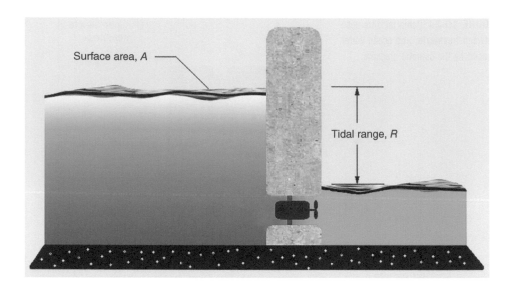

FIGURE 10.25 Schematic of a tidal barrage.

where A is the surface area of the tidal basin, ρ is the density of seawater, g is the gravitational acceleration, and R is the tidal range. Tidal barrages are mostly built in areas where semidiurnal tides exist. This means two high and two low tides are observed each lunar day. As one cycle is 12.4 hours, the potential energy that becomes available per solar day is:

$$E_{day} = \left(\frac{24 \dfrac{h}{day}}{12.4\ h}\right)\left(\frac{1}{2}A\rho g R^2\right) = 0.968\ A\rho g R^2 \tag{10.18}$$

The density of seawater varies in a range of 1021–1030 kg/m^3 and can be approximated as 1025 kg/m^3. With $g = 9.81$ m/s^2, daily available potential energy can be expressed as:

$$E_{day} = (\eta_{elec})(9733.5\ AR^2) \tag{10.19}$$

where η_{elec} is the conversion efficiency, which can be in a range of 20–35%. The unit of **Equation 10.19** is joules per day (J/day). The annual electricity production of a tidal barrage can then be calculated as:

$$E_{year} = (\eta_{elec})(9733.5\ AR^2)\left|\frac{1.1574 \times 10^{-14}\ GW}{1\dfrac{J}{day}}\right|\left|\frac{8760\ h}{1\ year}\right|$$
$$= (\eta_{elec})(0.987 \times 10^{-6}\ AR^2) \tag{10.20}$$

where E_{year} is in GWh/year.

As the energy generation rate is proportional to the square of the tidal height, it is important in tidal barrage projects to select a location that observes higher tidal ranges. In the United States, Anchorage, Alaska, observes tides of 12 m range. The highest tides in the world are seen in the Bay of Fundy, Canada, reaching 17 m in height. In addition to the height of the tide contributing to the tidal energy in this location, a physical phenomenon called *tidal resonance* can increase the energy content of the tides. This resonance occurs when the time it takes for a large wave to travel from a bay to an opposite shore and return back to the bay is about the same as the time between high and low tides.

Example 10.4 Determining Annual Tidal Power Generation

A tidal power plant has an average head of 6.1 m with a basin surface area of 35 km². The conversion efficiency of the turbine is 32%. Calculate the annual electricity production from this tidal barrage.

Solution

Annual electricity production is calculated employing **Equation 10.20**:

$$
\begin{aligned}
E_{\text{year}} &= (\eta_{\text{elec}})(0.987 \times 10^{-6}\, AR^2) \\
&= (0.32)(0.987 \times 10^{-6})(35 \times 10^6\ \text{m}^2)(6.1\ \text{m})^2 \\
&= 411.3\ \text{GWh/year}
\end{aligned}
$$

Note: *If the tidal range for the same construction and basin was 20% higher (7.32 m), this would reflect in an annual production of 592.3 GWh/year, which is a 44% increase. A 1.2 multiplier for the head yields a multiplier of (1.2)² = 1.44 due to the tidal height–power relation.*

10.2.4 Wave Energy

In practice, wave hydrodynamics is a complicated phenomenon. For theoretical purposes, the physics of the problem can be simplified such that the wave is considered to be in a sinusoidal form, as shown in **Figure 10.26**. The figure depicts a crest-to-crest schematic including a trough beneath the neutral (no wave) level. The wave has both potential energy due to the elevation difference between the crest and the trough, and also kinetic energy due to it moving in the horizontal direction. It has a wavelength of λ, amplitude of A, and a crest length of l which is orthogonal to the two-dimensional drawing. If we focus on the potential energy of a single wave per unit crest length, then:

$$
\frac{\text{PE}}{l} = \frac{m}{l} gh \tag{10.21}
$$

FIGURE 10.26 Schematic of a sinusoidal wave for approximating wave power.

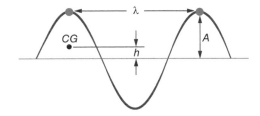

where mass of water in a single crest (or a trough) is given by:

$$m = \rho \left(\frac{A\lambda}{\pi} \right) l \tag{10.22}$$

In this expression, $A\lambda/\pi$ is the cross-sectional area of a single sine wave which is obtained by integration; h is the distance from the neutral line to the center of gravity of the wave which is:

$$h = \frac{\pi A}{8} \tag{10.23}$$

Substituting **Equations 10.22** and **10.23** into **Equation 10.21** gives:

$$\frac{\text{PE}}{l} = \rho \left(\frac{A\lambda}{\pi} \right) g \left(\frac{\pi A}{8} \right) \tag{10.24}$$

This is the potential energy equation for only a crest, which is half of a wavelength. Then, over one wavelength of a wave, potential energy would be:

$$\frac{\text{PE}}{l} = \frac{1}{4} \rho A^2 \lambda g \tag{10.25}$$

The kinetic energy of the wave is equal to the potential energy. This equality is obtained through a cumbersome process and is not derived here for the sake of focusing on the main point. The total energy per unit length of the wave, E/l, then is twice as much, yielding:

$$\frac{E}{l} = \frac{1}{2} \rho A^2 \lambda g \tag{10.26}$$

Determining the power capacity for a wave energy site requires a time quantity. For a wave, this is T, which is its period and is related to wavelength as:

$$\lambda = \frac{g T^2}{2\pi} \tag{10.27}$$

Then, total energy per unit crest length for one wavelength of a wave can be written in terms of time period by substituting **Equation 10.27** into **10.26**:

$$\frac{E}{l} = \frac{\rho A^2 T^2 g^2}{4\pi} \tag{10.28}$$

As power is energy per unit time, this equation can be converted into a power expression by dividing both sides by T, giving:

$$\frac{P}{l} = \frac{\rho A^2 T g^2}{4\pi} \tag{10.29}$$

This expression is derived in terms of the amplitude, A. For a sinusoidal wave, crest-to-trough wave height, H, is twice the amplitude. Hence, **Equation 10.29** can also be written in terms of H to yield:

$$\frac{P}{l} = \frac{\rho H^2 T g^2}{16\pi} \tag{10.30}$$

Since constants g and π are known, and the density of ocean water is approximated as 1025 kg/m^3, then the wave power per unit length can be expressed in a more compact form as:

$$\frac{P}{l} = 1.96\, H^2 T \tag{10.31}$$

where the unit is kW/m. As the wavelength and time period for the wave are known, its velocity can also be calculated:

$$V = \frac{\lambda}{T} = \frac{gT}{2\pi} \tag{10.32}$$

Example 10.5 Calculating Wave Power

A bay receives ocean waves with a height of 1.8 m and a crest length of 500 m. If the wave period is 8 s, determine

a. the wave power capacity (MW)
b. the wavelength (m)
c. the wave velocity (m/s)

Solution

a. Wave power available per 1 m of crest length can be calculated using **Equation 10.31**:

$$\frac{P}{l} = 1.96\, H^2 T = (1.96)(1.8\text{ m})^2(8\text{ s}) = 50.8\text{ kW/m}$$

Hence, P for 500 m of crest length will be:

$$P = \left(50.8\,\frac{\text{kW}}{\text{m}}\right)(500\text{ m})\left|\frac{1\text{ MW}}{10^3\text{ kW}}\right| = 25.4\text{ MW}$$

b. Wavelength of the wave is obtained with **Equation 10.27**:

$$\lambda = \frac{gT^2}{2\pi} = \frac{\left(9.81\,\frac{\text{m}}{\text{s}^2}\right)(8\text{ s})^2}{2\pi} = 99.9\text{ m}$$

c. Velocity of the wave is calculated from **Equation 10.32**:

$$V = \frac{\lambda}{T} = \frac{99.9\text{ m}}{8\text{ s}} = 12.49\,\frac{\text{m}}{\text{s}}$$

Example 10.6 Determining Wave Power Using Python

Consider an OWC project on the western shores of Cape Verde, an island country in the Atlantic Ocean. Annual average wave height and period for the application site are approximated as 1.1 m and 6 s, respectively. Twenty OWC units are planned to be deployed with each unit having a wave-receiving frontal length of 5 m. Energy conversion efficiency from wave-to-wire is 5.8%.

a. Determine the power capacity for this project.
b. Plot the total power output for 20 units for a wave height range of 0.4–1.5 m for the same energy conversion efficiency.
c. Simulate a scenario with 1000 waves of random wave heights using the Monte Carlo method with a normal distribution. Obtain the wave height and power generation plots over time (1 s period, 1000 s total time). Plot the histogram for the wave height.

Solution

a. Wave power available per unit crest length is obtained employing **Equation 10.31**.

```
import matplotlib.pyplot as plt
import numpy as np

H_wave_height=1.1
T_period=6
P_l=1.96*(H_wave_height**2)*T_period

P_l
14.229600000000001

units=20
frontal_length=5
total_crest_length=units*frontal_length

P_avail=P_l*total_crest_length

P_avail
1422.96

eff=5.8/100
P_gen= eff*P_avail

P_gen
82.53168
```

b. Power is proportional to the square of the wave height. Power generation for the same units being exposed to a wave height range of 0.4–1.5 m can be determined by introducing a wave height array.

```
H_wave_height_array=np.arange(40,150)/100
P_gen_array=1.96*eff*total_crest_length*(H_wave_height_array**2)
*T_period

plt.title("Power generation vs Wave Height")
plt.grid()
plt.xlabel("Wave Height(m)")
plt.ylabel("P_gen(kW)")
plt.plot(H_wave_height_array,P_gen_array)
```

```
[<matplotlib.lines.Line2D at 0x7ff2583862d0>]
```

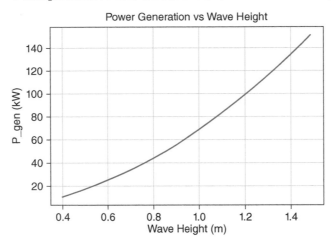

c. Monte Carlo methods can be used for simulating physical problems such as fluid systems (i.e., ocean waves) by utilizing repeated random sampling. For the given problem, we first generate 1000 random wave heights with a normal distribution of mean 1.5 meters and standard deviation of 1 meter. Although normally distributed values can be negative in this parameter setting, we will clip the negative values with the `clip(min=0)` command. It is assumed that the waves are sequential, and they follow each other by a second.

```python
mean=1.5
standard_deviation=0.5
H_wave_height_random=np.random.normal(mean, standard_deviation,
size=1000)
H_wave_height_random = H_wave_height_random.clip(min=0)

plt.title("Wave Height")
plt.grid()
plt.xlabel("time(t) ")
plt.ylabel("Wave height(m)")
plt.plot(H_wave_height_random)
```

```
[<matplotlib.lines.Line2D at 0x7ff25486d710>]
```

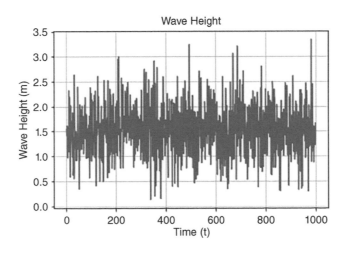

Hence, the wave heights over 1000 s are plotted. Next, we plot the power generation resulting from these random waves over the simulated time period.

```
P_gen_random=1.96*eff*total_crest_length*(H_wave_height_random**2)
*T_period
```

```
plt.title("Power generation")
plt.grid()
plt.xlabel("time(t)")
plt.ylabel("P_gen(kW)")
plt.plot(P_gen_random)
```

```
[<matplotlib.lines.Line2D at 0x7ff2543b2110>]
```

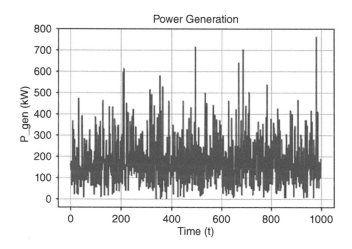

Now, let's plot power generation based on the random set of wave heights.

```
plt.title("Power generation")
plt.grid()
plt.xlabel("Wave Height(m)")
plt.ylabel("P_gen(kW)")
plt.scatter(H_wave_height_random, P_gen_random)
```

```
<matplotlib.collections.PathCollection at 0x7ff2542e8250>
```

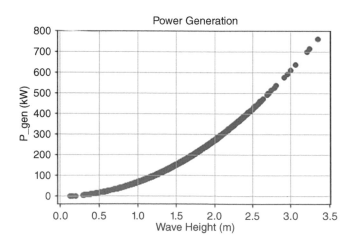

Finally, we plot the wave height histogram. In the simulation, 1000 ocean waves were produced with a normal distribution. By obtaining the histogram, the number of occurrences for each wave height can be seen.

```
plt.title("Wave Height Histogram")
plt.grid()
plt.xlabel("Wave Height(m)")
plt.ylabel("Frequency(number of occurences)")
plt.hist(H_wave_height_random,bins=100);
```

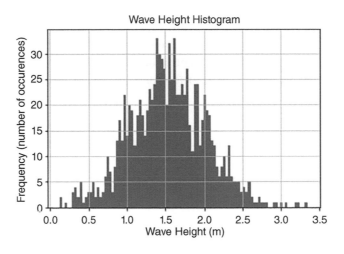

10.3 Applications and Case Study

In this section, real-world examples are explored through two applications and a case study. The applications include a closed-cycle OTEC plant in the United States, and a tidal power plant in South Korea. The case study is about an oscillating water column (OWC) application as part of a renewable energy integration project conducted in King Island, Australia.

Application 1 Closed-Cycle OTEC Plant (United States)

Makai OTEC Plant
Location: Kailua-Kona, Hawaii, United States
Operated by: Makai Ocean Engineering
Type: Closed-cycle, onshore
Working fluid: Ammonia
Capacity: 100 kW
www.d-maps.com

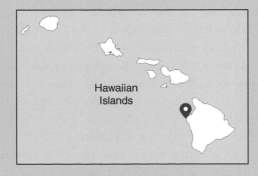

Makai OTEC Plant is a testing facility situated in Kailua-Kona on the west coast of Hawaii Island. It is an onshore application which works based on the closed-cycle principle, and employs ammonia (NH_3) as the working fluid. It has a capacity of 100 kW, which is sufficient to provide energy to about 120 Hawaiian homes. The facility is connected to the research camp microgrid at the HOST Park and is the first grid-connected closed-cycle OTEC plant in the United States. The system utilizes thin foil heat exchangers that provide enhanced heat transfer surface area within a compact unit. The volume flow rate of warm water is approximately 45,000 L/min [5]. The station is located in the *Hawaii Ocean Science and Technology (HOST) Park*, which houses a variety of renewable energy projects that are in R&D phase or in operation. Various groups working on designing, building, and testing the OTEC technology include University of Hawaii, Lockheed Martin, Makai Ocean Engineering, and the United States Navy. The projects have been funded by the National Science Foundation (NSF), Office of Naval Research (ONR), Pacific International Center for High Technology Research (PICHTR), U.S. Department of Energy, State of Hawaii, and some private sponsors [6].

Application 2 Tidal Power Plant (South Korea)

Sihwa Lake Tidal Power Plant
Location: Gyeonggi Province,
   South Korea
Operated by: Korea Water Resources
   Corp.
Type: Tidal barrage
Capacity: 254 MW
Capacity factor: 24.8%
www.d-maps.com

CLASSROOM DEBATE

Split in two groups and debate the idea of tidal barrage construction considering the advantages and disadvantages of these power stations.

FURTHER LEARNING

Natural Energy Laboratory of Hawaii Authority (NELHA)
https://nelha.hawaii.gov/

Sihwa Lake Tidal Power Plant is located in the northwestern part of South Korea at Gyeonggi Bay. It was the world's largest tidal power plant with a capacity of 254 MW when it was opened in 2010, surpassing La Rance Tidal Power Plant in France with a capacity of 240 MW, which had been the largest tidal plant since its completion in 1967, until the opening of Lake Sihwa station. The total cost for the project was $355.1 million. The plant has an annual generation of 552.7 GWh, with a head of 9.16 m. It houses 10 turbines with each turbine having a capacity of 25.4 MW, and eight sluice gates. Electricity production is achieved via single-effect flood generation. It is reported to substitute 862,000 barrels of oil in terms of the energy produced annually, with a reduction of 315,000 tons of CO_2 emission per year [7]. The barrage has a 12.7 km long seawall, embanking a tidal basin area of 43.8 km².

Annually 160 million tons of water flows through the gates. The circulation of water in and out of the basin has improved the water quality [8]. The Moon Observatory built next to the tidal barrage has become a tourist attraction, with a height of 75 m and a view of the barrage, the basin, and the Yellow Sea.

Case Study King Island Renewable Energy Integration Project (Australia)

King Island OWC System
Location: King Island, Australia
Led by: Hydro Tasmania
Supported by: Australian Renewable
  Energy Agency
Wave energy type: Oscillating water
  column
Brand/Model: Wave Swell/
  UniWave200
Capacity: 200 kW
www.d-maps.com

INTRODUCTION King Island is a small island located in the Bass Strait between Tasmania and mainland Australia. An initiative by Hydro Tasmania that was supported by the Australian Renewable Energy Agency (ARENA) resulted in implementation of a hybrid off-grid power system to provide 65% of the island's electricity needs using renewable energy. This initiative, *King Island Renewable Energy Integration Project* (KIREIP), has already been proven to cut use of diesel for power generation with a mix of wind, solar, and biodiesel energy supported by energy storage technologies. In late 2021, an oscillating water column unit with 200 kW capacity was also added to the mix, supplying renewable energy from the waves (**Figure 10.27**).

OBJECTIVE The objective of this case study is to learn about a wave energy converter (WEC) system deployed for a small island as part of a renewable energy integration project.

METHOD This case study was conducted through researching articles and reports.

RESULTS The OWC device, after extensive experiments under a variety of conditions simulated at the Australian Maritime College, has been anchored to the southern shore of King Island. The system is multifunctional with alternative uses of water purification and hydrogen production, besides electricity generation. The unit is at a demonstration stage, being tested for its efficiency, robustness, and accessibility. Capacities above 200 kW are considered necessary for the island with a population of about 1700 people to make the energy supply 100% renewable.

DISCUSSION The OWC application that harnesses wave energy contributing to the renewable energy mix of the island is a good example for small islands who rely on the mainland or diesel energy. Inclusion of energy storage technologies such as

FIGURE 10.27 Uni Wave200 oscillating water column wave energy converter. Source: Wave Swell Energy Ltd.

batteries and flywheels, a smart grid system, and an advanced control system provide a more efficient operation and assessment of the project. For these kinds of smart grid initiatives, the importance of such initiatives becomes more obvious during extreme weather conditions when the island may lose connection with the mainland.

RECOMMENDATIONS Hybrid technologies along with energy storage systems and ocean energy applications can provide more reliable energy production. Cost-effectiveness, accessibility of the renewable energy unit, and the environmental impacts are three important factors in assessing these projects. With smaller islands having limited population, qualified human resources are also critical for the O&M of the project site. Establishment of vocational schools and certificate training programs will address such needs.

10.4 Economics

The total cost of an ocean energy technology is mainly composed of the capital and the O&M costs. Capital costs include feasibility analysis, foundation and infrastructure, system design and construction, onshore and offshore components, and commissioning of the system. Feasibility analysis includes ocean bed morphology (if the system is to be built or anchored to the bottom of the ocean), sediment structure and behavior, and environmental impact assessment. Operational costs consist of marine and onshore operations, replacement parts, utilities, wages and salaries, social security, unemployment insurance, worker's compensation, medical insurance, and taxes. Maintenance costs include routine maintenance, troubleshooting, and repair costs.

In terms of levelized cost of energy (LCOE), ocean energy applications usually have higher LCOE than other renewable energy technologies. Depending on the location of the project site, weather conditions throughout the year, accessibility of the system, and measures taken for environmental compliance, the total cost can further increase. Conversely, advancement in marine energy technologies including components and materials and reduction in risk of investment over time will lower the costs. Combining ocean energy with wind or solar energy can also lower the risk and overall LCOE. Therefore, deployment of hybrid technologies is expected to increase globally.

10.5 Summary

Ocean energy in the form of temperature and salinity gradients, tidal barrages and streams, and waves was covered in this chapter. Depending on the type of ocean energy and the technologies under the umbrella of each type, power can be generated by means of converting thermal, chemical, potential, and kinetic energies into electrical energy. OTEC systems require a good understanding of thermal processes. Salinity gradient energy can involve chemical energy, as well as kinetic and potential energy. Tidal barrages, different than the other forms of tidal stream applications, can be treated as hydroelectric power during generation, and PHES if energy is being stored to accumulate more potential with the low tides on the other side. Wave energy technologies come in a variety, including onshore and offshore applications. The theory of OTEC, salinity gradient, tidal barrages, and wave energy were discussed using pertinent dynamics, thermodynamics, heat transfer, and fluid mechanics relations. A comprehensive engineering approach is not limited to theoretical computations. Evaluation of different marine energy technologies was performed based on several other criteria including efficiency, durability, accessibility, cost, and environmental impact. Feasibility analysis for an ocean energy project should take all these criteria and their constituent parameters into account for the project to be successful.

REFERENCES

[1] International Renewable Energy Agency, "Innovation Outlook: Ocean Energy Technologies," IRENA, Abu Dhabi, 2020.

[2] United Nations, "Conferences: Small Island Developing States." www.un.org/en/conferences/small-islands (accessed Nov. 17, 2021).

[3] United Nations, "List of SIDS." www.un.org/ohrlls/content/list-sids (accessed Nov. 17, 2021).

[4] L. Kilcher, M. Fogarty, and M. Lawson, "Marine Energy in the United States: An Overview of Opportunities," National Renewable Energy Laboratory, NREL/TP-5700-78773, Golden, CO, Feb. 2021.

[5] H. Kugeler, Makai Ocean Engineering, "Ocean Energy Expertise," United Nations, Jun. 2021. Accessed: Nov. 22, 2021 [Online]. Available: www.un.org/sites/un2.un.org/files/makai_ocean_engineering.pdf.

[6] Hawaii Ocean Science & Technology Park, Natural Energy Laboratory of Hawaii Authority, "Energy Portfolio and Projects," https://nelha.hawaii.gov/energy-portfolio (accessed Nov. 22, 2021).

[7] Korea Water Resources Corp., "Sihwa Tidal Power Plant," K-Water. www.kwater.or.kr/eng/busi/project03Page.do?s_mid=1192 (accessed Nov. 23, 2021).

[8] International Hydropower Association, "Technology Case Study: Sihwa Lake Tidal Power Station," Feb. 8, 2016. www.hydropower.org/blog/technology-case-study-sihwa-lake-tidal-power-station (accessed Nov. 23, 2021).

CHAPTER 10
EXERCISES

10.1 Fill in the blank circles with a technology related to each branching ocean energy group. Provide a global example for each application.

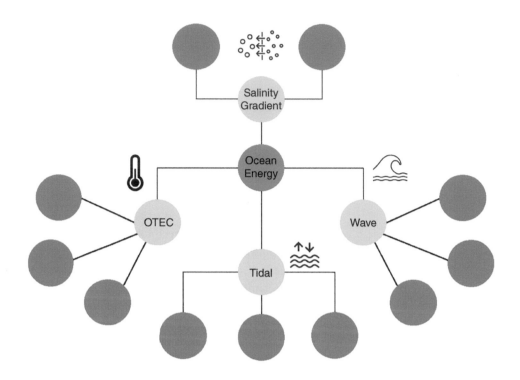

10.2 An open-cycle OTEC power station is supplied with ocean surface water at a temperature of 24 °C. Evaporation occurs at 22 °C and steam enters the turbine as saturated vapor. Isentropic efficiency of the turbine is 85%. The direct-contact condenser receives cold water at 7 °C. Condensation occurs at 9 °C, which is assumed as the discharge temperature of the condensate and cold water mixture. The station has a turbine power output of 125 kW and a total pump power input of 5 kW. Find:

  a. the mass flow rate of steam through the turbine (kg/s)
  b. the mass flow rate of warm water entering the flash evaporator (kg/s)
  c. the rate of heat addition at the evaporator (kW)
  d. the efficiency of the power plant (%)

10.3 A closed-cycle OTEC power plant uses propane (C_3H_8) in its closed loop. Warm surface water enters the evaporator at 27 °C and leaves at 23 °C. On the condenser side, cold water enters the heat exchanger at 5 °C and leaves at 7 °C. The evaporating temperature of propane in the evaporator is 18 °C, and the condensation temperature is 8 °C. Capacity of the power plant is 2 MW. Isentropic efficiency of the turbine is 88%. Overall heat transfer coefficient for both heat exchangers (evaporator and condenser) is 1250 W/m^{2}K. Calculate:

  a. the mass flow rate of propane (kg/s)
  b. the evaporator surface area (m^2)
  c. the condenser surface area (m^2)
  d. the warm water mass flow rate (kg/s)
  e. the cold water mass flow rate (kg/s)
  f. the thermal efficiency of the power plant (%)

10.4 A hybrid OTEC plant operates between 25 °C warm water and 5 °C cold water layers. Thermal efficiency of the plant is 4.7%. Compare the plant's overall efficiency to its Carnot efficiency.

10.5 *Garabogazkol* (*Black Strait Lake*) is a highly saline lagoon of the Caspian Sea in the northwestern part of Turkmenistan. It is one of the saltiest water bodies in the world with a salinity of 350 g/kg. Determine:

    a. the molarity of the water assuming it is mostly NaCl (mol/L)

    b. the height difference of the water columns due to osmotic pressure if the water in the lagoon was introduced to freshwater with a mean water temperature of 12 °C

10.6 Consider the salinity gradient energy facility in Example 10.3. Obtain a height difference versus temperature plot in the form of $h = f(T)$ for saltwater and freshwater that are separated by a semipermeable membrane for a mean temperature range of 10–20 °C.

10.7 Using NOAA's tide and current prediction tool, obtain the monthly tidal range plots for the locations and months listed below.

Location	Station	Month
Seattle, WA	Seattle	May 2025
Miami, FL	Virginia Key	February 2028
San Diego, CA	San Diego	August 2033
Eastport, ME	Eastport	September 2026

10.8 Research on–off control and modulated control for HVAC systems. If you were to make an analogy between these control systems and the two tidal barrage application methods, one-way generation (flood or ebb) and two-way generation, how would you match a control method with a tidal power generation method? How are they similar in terms of the operating principles?

10.9 A tidal barrage has a basin that is 8 km wide and 5.5 km long. Average tidal height is 9.4 m. If the electricity generation efficiency of the turbine is 28%, determine the annual production of this tidal power station.

10.10 A tidal power plant has a daily energy production of 1200 MWh. If the tidal height is 7.8 m and the conversion efficiency of the turbine is 33%, calculate the surface area of the tidal basin.

10.11 Calculate the tidal energy generated in one cycle for a barrage in Rio Gallegos estuary, Argentina, if the tidal height is 8.2 m and the surface area of the tidal basin is approximately 80 km^2. Assume the turbine generation efficiency to be 27%.

10.12 Research tidal kite and tidal fence technologies. Make a list of strengths and weaknesses of both applications.

10.13 The velocity of the ocean waves at a pilot site is measured to be 11 m/s. Determine the period and wavelength of the waves.

10.14 Consider an area with ocean waves having a height of 1.6 m and a period of 10 s. For this location, determine:

    a. the wave power for a wave crest length of 700 m (MW)

    b. the wavelength (m)

    c. the wave velocity (m/s)

10.15 Consider a pilot OWC application for Micronesia as supplementary energy generation. The island has an installed electricity capacity of approximately 20 MW. Annual average wave height and period for the selected site are recorded as 1.8 m and 12 s, respectively. Frontal length of the chamber is 6 m for each OWC unit. If the wave-to-wire efficiency for the system is 6.5%, determine what percentage of the island's installed electricity generation capacity could be supplied with 50 OWC units.

10.16 Explain the difference between tides, waves, and currents. What are the driving forces for each term?

10.17 Wave speed is also a function of ocean depth and is given by:

$$V = \sqrt{gD}$$

where D is depth (vertical distance between the neutral surface line and the ocean bed). This equation is valid for shallow-water waves where the ratio of depth to wavelength is very small. Tsunamis can have wavelengths exceeding 500 km, hence resulting in a small depth-to-wavelength ratio and making them shallow-water waves. It can be interpreted from the given velocity equation that the tsunami slows down as the water gets shallower. However, as the total energy remains conserved and the losses are very small, all that energy results in a growing wave height. This phenomenon is called *shoaling*. Wave heights as a function of water depths for shallow and deep waters are expressed by *Green's law* as:

$$\frac{H_s}{H_d} = \left(\frac{D_d}{D_s}\right)^{0.25}$$

Based on these two equations, plot a graph of wave speed and wave height as a function of depth (0–3000 m, 500 m increments) with the speed and height being on the left and right vertical axes, respectively. Assume the wave height at 3000 m depth is 1 m.

10.18 Research the following hybrid renewable energy applications involving ocean energy. Describe each and provide an example project that has been applied or is in progress.

a. Wave–solar energy

b. Wave–wind energy

CHAPTER 11
Solar Energy

11.1 Introduction

Energy from the Sun can be utilized in many ways. The hydrologic cycle is driven by the Sun. It is this cycle that provides water to gain potential energy and be used in harvesting energy at the dams built on rivers for hydropower generation. It is again the sunlight that is absorbed by chlorophyll pigments in plants to initiate photosynthesis which yields energy in the form of carbohydrates. The plants eventually become biomass which is another source of energy. Other forms of energy, which are directly or indirectly related to the energy from the Sun, also exist.

In this chapter, the use of energy from the Sun is discussed based on solar energy fundamentals and applications. There are two main categories in utilizing solar energy: direct use and electricity generation. Direct use of solar energy is associated with space heating or hot-water production. These can be done by active and passive means. Heliostats belong to a specific category in direct use applications where higher operating temperatures can be achieved for various purposes. Active solar systems require electrical energy input to power equipment (e.g., a turbomachine such as a pump or a fan) in utilizing the energy from the Sun. In passive solar heating systems, energy is harvested without the need for an external means of electric power. The other category in solar energy applications is electricity generation. Parabolic troughs, parabolic dishes, central receivers, and photovoltaic (PV) systems are discussed pertaining to the power generation category.

11.1.1 Global Overview

There has been a steady and significant increase in the capacities of solar energy applications including solar PV, concentrating solar power (CSP), thermal energy storage (TES), solar water heating, and solar district heating. Changes in capacities, annual additions, analyses based on countries and regions, and top countries in specific applications can be seen in **Figures 11.1–11.8**. According to the Renewables 2021 Global Status Report of REN21 [1], global PV capacity reached approximately 760 GW by the end of 2020. China has the largest capacity followed by the European Union (sum of all EU countries), and the United States. In terms of CSP capacity, Spain and the United States were the top two countries. Thermal energy storage capacity by means of solar energy reached 21.1 GWh, which is an important indicator of increasing global investments in energy storage technologies. Global solar water heating collector capacity reached 501 GW-thermal, with the majority coming from glazed collectors. On a country basis, China, Turkey, and India had the highest total collector capacities.

FIGURE 11.1 Solar PV global capacity and annual additions. Source: REN21 Renewables 2021 Global Status Report (Paris: REN21 Secretariat), with data from Becquerel Institute and IEA PVPS.

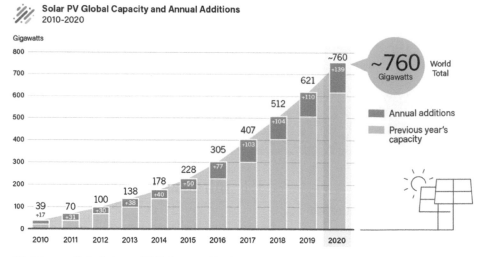

FIGURE 11.2 Solar PV global capacity by country and region. Source: REN21 Renewables 2021 Global Status Report (Paris: REN21 Secretariat).

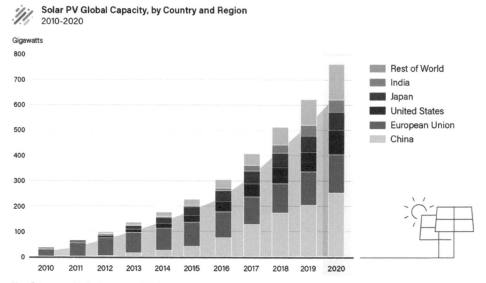

In terms of direct use applications of solar energy, more buildings have been introduced to active or passive solar designs. These are either new constructions being built or existing buildings being renovated. There are numerous coastal Mediterranean cities that have been promoting active solar systems. Some examples include Barcelona (Spain), Casablanca (Morocco), Marseille (France), Naples (Italy), Athens (Greece), and Antalya (Turkey). More cities have been advocating for and supporting sustainable architectural designs for buildings having passive solar applications as well. Policies, regulations, standards, and codes have had an impact on this environmentalist approach. Improved communication and an interdisciplinary working culture between the architects and engineers have also played an important role in this progress.

FIGURE 11.3 Top 10 countries in solar PV capacity and additions. Source: REN21 Renewables 2021 Global Status Report (Paris: REN21 Secretariat).

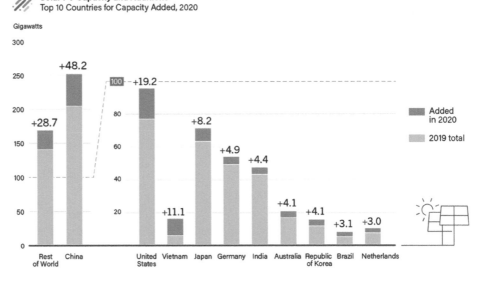

FIGURE 11.4 Global capacity for concentrating solar power (CSP) by country and region. Source: REN21 Renewables 2021 Global Status Report (Paris: REN21 Secretariat).

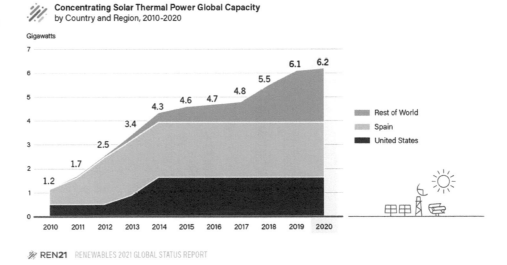

FIGURE 11.5 Global capacity and annual additions in thermal energy storage (TES). Source: REN21 Renewables 2021 Global Status Report (Paris: REN21 Secretariat).

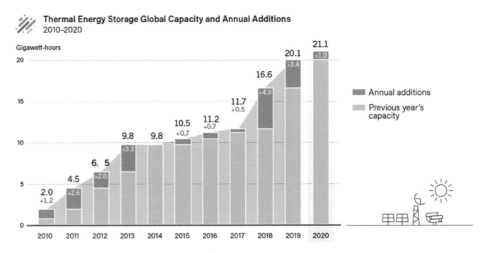

FIGURE 11.6 Global solar water heating collector capacity. Source: REN21 Renewables 2021 Global Status Report (Paris: REN21 Secretariat), with data from IEA SHC.

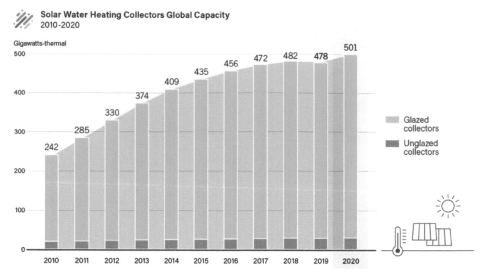

Solar Water Heating Collectors Global Capacity
2010-2020

Note: Data are for glazed and unglazed solar water collectors and do not include concentrating, air or hybrid collectors. The drop in 2019 was caused by revised annual additions for China in 2019 and new assumptions for projecting total capacity in operation for 2019 and 2020.
Source: IEA SHC.

REN21 RENEWABLES 2021 GLOBAL STATUS REPORT

FIGURE 11.7 Top 20 countries for collector capacity additions in solar water heating in 2020 . Source: REN21 Renewables 2021 Global Status Report (Paris: REN21 Secretariat).

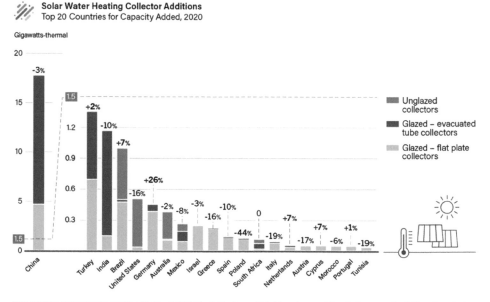

Solar Water Heating Collector Additions
Top 20 Countries for Capacity Added, 2020

Note: Additions represent gross capacity added. For the Netherlands, the shares of flat plate and vacuum tube collectors were estimated based on actual shares in 2019. For Morocco, the share of collector types was not available.

REN21 RENEWABLES 2021 GLOBAL STATUS REPORT

11.1.2 Properties of Sunlight and Solar Radiation

In engineering analysis, we approach most of the problems with conservation equations. These equations are for conservation of mass, momentum, and energy. In energy-related problems, the law of conservation of energy is frequently used. Energy cannot

FIGURE 11.8 Global annual additions and total area in operation in solar district heating. Source: REN21 Renewables 2021 Global Status Report (Paris: REN21 Secretariat).

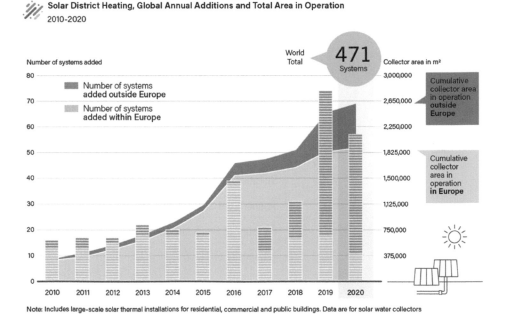

Solar District Heating, Global Annual Additions and Total Area in Operation 2010-2020

Note: Includes large-scale solar thermal installations for residential, commercial and public buildings. Data are for solar water collectors and concentrating collectors.

REN21 RENEWABLES 2021 GLOBAL STATUS REPORT

be created, nor can it be destroyed. It can, however, be converted from one form of energy to another form. In the Sun, nuclear energy transforms into heat or light. When light is received by plants, it is again converted into chemical energy during photosynthesis. In solar energy applications, thermal and light energy can be utilized either to heat a solid (e.g., a house) or a fluid (e.g., water), or to generate electricity from photons striking a solar cell.

To better understand solar energy fundamentals, it is essential to first understand the origin of this energy. The Sun is the largest star within our solar system. It comprises 99.7% of the total mass of the system. The Earth orbits approximately 149.6×10^6 km from the Sun. The mean radius of the Sun is 696,000 km. Despite its massive size, there are other stars that are much larger than the Sun. *Rigel* has a radius that is about 75 times the radius of the Sun. Its luminosity is more than 100,000 times that of the Sun. *Betelgeuse* is 700 times bigger than the Sun with a luminosity that is 14,000 times that of the Sun. These are just two simple comparisons to give a feeling of how large the universe is and how enormous the energy capacity of it is. It is beyond our imagination when we try to think of quantitative examples from our earthly experiences and knowledge.

The Sun consists mainly of hydrogen and helium. Some of its properties and facts about it are given in **Table 11.1** [2]. The layers of the Sun are illustrated in **Figure 11.9**.

In **Table 11.1**, luminosity is the energy radiation of the Sun in watts. It has a magnitude of about 3.83×10^{26} W. Energy received by Earth from this source is 1.7×10^{18} W. Electromagnetic radiation is the propagation of electromagnetic waves through space, carrying electromagnetic radiant energy. It consists of gamma rays, X-rays, ultraviolet light, visible light, infrared light, microwaves, and radio waves. The electromagnetic spectrum can be seen in **Figure 11.10**. The illustrations beneath the

Table 11.1

Properties of the Sun and some facts about it [2]	
Quantity	**Value**
Volumetric mean radius (km)	695,700
Volume (10^{12} km^3)	1,412,000
Mean density (kg/m^3)	1408
Mass (10^{24} kg)	1,988,500
Distance from Earth (km)	149,600,000
Temperature at the center (K)	1.57×10^7
Pressure at the center (bar)	2.477×10^{11}
Luminosity (10^{24} W)	382.8

FIGURE 11.9 Layers of the Sun. Source: NASA.

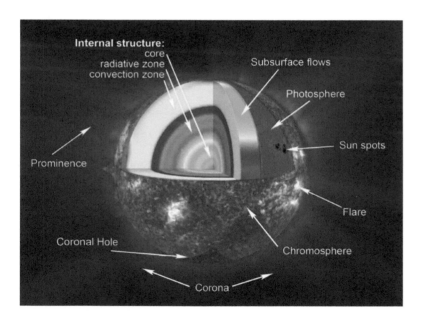

spectral chart help to give a feeling of the order of magnitude of wavelengths from high-energy gamma rays to low-energy radio waves. Solar radiation is the electromagnetic radiation emitted by the Sun. **Figure 11.11** illustrates the solar spectrum. The amount of solar radiation received by a given point on Earth's surface depends mainly on the geographic location, time of year, and time of day. These parameters are quantified in terms of latitude (l) for geographic location, Sun's declination angle (δ) for time of year, and hour angle (h) for time of day. The Earth rotates around the Sun on an elliptical orbit. This causes the planet to be closer to the Sun at certain times, resulting in higher solar energy received on the surface. Besides the changing distance due to the elliptical path, there is another parameter that affects the amount of solar radiation even more. This is the 23.44° axial tilt of the Earth. This tilt results in longer days between the vernal and autumnal equinoxes in the northern hemisphere, and vice versa in the southern hemisphere. The hourly changes result from the rotation of Earth

FIGURE 11.10 Electromagnetic spectrum. Source: NASA.

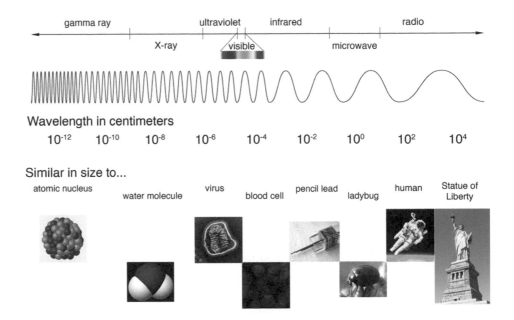

FIGURE 11.11 Solar spectrum. Source: NASA.

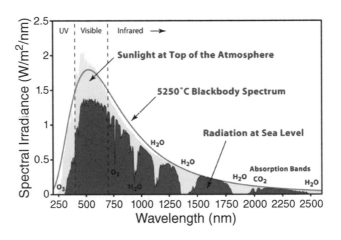

about its axis. During sunrise and sunset, the Sun makes a lower angle with the Earth's surface and the rays must travel a longer distance through the atmosphere, which results in loss of energy received on the surface. When the Sun's rays are normal to the Earth's surface, this is when the maximum amount of solar radiation is received. This occurs usually around noon; however, the time may vary due to the geographic location and time of year. A two-axis solar tracking system uses this information to adjust its tilt and rotation to receive the sunlight at 90°. A linear actuator can control the tilt of the panel to adjust to seasonal changes, while a motor can rotate the panel to account for the hourly changes within the day from sunrise to sunset. There are also other factors that can affect the incident radiation on a surface, such as reflected light from surrounding objects (e.g., glass structures), weather, and surrounding landscape.

Solar radiation can be direct, diffuse, or reflected. *Direct radiation* (or beam radiation) is the solar radiation that travels in a straight line from the Sun to the surface of the Earth. Some of the sunlight can be absorbed, scattered, or reflected by molecules and particles while passing through the atmosphere. This is the *diffuse radiation*. Air molecules, clouds, water vapor, and pollutants are some examples of molecules and particles causing diffuse radiation. The ratio of direct-to-diffuse radiation can vary depending on the time of day, cloudiness, and pollution. On a day with clear sky when the Sun is high, direct radiation can make up about 85% of the total radiation, while the remaining 15% would be the diffuse radiation. The percentage of diffuse radiation will be higher in the morning or later in the afternoon when the sun is lower, or if the sky is cloudy, or if there are pollutants in the air. Direct radiation has a certain path it follows; however, diffuse radiation can travel in irregular paths and can come from any direction. On a sunny day, our shadow on the ground is a result of us blocking the direct radiation as all the sunrays blocked are traveling in the same direction. On an overcast day, nearly all of the solar radiation will be diffuse radiation as the direct radiation will almost be fully blocked. *Reflected radiation* is the sunlight that is reflected from surrounding surfaces such as roads, buildings, and landscapes. Gray surfaces such as concrete or asphalt absorb radiation, hence they are considered to be non-reflective. This causes a microclimate problem called the *urban heat island effect*, which is the overheating of urban areas in summer months due to more coverage of non-reflective surfaces. Lighter colors on the other hand are more reflective. A building with a white roofing membrane would reflect more radiation in summertime, hence reducing the cooling load of the building and saving air-conditioning energy. Snow is also a good reflector. It can reflect about 80–90% of the sunlight falling on it. That is why ultraviolet protection during skiing or snowboarding is as important as protecting the skin from ultraviolet radiation in summer. Even though the ultraviolet index is lower in the winter, the enhanced effect due to the high reflectivity of snow can make it on par with summer values.

Global direct normal irradiation (DNI) and global horizontal irradiation (GHI) for the world can be seen in **Figures 11.12** and **11.13**. The GHI for the United States is also illustrated in **Figure 11.14**. *Global horizontal irradiation* is the total solar radiation incident on a horizontal surface. It is the sum of direct normal irradiation, diffuse horizontal irradiation, and ground-reflected radiation.

Irradiance is the measure of instantaneous solar energy on a unit surface area (i.e., in W/m^2), whereas another solar energy term, insolation, is the measure of cumulative solar energy on a unit surface area over a unit amount of time (i.e., in kWh/m^2). Hence,

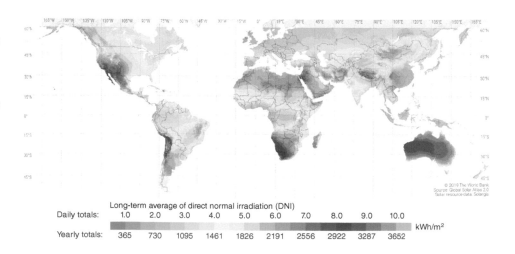

FIGURE 11.12 Global direct normal irradiation (DNI). Source: Global Solar Atlas 2.0, a free, web-based application developed and operated by Solargis on behalf of the World Bank Group, utilizing Solargis data, with funding provided by the Energy Sector Management Assistance Program (ESMAP).

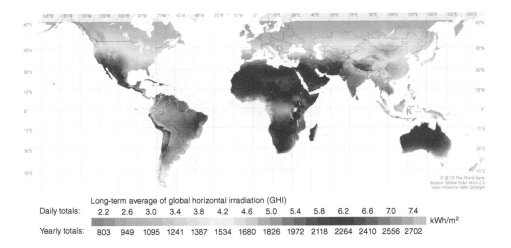

FIGURE 11.13 Global horizontal irradiation (GHI). Source: Global Solar Atlas 2.0, a free, web-based application developed and operated by Solargis on behalf of the World Bank Group, utilizing Solargis data, with funding provided by the Energy Sector Management Assistance Program (ESMAP).

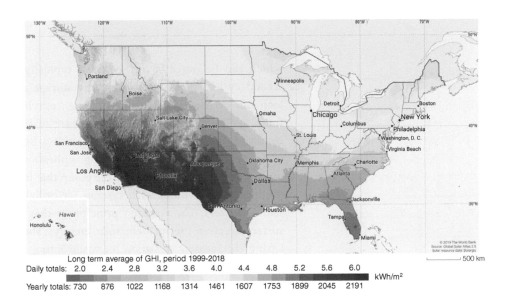

FIGURE 11.14 Global horizontal irradiation (GHI) for the United States. Source: Global Solar Atlas 2.0, a free, web-based application developed and operated by Solargis on behalf of the World Bank Group, utilizing Solargis data, with funding provided by the Energy Sector Management Assistance Program (ESMAP).

irradiation can be considered as power, and insolation can be thought of as energy. The insolation value for a project depends on several factors, some of which can be determined upfront. These factors are:

- Geographic location
- Time of year
- Time of day

The theoretical significance of these aspects is discussed in **Section 11.2.1**. Another factor affecting insolation is the weather conditions. As this parameter cannot easily be predicted, insolation can be obtained based on historical weather data to get average values.

11.1.3 Active Solar Thermal Applications

Solar energy is converted into thermal energy by active solar thermal systems for hot water or space heating purposes. The transfer of heat is achieved with a working fluid

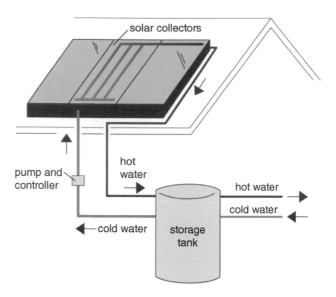

FIGURE 11.15 Basic components of a solar thermal residential heating system. Source: U.S. Energy Information Administration, www.eia.gov/energyexplained/solar/solar-thermal-collectors.php

which can be water or air. Water is typically used for hot-water applications. If a secondary loop along with an energy storage unit is involved, water in the primary loop can be mixed with antifreeze (e.g., ethylene glycol) in colder climates to avoid freezing. In such cases, the dilution rate of the antifreeze depends on the external conditions. Air is a more suitable working fluid for space heating in residential or commercial buildings, or crop drying applications in agriculture. The active component in these systems is a fan, unlike the pump that is used with systems employing water as the heat transfer fluid. Such systems for utilizing the energy from the Sun are called *solar collectors*. The working principle is that a dark color (for higher absorptivity) surface receives the incident solar energy. There might be glazing systems used for letting the incident radiation in, while acting as a barrier for heat dissipation to the surroundings which can take place by means of radiation and convection. The idea here is analogous to check valves in fluid systems, where the valve allows the fluid to flow in one direction only. Beneath the absorber plate, a fluid flow system is placed. If the heat transfer fluid is water, then tubing will be used. If the working fluid is air, then air-flow channels are implemented underneath the absorber. The sides and bottom of the whole system are insulated so that the heat leakage is minimized. Basic components used in a residential solar thermal heating system are given in **Figure 11.15**. The schematic includes the solar collector, a circulation pump and controller, supply and return lines, and the energy storage tank. The storage tank can utilize enhanced applications such as use of phase change materials (PCMs). A wide variety of energy storage options are discussed in Chapter 13.

Solar collectors can be categorized based on how they receive the sunlight and how the heat transfer mechanism works. Five common types of collectors are unglazed, flat-plate, evacuated tube, transpired, and concentrating solar collectors. These configurations are illustrated in **Figure 11.16**.

Unglazed solar collectors do not have a glazing system. They are used in applications that do not require high temperatures such as swimming pools and showers. As the temperatures are not high, heat loss is not a concern. In fact, on hot summer days, the surrounding air temperature can be higher than the collector temperature, hence contributing to heat input by convection. Unglazed systems are more cost-effective than other types.

FIGURE 11.16 Collector types used in solar thermal applications. Source: U.S. Environmental Protection Agency, www.epa.gov/rhc/solar-heating-and-cooling-technologies.

Unglazed Solar Collector

1. **Sunlight:** Sunlight hits the dark material in the collector, which heats up.

2. **Circulation:** Cool fluid (water) or air circulates through the collector, absorbing heat.

3. **Use:** The warmer fluid is used for applications such as pool heating.

Flat-Plate Solar Collector

1. **Sunlight:** Sunlight travels through the glass and hits the dark material inside the collector, which heats up.

2. **Heat reflection:** A clear glass or plastic casing traps heat that would otherwise radiate out. This is similar to the way a greenhouse traps heat inside.

3. **Circulation:** Cold water or another fluid circulates through the collector, absorbing heat.

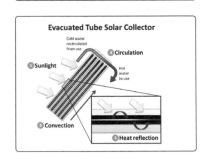

Evacuated Tube Solar Collector

1. **Sunlight:** Sunlight hits a dark cylinder, efficiently heating it from any angle.

2. **Heat reflection:** A clear glass or plastic casing traps heat that would otherwise radiate out. This is similar to the way a greenhouse traps heat inside.

3. **Convection:** A copper tube running through each cylinder absorbs the cylinder's stored heat, causing fluid inside the tube to heat up and rise to the top of the cylinder.

4. **Circulation:** Cold water circulates through the tops of the cylinders, absorbing heat.

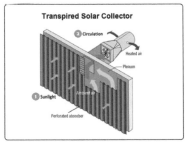

Transpired Solar Collector

1. **Sunlight:** Sunlight hits the dark perforated metal cladding, which heats up.

2. **Circulation:** A circulation fan pulls air through the perforations behind the metal cladding, heating the air, which is then pulled into the building for distribution.

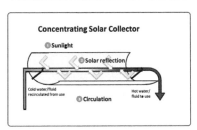

Concentrating Solar Collector

1. **Sunlight:** Sunlight hits a reflective material (i.e., a mirrored surface), usually in the shape of a trough (shown here) or a dish.

2. **Solar reflection:** The reflective material redirects the sunlight onto to a single point (for a dish) or a pipe (for a trough).

3. **Circulation:** Cold water or a special heat transfer fluid circulates through the pipe, absorbing heat.

Flat-plate solar collectors are similar to unglazed collectors, plus a glazing material atop the absorber plate. The glazing material can be glass or plastic. Glass comes with higher cost and damage risk under certain external conditions such as hail. It does, however, retain its optical properties unlike plastic. Although plastic material is less expensive, discoloration (yellowing) and degradation of plastic glazing results in reduced transmissivity. Another analogy to fluid systems in this case would be the fouling of pipes which impedes the flow of the fluid.

FIGURE 11.17 An evacuated tube solar collector residential application in Alibey village atop Mt. Ilgaz in Central Turkey. Another solar collector system can be seen on top of a neighboring house in the background. Source: Serdar Celik.

Evacuated tube solar collectors utilize a heat transfer mechanism similar to one in tube-in-tube heat exchangers. The outer tube is made of glass, hence transmitting light in. The inner tube, which is concentrically placed inside the outer, can be made of a metal such as copper. These tubes can include fins and a selective absorber coating for enhancing heat transfer. The space between the outer glass tube and the inner tube is a vacuum, which reduces heat loss by means of convection. Efficiency of these systems is typically higher than that of flat-plate solar collectors. An evacuated tube solar collector application in Alibey, a village of Ilgaz in Central Turkey, can be seen in **Figure 11.17**.

Transpired solar collectors involve a perforated absorber through which air can flow. A fan placed behind the absorber plate pulls the air through the perforations of the heated surface into the plenum, and then directs it to the space to be heated. This is similar to a draw-through type air handling unit in HVAC (heating, ventilating, and air-conditioning) applications. The absorber plate can have fins to provide an extended surface area for increased heat transfer rates.

Concentrating solar collectors can provide higher temperatures than the flat-plate collectors, which are bounded by temperatures below the boiling temperature of water and provide a limited variety of applications. Concentrating solar collectors can enable applications such as solar cooking, chemical production, water desalination, and electricity generation. They comprise a *receiver*, where the radiation is absorbed, and a *concentrator*, which directs the beam radiation onto the receiver. Three concentrating solar collector configurations are parabolic troughs, parabolic dishes, and central receivers. These systems are discussed in **Section 11.1.5** as they can be used for electricity generation.

11.1.4 Passive Solar Thermal Applications

Passive solar thermal systems provide heat transport by passive means only, meaning transfer of heat without any turbomachines such as pumps or fans, or any other device that requires energy input. Heat transfer occurs in the modes of thermal radiation, natural convection, and conduction.

FIGURE 11.18 Convex glass domes called *elephant eyes* have been used in Turkish baths for centuries. These domes were used to provide sufficient lighting into the bath chamber while covering up the bathhouse from the outside. Source: Izzet Keribar/Stone via Getty Images.

Passive solar buildings have been designed and used by several civilizations in the past. Neolithic Chinese built their houses facing south to combat the harsh winter conditions. Romans had their *heliocaminus* (sun furnace) baths. Inspired by the first-century architect and author *Vitruvius*, buildings in the Roman Empire were designed to benefit from solar heating and reduce fuel consumption. Ancient Greeks benefitted from solar energy by passive means in their houses and baths. Socrates, the famous Greek philosopher, was also an observer of the Sun's motion and concluded that use of protruding structures (overhangs in modern architecture) can benefit buildings in summer and winter. Turkish baths, called *hammams*, have also been utilizing solar energy by passive means for heating and daylighting (**Figure 11.18**) for centuries.

In terms of space heating, passive solar thermal applications can be categorized based on the way heat is stored and used. The three methods of harvesting solar energy by passive means to heat a space are direct gain, indirect gain, and isolated gain. Schematics for each method of heating are illustrated in **Figure 11.19**.

Direct gain refers to the penetration of the Sun's rays through fenestration systems and absorption of the heat by the space directly. Energy in the form of heat dissipates within the space by means of natural convection and radiation. Conduction heat transfer occurs within the thermal mass which is the flooring. It is important for the floors to be uncarpeted for the direct heat gain method to work effectively. The advantages of direct heat gain applications include lower costs and faster heating early in the day. A disadvantage of these applications is that sometimes the space can overheat during peak hours. Another drawback is the flooring material and furniture being exposed to ultraviolet radiation, which is a major damaging factor alongside humidity and artificial light.

Indirect heat gain occurs when solar energy is absorbed by a thermal mass initially, and then dissipated into the space to be heated. This is a delayed heating process and can heat the surface for hours even after the Sun is down. A common application of indirect heat gain systems is a Trombe wall, which is alternatively called a solar wall or thermal storage wall as it operates as a thermal energy storage unit. These walls are dark colored structures which are placed on the Equator-facing side of the building envelope. They are placed about 15 cm apart from the glazing unit and they have vents to enable air flow through them. The narrow gap between the glazing and the wall allows the thermosyphon effect to circulate air through the gap into the space to be heated. Based on the thickness of the wall (thermal mass), the stored energy can be radiated into the space 10–12 hours after the peak heat gain hours. The vents can be closed at night to prevent the reverse heat leak from the space to the outside.

FIGURE 11.19 Schematics of (a) direct gain, (b) indirect gain, and (c) isolated gain solar buildings.

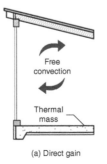

(a) Direct gain

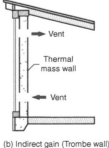

(b) Indirect gain (Trombe wall)

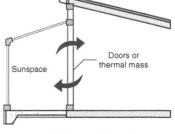

(c) Isolated gain (Sunspace)

Isolated heat gain structures store energy in a separate section. Some of the terminology to define such spaces includes *sunspaces, sunrooms, solariums,* or *solar greenhouses.* The stored energy is transferred into the house through doors, windows, or walls. These are sections added to the living space. Sunspaces act as auxiliary heating media for the buildings, hence yielding reduced energy consumption during heating seasons. Other benefits of these spaces are that they provide an atmosphere suitable for growing plants, and they can be relaxing living areas for the occupants. Depending on the season and solar radiation received, the indoor temperature in sunspaces can exceed comfort levels for people or plants. Evapotranspiration of plant canopies within the sunspaces can also impact the humidity levels in these chambers. Addition of ventilation units can help regulate temperature and relative humidity values within these spaces.

The seasonal heating needs of a building depend on several factors including the geographic location and structural components of the building. One of the methods for determining how much heating would be required for a building is use of the *heating degree days* (HDD) method. The basic idea of this approach is determining the number of degrees below a reference level on each day of the heating season. This reference level is considered as an outdoor design condition that requires no heating. In the United States, HDD is mostly tabulated based on a 65 °F outdoor temperature, which corresponds to 18.3 °C. Heating load calculations formulated on HDD are discussed in the theoretical section of this chapter. Heating degree days data for selected US and Canadian cities are listed in **Table 11.2** [3].

11.1.5 Electricity Generation from Solar Energy

Uses of solar energy in a direct manner for various needs including space heating and hot water were discussed under the active and passive solar thermal sections. There is yet another benefit from solar energy that can serve an endless number of needs in buildings, transportation, and industry: *electricity generation.* Generating power through solar energy can be achieved by two approaches. One method is conversion of energy through solar radiation into heat which is then used to achieve mechanical energy to produce electricity. This is no different than a thermal power plant in terms of working principle except for the heat source being solar radiation rather than other fuel types such as coal, natural gas, or nuclear energy. Three common applications that harvest solar energy employing heat engines are *parabolic troughs, parabolic dishes,* and *central receivers.* The second approach in generating electricity from solar energy is *photovoltaics,* which operates based on the idea of conversion of light into electricity using semiconducting materials. In this section, electricity generation applications pertaining to both heat engine and PV approaches are reviewed.

11.1.5.1 PARABOLIC TROUGHS

Parabolic trough solar collectors are formed of concave mirrors that concentrate solar rays onto a pipe that is located at the focus. The name is associated with the fact that the reflecting surfaces are curved as a parabola. Radiation is concentrated on the linear pipe through which the primary working fluid runs. This fluid is typically an oil, which can be heated up to high temperatures to generate steam by employing a heat exchanger through which water flows as the secondary fluid. The steam is then sent to the turbine to produce

Table 11.2

Average number of heating degree days and insolation for selected US and Canadian cities. Data converted into SI units [3]					
City	State	Latitude (°N)	Longitude (°W)	HDD_18.3 °C (65 °F)	Average daily insolation (kWh/m^2)
United States					
Anchorage	AK	61.22	149.9	5775	2.44
Birmingham	AL	33.52	86.8	1569	4.50
San Diego	CA	32.72	117.2	598	5.09
Key West	FL	24.56	81.8	34	5.13
Honolulu	HI	21.31	157.9	0	5.34
Chicago	IL	41.88	87.6	3583	3.92
Lexington	KY	38.04	84.5	2774	4.10
Detroit	MI	42.33	83.0	3738	3.78
Duluth	MN	46.79	92.1	5674	3.70
St. Louis	MO	38.63	90.2	2789	4.24
Raleigh	NC	35.78	78.6	1970	4.46
Concord	NH	43.21	71.5	4264	3.90
Albuquerque	NM	35.08	106.7	2423	5.61
Buffalo	NY	42.89	78.9	3734	3.68
Akron	OH	41.08	81.5	3446	3.80
Charleston	SC	32.78	79.9	1227	4.61
Memphis	TN	35.15	90.0	1726	4.63
San Antonio	TX	29.42	98.5	933	4.97
Seattle	WA	47.61	122.3	2704	3.35
Madison	WI	43.07	89.4	4169	3.94
Canada					
Calgary	Alberta	51.04	114.1	5108	3.85
Vancouver	B. Columbia	49.28	123.1	2926	3.26
Toronto	Ontario	43.65	79.4	4066	3.88
Montreal	Quebec	45.5	73.6	4575	3.76

FIGURE 11.20 SkyTrough® parabolic solar collector site near Barstow, California. Source: U.S. Department of Energy, Office of Energy Efficiency and Renewable Energy.

FIGURE 11.21 Parabolic dishes at the White Cliff Solar Thermal Power Station in Australia. The station was in operation from 1982 to 2004. It was Australia's first solar power station and one of the world's first commercial solar power stations. Electricity was generated with a uniflow steam engine, reaching up to a capacity of 25 kWe. Source: Andrew Merry/Moment via Getty Images.

mechanical work and hence generate electricity. The parabolic troughs can rotate to track the Sun and focus the radiation on the fluid pipe. **Figure 11.20** shows the concave troughs and the pipe at the SkyTrough® parabolic solar collector site in California. This system employs weatherproof, high-reflectivity polymer films instead of conventionally used glass-based mirrors, which reduces the cost and weight of the project.

11.1.5.2 PARABOLIC DISHES

Parabolic dishes operate similarly to parabolic troughs with the geometric difference of being in a bowl-shape rather than a concave surface. Therefore, instead of focusing solar radiation on a line like parabolic troughs do, parabolic dishes concentrate sunlight on a single point. At this point, the received energy can be utilized in two different ways to generate electricity. The first approach is generating steam with a working fluid and then sending it to a turbine to produce electricity, as in parabolic trough applications. The second approach is a more direct method. A Stirling engine that is placed at the focal point of the dish receives the heat. The working fluid sealed within the cylinders converts thermal energy into kinetic energy, which in turn is used for electricity generation. Some of the parameters that affect the power output of these systems are received solar radiation at the reflector concentrator, and material, diameter, aperture area, and focal length of the concentrator. Parabolic dishes at the White Cliff Solar Thermal Power Station in Australia are illustrated in **Figure 11.21**. Each dish comprises more than 2000 mirrors. Each bowl-shaped concentrator has 5 m diameter, giving a surface area of approximately 20 m^2 per dish.

FURTHER LEARNING

Solar Energy Factsheet Archives
www.seia.org/seia-factsheet-archive

FURTHER LEARNING

Photovoltaic Geographical Information System (PVGIS)
https://ec.europa.eu/jrc/en/pvgis

FIGURE 11.22 An aerial view of a heliostat in the desert in Nevada. Source: Mlenny/E+ via Getty Images.

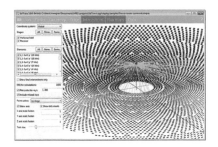

FIGURE 11.23 SolTrace is software developed by NREL for modeling concentrating solar power (CSP) systems and analyzing their performance. Source: NREL.

11.1.5.3 CENTRAL RECEIVERS

Central receivers are similar to parabolic dishes. The only difference is that instead of having small mirrors or reflectors on a bowl-shaped dish, there are mirrors positioned on a planar surface (ground) and each can be individually controlled to track the Sun. Solar energy is concentrated by the heliostat to the central receiver unit which is located at the top of a power tower. That is why these systems can also be referred to as power tower concentrating solar collectors. The heliostat field consists of many individual mirrors that can track the Sun to compensate for the Sun's apparent motion, and hence direct the sunlight to the central receiver. Besides power generation applications, heliostats can also be used for daylighting or solar cooking purposes where the system is simpler, and the number of mirrors used is much smaller than in those used for power generation. An aerial view of a heliostat field and a power tower in Nevada can be seen in **Figure 11.22**.

FIGURE 11.24 A solar cooking stove in a village in Nepal. Source: John Elk III/The Image Bank via Getty Images.

11.1.5.4 PHOTOVOLTAIC SYSTEMS

The term *photovoltaic* (PV) is a combination of two words, *photo* which comes from the Greek word *phōs* (meaning light), and *voltaic* which is named after the Italian physicist *Alessandro Volta*, who is known as the developer of the first modern electric battery. (The world's first electrochemical battery is believed to be the *Baghdad Battery* which was discovered in a village near Baghdad in the Mesopotamia region and is claimed to be a Sumerian artifact dating back to 250 BC. The clay jar had a copper disk at the bottom which was sealed with bitumen or asphalt.) A PV cell is a system that converts photon energy into electrical energy by use of semiconducting materials. Photovoltaic systems can be used in a wide range of applications, from small-scale electrical uses to large-scale power generation. The main categories of PV systems are illustrated in **Figure 11.25**.

Stand-alone (or *off-grid*) systems operate without a connection to the utility grid. These systems may or may not need energy storage components (batteries). Applications where excess energy is stored when there is no need for it and used at later times when sunlight is not available require batteries. These systems can be connected either directly to DC loads, or to an AC load through *inverters*. If the supply and demand are close to each other, then there will not be a need for batteries. Such systems are directly connected to the load and are called *direct-coupled* PV systems.

They cost less than the systems with storage as the need for batteries and *charge controllers* is eliminated with this option.

Grid-connected systems, on the other hand, are coupled to the grid. These applications mostly do not need batteries for large-scale production. Clients can reduce their bills by relying on solar energy and utilize electricity from the grid when energy from the PV system is not sufficient. Systems that are connected to the grid can be a piece of a bigger picture where other renewable and non-renewable energy sources are involved along with system controllers, transmission and distribution lines, and energy storage technologies which make up *smart grids*.

Hybrid PV systems combine PV systems with another electricity generating source. Some of these alternatives include diesel generators, micro-hydropower systems, wind turbines, and fuel cells. The main advantage of hybrid applications is that the combination of solar energy along with the additional energy source helps the two sources complement each other. Hence, the total power generation from both systems experiences less fluctuation and is more stable over time.

Electricity generation by a PV cell occurs due to the *photovoltaic effect*, which was first demonstrated by *Edmund Becquerel*, a French physicist, in 1839. The PV effect is the generation of electric current when light hits a material. Absorption of light results in excitation of an electron to a higher energy state. Electrons are positioned in orbits around the nucleus. The inner orbits should be filled for electrons to be positioned on the outer bands. The outermost orbit that has an electron is called the *valence orbit* or *valence band*. Electrons in the valence band can jump into a further band if they are given sufficient energy. This further band is called the *conduction band*. The amount of energy required to excite an electron from the valence band to the conduction band is the *band gap energy*. Metals, as conductors, have overlaps in terms of electrons in both bands. Insulators on the other hand have high band gap energies, which makes the excitation of the electrons difficult. Semiconductors exhibit characteristics between conductors and insulators. They have fuller valence bands with lower band gap energy which makes them ideal for PV applications. Silicon (Si) is the most frequently used material in semiconductors. Being the second most abundant element on Earth after oxygen, it provides logistic and economic feasibility for PV cell manufacturing. Silicon has four electrons in its valence band. To enable a flow of electrons, two adjacent silicon-based layers, one with extra electrons and one with "holes" (fewer electrons) can be fabricated. One layer is doped with phosphorus to have five valence electrons (n-type silicon), while the other layer is doped with boron to have three valence electrons (p-type silicon). Hence a *p-n junction* is formed. As a photon striking the top surface of the cell breaks loose an electron, the electron moves towards

FIGURE 11.25 Classification of PV systems.

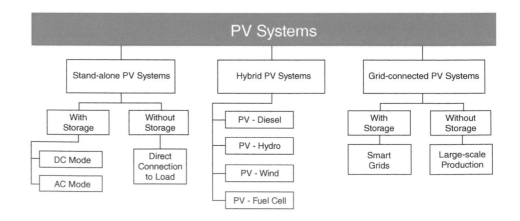

FIGURE 11.26 Working principle of a PV cell.

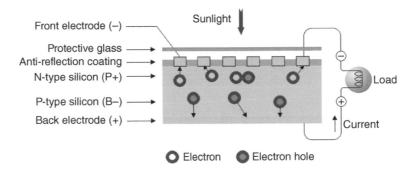

Front electrode (−)
Protective glass
Anti-reflection coating
N-type silicon (P+)
P-type silicon (B−)
Back electrode (+)

Sunlight

Load

Current

◯ Electron ● Electron hole

FIGURE 11.27 Photovoltaic cells, modules, and arrays.

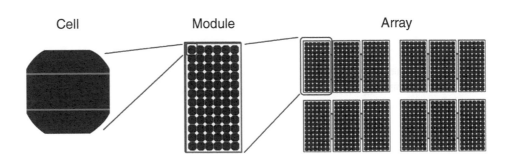

Cell   Module   Array

FURTHER LEARNING

PVWatts® Calculator by NREL
https://pvwatts.nrel.gov/

the front electrode while the electron hole travels towards the back electrode. This results in an electric current through a load that is connected to the cell. The working principle of the PV cell is illustrated in **Figure 11.26**. A PV cell is the building block of a PV system. When multiple cells are brought together in an organized manner, a module is formed. Grouping of multiple modules that are connected to each other forms arrays (**Figure 11.27**).

The performance of a solar cell when illuminated can be demonstrated by an *I–V* curve. Experimenting with a single solar cell or multiple cells interconnected in series or parallel to get the *I–V* curve does not change the characteristic of the curve. Therefore, a single cell can be tested to obtain the curve. The characteristic of the *I–V* curve depends mainly on two parameters: the irradiance and the cell temperature.

As light is the source and photons are the fuel for solar energy in photovoltaics, increasing irradiance will increase the current as well as the voltage, although the increase in voltage will be modest. Increase in cell temperature on the other hand results in a slight increase in current while the voltage rises noticeably. This increase is proportional to the temperature difference between the *nominal operating cell temperature* (NOCT) and the actual temperature of the solar cell. This fact highlights how photovoltaics and thermal management are interrelated. Cooling of a solar panel can increase its performance remarkably. The *I–V* curve characteristic changing with selected irradiance and cell temperature values can be seen in **Figures 11.28** and **11.29**, respectively.

During the daytime, the output of a PV module can change instantaneously with changes in irradiance and cell temperature, as depicted in **Figures 11.28** and **11.29**. Irradiance will change due to the position of the sun changing constantly with respect to the module. Cloudiness and possible objects in the surroundings that may cause shading or reflection (buildings, trees, etc.) can also impact the irradiance on the module.

FIGURE 11.28 *I–V* curve characteristic with increasing irradiance.

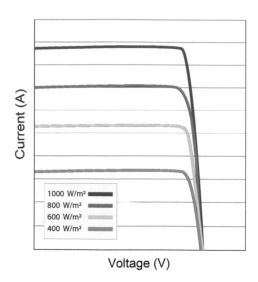

FIGURE 11.29 *I–V* curve characteristic with increasing cell temperature.

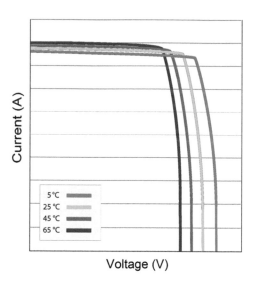

FIGURE 11.30 *I–V* and power curves for a photovoltaic cell.

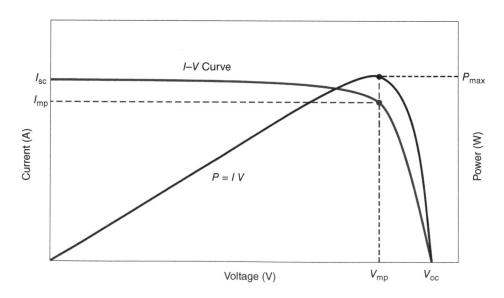

Cell temperature is another factor affecting the instantaneous performance of a solar panel. Temperature of the module can change due to similar reasons as for irradiance, including time of day, sky clearness, or objects in the immediate vicinity. Other factors such as ambient temperature and wind can also affect the cell temperature.

At a given time, under existing environmental conditions, the characteristic of the PV module is defined by its *operating point*. This point corresponds to a specific current and voltage value on the I–V curve. The desired scenario is that the operating point on the I–V curve falls onto the spot that gives the highest power output. This peak is called the *maximum power* point and is illustrated in **Figure 11.30**. The current and voltage values corresponding to the maximum power point are denoted as I_{mp} and V_{mp}. There are also two other terms that are presented on the graph: I_{sc} is the short-circuit current, which is the current that flows when there is no voltage difference between the terminals; V_{oc} is the open-circuit voltage, which is the voltage difference between the terminals recorded when there is no current flowing.

FIGURE 11.31 Engineer working on a solar power plant wearing a virtual reality (VR) headset. VR technology has multiple uses in the power sector including training, education, planning, and site management. Source: Ratchaneeyakorn Suwankhachasit/ Moment via Getty Images.

FIGURE 11.32 Thin-film solar panels being tested on reflective (TPO) and non-reflective (EPDM) roofing membranes. Thermal images of the PV panel–roofing membrane assemblies are being captured for studying the relation between surface temperature and panel performance. Source: Serdar Celik.

11.2 Theory

Theoretical analysis of solar energy involves five important topics which we discuss in this section:

- Sun's motion
- Solar radiation
- Active solar thermal applications: Solar collectors
- Passive solar thermal applications: Solar heat gain
- Photovoltaic systems

11.2.1 Sun's Motion

The power output of a PV application is a function of several parameters including total module area, efficiency, orientation, sky clearness, irradiation, location, and time. Some of these parameters can be controlled; however, some of them are uncontrollable external conditions. Understanding the Sun's motion, time, and solar angles is an important aspect of a solar energy project for estimating system performance. Sun path (or day arc) is the seasonal path of the Sun observed across the sky while the Earth rotates about its axis and orbits the Sun. The National Oceanic and Atmospheric Administration (NOAA) offers a user-friendly online solar position calculator which is a helpful tool for exploring SUN path. Time is another parameter in solar energy calculations. There are two key time terms in such calculations: standard time and local solar time. Standard time is an artificial time that has been standardized for split regions of the world to provide convenience and avoid confusion. Local solar time, on the other hand, is the exact time for a specific location that is determined based on the position of the Sun in the sky with respect to an observer on the ground at that specific location. Standard time in terms of local solar time (LST) is expressed as:

$$\text{Standard time} = \text{LST} + \left(4\frac{\text{min}}{\text{degW}}\right)(L_{\text{loc}} - L_{\text{std}}) - \text{EoT} \tag{11.1}$$

where L_{loc} and L_{std} are the local and standard longitudes and EoT is the equation of time. Standard longitudes correspond to standard time zones. As Earth completes one full rotation of $360°$ in 24 hours, then each time zone is equivalent to $15°$. Greenwich,

United Kingdom, being the reference ($0°$), any location to the east of it has longitude assigned in degrees E, and any location to the west of it is defined by degrees W longitude. Standard longitudes for US time zones are $75°$, $90°$, $105°$, and $120°$ for Eastern, Central, Mountain, and Pacific time zones, respectively.

The EoT can be calculated in different ways depending on how much accuracy is expected. Astronomical methods are very accurate (i.e., ± 2 seconds); however, they are more involved. If there is no need for such accuracy, Fourier's method is still a good alternative with competitive accuracies (i.e., ± 3 seconds) if the expansion has at least four terms. Omitting each term adds time to the errors in calculation. Using Fourier series, the EoT can be calculated as [4]:

$$\text{EoT} = a_1 \sin(\tau) + b_1 \cos(\tau) + a_2 \sin(2\tau) + b_2 \cos(2\tau) + a_3 \sin(3\tau) + b_3 \cos(3\tau)$$
$$+ a_4 \sin(4\tau) + b_4 \cos(4\tau) \tag{11.2}$$

where τ is the scaled time of the year and the coefficients of each term in the series are:

$a_1 = -7.3412$	$b_1 = +0.4944$
$a_2 = -9.3795$	$b_2 = -3.2568$
$a_3 = -0.3179$	$b_3 = -0.0774$
$a_4 = -0.1739$	$b_4 = -0.1283$

Scaled time of the year can be obtained by:

$$\tau = (360°)\left(\frac{N}{365}\right) \tag{11.3}$$

where N is the day number of the year. To determine the direction of the Sun's rays, three quantities need to be known:

- Geographic location $\rightarrow$ Latitude (l)
- Time of year $\rightarrow$ Sun's declination angle (δ)
- Time of day $\rightarrow$ Hour angle (h)

Once the latitude for the location of interest is obtained, the Sun's declination and hour angles need to be determined. Declination angle (δ) is calculated by:

$$\delta = 23.45° \sin\left[360°\left(\frac{284 + N}{365}\right)\right] \tag{11.4}$$

Hour angle (h) converts the local solar time into the number of degrees the Sun is observed to be moving across the sky during the day. It is given by:

$$h = \left(\frac{15°}{\text{hour}}\right)(\text{LST} - 12{:}00 \text{ noon}) \tag{11.5}$$

Hence, h has a negative sign in the morning, positive sign in the afternoon, and is equal to zero at noon. There are a number of solar angles that are utilized in solar energy analyses. These solar angles are listed, along with their illustrations in **Figures 11.33** and **11.34.**

- Solar altitude angle (β): angle between Sun's rays and horizontal
- Solar azimuth angle (φ): angle between north and Sun's rays' projection on the horizontal plane in the clockwise direction
- Solar zenith angle (θ_z): angle between Sun's rays and vertical ($90° - \beta$)
- Surface solar azimuth angle (γ): the angle between the solar azimuth angle (φ) and the panel surface azimuth angle (ψ), with both angles measured clockwise from north

FIGURE 11.33 Solar altitude, azimuth, and zenith angles. Source: Modified from Sandia National Laboratories, PV Performance Modeling Collaborative (PVPMC).

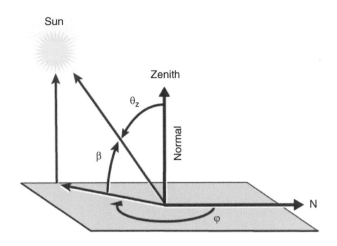

FIGURE 11.34 Incidence and tilt angles for a solar panel exposed to the Sun's rays.

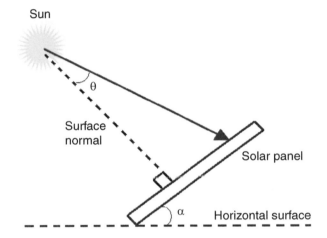

- Angle of incidence (θ): angle between Sun's rays and normal to the tilted surface
- Tilt angle (α): angle between tilted surface and horizontal

Solar altitude, solar azimuth, solar zenith, and surface solar azimuth angles as functions of pertinent angles are given in **Equations 11.6–11.9**.

$$\sin \beta = \cos l \, \cos \delta \, \cos h + \sin l \, \sin \delta \tag{11.6}$$

$$\cos \varphi = \frac{\sin \delta \, \cos l - \cos \delta \, \sin l \, \cos h}{\cos \beta} \tag{11.7}$$

$$\beta + \theta_z = 90° \tag{11.8}$$

$$\gamma = |\varphi - \psi| \tag{11.9}$$

The relation between the angle of incidence and solar altitude, tilt, and surface solar azimuth angles is given by:

$$\cos \theta = \cos \beta \sin \alpha \cos \gamma + \sin \beta \cos \alpha \tag{11.10}$$

If the panel is placed on a flat surface (e.g., thin-film solar panels on a flat roof), then the tilt angle, α, will be zero. Another special case is panels mounted parallel on vertical walls. In this scenario, $\alpha = 90°$.

Example 11.1 Determining Local Solar Time

Calculate the local solar time on February 15, 2024, in Duluth, Minnesota, when standard time is 3:30 PM.

Solution

Duluth is in Central Time Zone. In February, daylight savings time is not yet being observed. Hence, without a need to convert Central Daylight Time (CDT) to Central Standard Time (CST):

$$CST = LST + \left(4\frac{min}{degW}\right)(L_{loc} - L_{std}) - EoT$$

$L_{std} = 90°$ (Central Time Zone)
$L_{loc} = 92.1°$ (**Table 11.2**)
EoT $= -14.14$ min

EoT for February 15, 2024 can be obtained from online calculators or it can be approximated by using Phyton and solving **Equation 11.2** for $N = 46$ (day number of the year).
   Then,

$$3:30\ PM = LST + \left(4\frac{min}{degW}\right)(92.1° - 90°) - (-14.14\ min)$$

Rearranging,

$$LST = 3:30\ PM - \left(4\frac{min}{degW}\right)(2.1°) - 14.14\ min$$
$$= 3:30\ PM - 8.4\ min - 14.14\ min \cong 3:07\ PM$$

Note: *If the given date was during the time when daylight savings is observed, an hour would be subtracted from the given daylight time (CDT) to get the standard time (CST).*

Example 11.2 Finding Hour Angle for Given Times

Calculate the hour angles for the given times below.

a. 8:25 AM
b. 4:40 PM

Solution

a. At 8:25 AM:

$$h = \left(\frac{15°}{hour}\right)(8:25\ AM - 12:00\ noon) = \left(\frac{15°}{hour}\right)(-215\ min)\left|\frac{1h}{60\ min}\right|$$
$$= \left(\frac{15°}{hour}\right)(-3.583\ hr) = -53.75°$$

b. At 4:40 PM:

$$h = \left(\frac{15°}{\text{hour}}\right)(4\text{:}40\text{ AM} - 12\text{:}00\text{ noon}) = \left(\frac{15°}{\text{hour}}\right)(280\text{ min})\left|\frac{1h}{60\text{ min}}\right|$$

$$= \left(\frac{15°}{\text{hour}}\right)(4.667\text{ hr}) = 70°$$

Example 11.3 Obtaining Solar Angles

Calculate the solar altitude (β) and azimuth (φ) angles at 10:00 AM local solar time on May 21 for Dakar, Senegal.

Solution

Latitude (l) for Dakar is 14.7° N.

$$h = \left(\frac{15°}{\text{hour}}\right)(10\text{:}00\text{ AM} - 12\text{:}00\text{ noon}) = -30°$$

Sun's declination angle can be calculated using **Equation 11.4**:

$$\delta = 23.45° \sin\left[360°\left(\frac{284 + N}{365}\right)\right]$$

where May 21 is day 141 of the year ($N = 141$)

$$\delta = 23.45° \sin\left[360°\left(\frac{284 + 141}{365}\right)\right] \cong 20°$$

Now, we use **Equation 11.6** to calculate solar altitude angle:

$$\sin\beta = \cos l \cos \delta \cos h + \sin l \sin \delta$$
$$= \cos(14.7°)\cos(20°)\cos(-30°) + \sin(14.7°)\sin(20°) = 0.70$$

Then,

$$\beta = \sin^{-1}(0.70) = 44.4°$$

Next, **Equation 11.7** is used to find the solar azimuth angle:

$$\cos\varphi = \frac{\sin\delta\cos l - \cos\delta\sin l\cos h}{\cos\beta}$$
$$= \frac{\sin(20°)\cos(14.7°) - \cos(20°)\sin(14.7°)\cos(-30°)}{\cos(44.4°)} = 0.174$$

Hence,

$$\varphi = \cos^{-1}(0.174) \cong 80°$$

11.2.2 Solar Radiation

In the design and calculation phases of PV system projects, the daily irradiation values are crucial for approximating the capacity of the application. Solar irradiation as a function of location, time of day, and day of year needs to be known for these calculations. Tilt angle of the PV panel surface is another important parameter. In the earlier sections, we also discussed other factors such as weather conditions. Humidity, air pollution, suspended particles in the atmosphere, and cloudiness are also factors affecting PV panel performance. First, let's start with some definitions of irradiation values. These quantities have units of W/m^2.

- G_o: extraterrestrial radiation (power of the Sun at the top of Earth's atmosphere)
- G_{SC}: solar constant (amount of solar energy per unit time over a unit surface area at a mean distance of the Earth from the Sun, 1367 W/m^2)
- G: global daily radiation
- G_B: direct (or beam) radiation
- G_D: diffuse radiation
- G_R: reflected radiation

Average attenuation of solar radiation through the atmosphere depends on several parameters which are factored in a term, K_T, which is the *clearness index* expressed as:

$$K_T = \frac{G}{G_o} \tag{11.11}$$

Extraterrestrial radiation can be determined by using the solar constant value for a given location, time of day, and day of year:

$$G_o = \frac{24}{\pi} G_{SC} \left[1 + 0.033 \left(\frac{360\,N}{365} \right) \right] \left(\cos l \cos \delta \sin h_s + \frac{\pi}{180} h_s \sin l \sin \delta \right) \tag{11.12}$$

where N is the number of the day of the year, and h_s is the mean sunset hour angle in degrees. This value needs to be obtained based on the time of the year. If the given date is between fall and spring equinoxes, then:

$$h_{s1} = \cos^{-1} [-\tan l \tan \delta] \tag{11.13a}$$

If the day falls between spring and fall equinoxes, then the smaller of the two values below is used as the hour angle:

$$h_{s2} = \min \left\{ h_{s1},\ \cos^{-1}[-\tan(l-\alpha)\tan\delta] \right\} \tag{11.13b}$$

For known values of K_T and G_o, global daily radiation, G, can be determined. This value is equal to the sum of direct, diffuse, and reflected irradiances:

$$G = G_B + G_D + G_R \tag{11.14}$$

For studying solar irradiation on a PV panel with a tilt angle of α, all components at the given tilt angle should be accounted for. Then, specifying **Equation 11.14**, we get:

$$G(\alpha) = G_B(\alpha) + G_D(\alpha) + G_R(\alpha) \tag{11.15}$$

Diffuse and reflected radiation values can be calculated employing **Equations 11.16** and **11.17**. In **Equation 11.17**, ρ_g is the ground reflectivity (or *albedo*), which is the ratio of diffuse reflection of solar radiation to total solar radiation. This value varies from 0 to 1, where a black body that absorbs all light has a value of zero, and a body that reflects all incident solar radiation has a reflectivity of unity. Once the diffuse and reflected radiation values are calculated, direct radiation can be determined by subtracting the first two from the total radiation, as given in **Equation 11.18**:

$$G_D(\alpha) = G_D\left(\frac{1 + \cos\alpha}{2}\right) \tag{11.16}$$

$$G_R(\alpha) = G\,\rho_g\left(\frac{1 - \cos\alpha}{2}\right) \tag{11.17}$$

$$G_B(\alpha) = G(\alpha) - [G_D(\alpha) + G_D(\alpha)] \tag{11.18}$$

The diffuse radiation expression in **Equation 11.17** is obtained based on the isotropic sky model, which assumes that the diffuse radiation from the sky dome is uniform across the sky.

11.2.3 Active Solar Thermal Applications: Solar Collectors

The energy balance equation and its components for a flat-plate solar collector as depicted in **Figure 11.35** are provided, in order, in **Equations 11.19–11.22**:

$$\dot{Q}_u = \dot{Q}_{abs} - \dot{Q}_{loss} \tag{11.19}$$

$$\dot{Q}_u = \dot{m}c_p(T_e - T_i) \tag{11.20}$$

$$\dot{Q}_{abs} = \alpha\tau I_T A_{sc} \tag{11.21}$$

$$\dot{Q}_{loss} = UA_{sc}(T_{sc} - T_\infty) \tag{11.22}$$

In **Equation 11.22**, U is the overall heat transfer coefficient for the collector, T_{sc} is the solar collector (absorber plate) temperature, and T_∞ is the surrounding ambient temperature. Plugging the last three equations into **Equation 11.22** gives:

$$\dot{m}c_p(T_e - T_i) = (\alpha\tau I_T A_{sc}) - [UA_{sc}(T_{sc} - T_\infty)] \tag{11.23}$$

Efficiency for the solar collector is defined as the ratio of the rate of useful energy extraction to the rate of energy received by the collector:

$$\eta = \frac{\dot{Q}_u}{\dot{E}_{in}} = \frac{(\alpha\tau I_T A_{sc}) - [UA_{sc}(T_{sc} - T_\infty)]}{I_T A_{sc}} = \alpha\tau - U\frac{(T_{sc} - T_\infty)}{I_T} \tag{11.24}$$

FIGURE 11.35 Control volume analysis for a flat-plate solar collector.

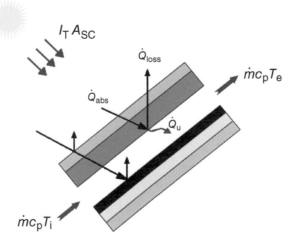

In **Equation 11.23**, both sides of the expression quantify rate of useful energy. This equation uses the absorber plate temperature. There is another way of expressing useful energy flow rate in terms of fluid inlet temperature rather than utilizing the absorber plate temperature. This is done by introducing a new term, F_R, which is the *heat removal factor*. Then:

$$\dot{Q}_u = (F_R \alpha \tau I_T A_{sc}) - [F_R U A_{sc}(T_i - T_\infty)]$$

(11.25)

The efficiency equation can be modified to include the F_R term in it as:

$$\eta = F_R \alpha \tau - F_R U \frac{(T_i - T_\infty)}{I_T}$$

(11.26)

Example 11.4 Flat-Plate Solar Collector Analysis

A residential building has a flat-plate solar collector atop its roof. Absorptivity and transmissivity values for the collector are 0.92 and 0.88, respectively. The graph provided by the manufacturer shows efficiency as a function of temperature difference per irradiance. Determine:

a. the heat removal factor, F_R (–)
b. the overall heat transfer coefficient for the collector, U (W/m^{2}K)
c. the efficiency if the water inlet temperature is 24 °C and the outside ambient temperature is 12 °C, with an irradiance value of 480 W/m^2
d. the exit temperature of water, T_e (°C), if the mass flow rate of water is 0.09 kg/s, and the solar collector area, A_{sc}, is 6 m^2 ($c_{p,\text{water}} \approx 4.18$ kJ/kg K)

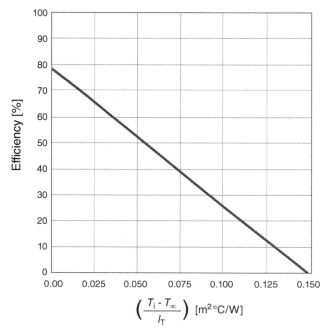

$$\left(\frac{T_i - T_\infty}{I_T}\right) \, [\text{m}^2\,°\text{C/W}]$$

Solution

a. Heat removal factor can be calculated using **Equation 11.26** for selected values from the graph:

$$\eta = F_R \alpha \tau - F_R U \frac{(T_i - T_\infty)}{I_T}$$

When the abscissa is chosen as zero, the ordinate (efficiency) value is read from the graph as 0.78.

Then,

$$0.78 = F_R\alpha\tau = F_R(0.92)(0.88) \text{ which yields } F_R = 0.96.$$

b. Overall heat transfer coefficient, U, is also determined using **Equation 11.26**. With the calculated F_R value,

$$\eta = 0.777 - 0.96U\frac{(T_i - T_\infty)}{I_T}$$

which is in the form of $y = b + mx$. The slope of this line is $-0.96U$.

By definition, the slope of the line on the graph can be expressed as:

$$\tan \alpha = \frac{\Delta\eta}{\Delta\left[\dfrac{(T_i - T_\infty)}{I_T}\right]} = -0.96U$$

Selecting two arbitrary points on the line:

$$\frac{0.78 - 0.40}{0 - 0.075} = -0.96U$$

which gives the U-value as 5.28 W/m²K.

c. Now that we know F_R and U, we can solve for the efficiency for the given values of water inlet temperature, ambient temperature, and irradiance:

$$\eta = (0.96)(0.92)(0.88) - (0.96)(5.28)\left[\frac{(24 - 12)}{480}\right] = 0.653 (65.3\%)$$

d. To find the exit temperature of water, **Equation 11.20** is used:

$$\dot{Q}_u = \dot{m}c_p(T_e - T_i)\eta = \frac{\dot{Q}_u}{\dot{E}_{in}} = \frac{\dot{m}c_p(T_e - T_i)}{I_T A_{sc}}$$

Then,

$$0.653 = \frac{\left(0.09\ \dfrac{kg}{s}\right)\left(4180\ \dfrac{J}{kg\ K}\right)(T_e - 24\,^\circ C)}{\left(480\ \dfrac{W}{m^2}\right)(6\ m^2)}$$

$$T_e = 29\,^\circ C$$

ƒ-Chart Method

There is another approach in estimating the annual performance of solar collectors. This method relies on a correlation that is developed based on large hourly data and is called the ƒ-chart method. The analysis takes two non-dimensional parameters into account. One of these parameters is the ratio of collector losses to the total heating load (space heating and hot water), and is expressed as:

$$X = \frac{\text{Total collector losses}}{\text{Total heating load}} = \frac{F'_R\,U\,A_{sc}\Delta t(T_{ref} - \overline{T}_\infty)}{L} \tag{11.27}$$

where X is the normalized heat loss parameter, F_R' is the corrected heat removal factor, Δt is the total number of seconds in the month, T_{ref} is the empirical reference temperature (100 °C), $\overline{T}_\infty$ is the monthly average ambient temperature, and L is the total heating

FIGURE 11.36 Comparative f-chart for solar collectors utilizing liquid or air as the heat transfer fluid.

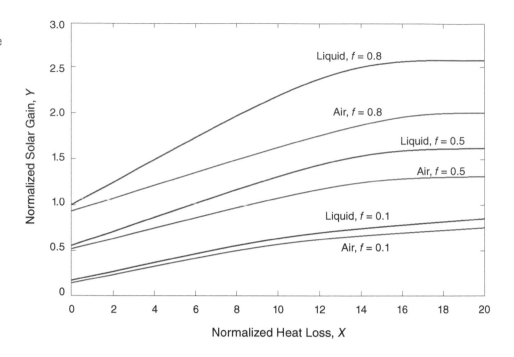

load (J). The other non-dimensional parameter is the ratio of total energy absorbed by the solar collector to the total heating load:

$$Y = \frac{\text{Total energy absorbed by collector}}{\text{Total heating load}} = \frac{F'_R \, \overline{\alpha\tau} \, A_{sc} \overline{H}_T \, N}{L} \qquad (11.28)$$

where $\overline{\alpha\tau}$ is the monthly average absorptivity–transmissivity product, $\overline{H}_T$ is the monthly average daily insolation on the collector (J/m^2), and N is the number of days in the month.

These two parameters are used to calculate f which is the monthly fraction of the total heating load supplied by solar energy. The empirical equations using the two non-dimensional parameters to provide f differ for liquid and air solar collectors.

For liquids:

$$f = 1.029Y - 0.065X - 0.245Y^2 + 0.0018X^2 + 0.0215Y^3 \qquad (11.29)$$

For air:

$$f = 1.040Y - 0.065X - 0.159Y^2 + 0.00187X^2 + 0.0095Y^3 \qquad (11.30)$$

Figure 11.36 illustrates the f-chart for different values of f-values for both liquid and air. The plots are presented on the same chart for comparison. It can be seen from the graph that, at a given f-value, the heat loss parameter is larger for air than it is for liquid heat transport media for the same amount of normalized solar heat gain.

11.2.4 Passive Solar Thermal Applications: Solar Heat Gain

The amount of heating required for a building can be calculated through different methods. This amount differs based on the outdoor and indoor design conditions, as well as on the structural properties of the building. These are discussed in Chapter 14 under Load Calculations (Section 14.2.3). As for the outdoor conditions, there is a concept that helps determine the magnitude of heating supply needed for a reference indoor design temperature. This concept is called *heating degree days* (HDD) and is determined based on the average daily outside temperature and a reference indoor temperature. Typically, the reference temperature is selected as 18.3 °C (65 °F), which is assumed to be a standard comfort temperature. Then:

$$\text{HDD}_{18.3°C} = 18.3 - T_{ave} \tag{11.31}$$

where average daily outside temperature, T_{ave}, is:

$$T_{ave} = \frac{T_{max} - T_{min}}{2} \tag{11.32}$$

Heat required by a building on a daily basis depends on its structural properties as well. These properties in terms of heat transmission are factored in an overall heat transfer coefficient:

$$Q = UA(\text{HDD}) \tag{11.33}$$

U: overall heat transfer coefficient (W/m^2 °C)
A: surface area (m^2)
HDD: heating degree days (°C day)

A building envelope consists of various surfaces with different U-values and areas. For simplicity, a global UA value for a building can be obtained. This is called the overall average heat transfer of a gross area of the exterior building envelope. The same method is applied to refrigerators to determine the heat leakage into the cabinet at given exterior (kitchen) and interior (cabinet) temperatures. The UA value used in these calculations is usually referred to as the uniform global heat transfer coefficient between external air and cabinet air. The UA value has the SI units of W/°C.

Equation 11.33 can be rearranged to yield a total heat transfer of Q as:

$$Q\,[\text{W h}] = UA\left[\frac{\text{W}}{°\text{C}}\right]\text{HDD}\,[°\text{C day}]\left|\frac{24\,\text{h}}{\text{day}}\right| = 24(UA)(\text{HDD}) \tag{11.34}$$

Load Collector Ratio (LCR) Method

A common method in investigating the performance of passive solar systems is the *load collector ratio* (LCR) method. This method relies on a non-dimensional energy ratio, *solar savings fraction* (SSF), which is retrieved from LCR tables that are developed for different cities and passive solar configurations. These configurations are direct gain (DG), vented or unvented Trombe wall (TW), waterwall (WW), and sunspace (SS).

The total heating load of a building can be determined by:

$$Q_{total} = (\text{NLC})(\text{HDD}) \tag{11.35}$$

where NLC is the *net building load coefficient*, which is specific for each building and is given by:

$$\text{NLC} = (\text{LCR})(A_f) \tag{11.36}$$

In **Equation 11.36**, A_f is the fenestration area that lets sunlight into the space. The total heating load can also be determined by summation of heating supplied through utilities (natural gas, electricity, or other non-solar means) and the heating received through solar gain by passive means:

$$Q_{total} = Q_{supply} + Q_{solar} \tag{11.37}$$

Solar savings fraction (SSF) is an informative ratio which reflects the amount of energy that can be saved by solar heat gain. SSF is expressed as:

$$SSF = \frac{Q_{solar}}{Q_{total}} = 1 - \frac{Q_{supply}}{Q_{total}} \tag{11.38}$$

Hence, if a building received no solar heat gain, it would have an SSF of zero, and a building that is solely heated by solar would have an SSF equaling 1.0. Within this range, the SSF value of a building will vary for different geographic locations and building envelopes.

Example 11.5 Determining Solar Savings Fraction

Consider a house in Albuquerque, New Mexico. The house benefits from indirect heat gain utilizing a Trombe wall. Net building load coefficient is 9000 kJ/°C day. Determine:

a. total heating load (kWh)
b. SSF if $Q_{supply} = 3150$ kWh

Solution

a. For Albuquerque HDD = 2423 °C day (**Table 11.2**)

$$Q_{total} = (NLC)(HDD) = \left(9000 \frac{kJ}{°C \ day}\right)(2423 °C \ day)\left|\frac{1 \ kWh}{3600 \ kJ}\right| = 6057.5 \ kWh$$

b. SSF can be calculated using **Equation 11.38**:

$$SSF = \frac{Q_{solar}}{Q_{total}} = \frac{Q_{total} - Q_{supply}}{Q_{total}} = \frac{6057.5 - 3150}{6057.5} = 0.48$$

Hence, 48% of the total space heating needs of this house are provided by the passive solar thermal application. This percentage can increase with modifications to the fenestration and Trombe wall, as well as overall enhancement of building insulation.

11.2.5 Photovoltaic Systems

Theoretical analysis of PV systems focuses on two topics in this section. The first topic of interest is band gap analysis. The amount of energy that a photon has is given by:

$$E = hv \tag{11.39}$$

where $h = 6.626 \times 10^{-34}$ J s is the *Planck constant*, and v is frequency, which is expressed as:

$$v = \frac{c}{\lambda} \tag{11.40}$$

In this equation, $c = 2.998 \times 10^8$ m/s is the speed of light, and λ is the wavelength of light. Combining **Equations 11.39** and **11.40**, we can get the photon energy in terms of its wavelength as:

$$E = h\frac{c}{\lambda} \tag{11.41}$$

The energy that a photon requires to be able to excite an electron from the valence band should have a minimum value of the band gap. If the photon has excess energy, the energy is wasted. Therefore, the ideal case is that the photon energy is equal to the band gap, where the photon is efficiently absorbed. Photon energy is inversely proportional to wavelength. Hence, the smaller the wavelength, the more energy the photon will have, and the longer the wavelength, the less energetic the photon will be.

Example 11.6 Maximum Wavelength to Excite an Electron from the Valence Band

Cadmium telluride (CdTe) is an alternative semiconductor material for solar cells. It has a band gap energy of 1.45 eV. Find the maximum wavelength for CdTe that can excite an electron from the valence band to the conduction band.

Solution

Rearranging **Equation 11.41**, we get:

$$\lambda = \frac{hc}{E}$$

Then we apply the values and conversion factors to get the result in nanometers (nm):

$$\lambda = \frac{(6.626 \times 10^{-34}\ \text{J s})(2.998 \times 10^8\ \text{m/s})}{1.45\ \text{eV}} \left| \frac{1\ \text{eV}}{1.602 \times 10^{-19}\ \text{J}} \right| \left| \frac{10^9\,\text{nm}}{1\ \text{m}} \right| = 855\ \text{nm}$$

Hence, the striking photon needs to have a maximum wavelength of 855 nm to be able to excite an electron. Any wavelength greater than this value will result in lower energy values which will not be able to provide the band gap energy.

Another topic under theoretical analysis of PV systems is at a more macroscopic level. Current flow, voltage, fill factor, and power output of solar panels are discussed in this section. Remembering the I–V curve plotted in **Figure 11.30**, the relation between the current flow and voltage output determines the capacity of the PV system. The net current flow through the system is equal to the difference between the cell current that is generated under sunlight and a backward current, which is called the *dark current*. Net and dark current expressions are listed in **Equations 11.42–11.44**.

$$I = I_c - I_D \tag{11.42}$$

$$I_D = I_o \left[\exp\left(\frac{eV}{kT}\right) - 1 \right] \tag{11.43}$$

$$I = I_c - I_o \left[\exp\left(\frac{eV}{kT}\right) - 1 \right] \tag{11.44}$$

In these equations, I is the net current, I_C is the cell current, I_D is the *dark current*, and I_o is the *saturation current* for the cell. Dark current is a reverse leakage of electrons which can exist when there is sunlight or there are no photons hitting the semiconductor surface. Saturation current is the portion of the reverse current that is caused by the diffusion of charge carriers from the neutral layer to the junction layer. e is the charge per electron $(1.602 \times 10^{-19}$ J/V$)$, V is the voltage, k is the Boltzmann constant $(1.381 \times 10^{-23}$ J/K$)$, and T is the operating cell temperature in kelvin.

Equation 11.44 can be examined for two special cases. If $V = 0$, then $I = I_C = I_{sc}$, where I_{sc} is the *short-circuit current*. If $I = 0$, then

$$V = \frac{kT}{e} \ln \left[\frac{I_c}{I_o} \right] = V_{oc} \qquad (11.45)$$

where V_{oc} is the *open-circuit voltage*.

For known values of current and voltage, the power output of the panel is:

$$P = IV \qquad (11.46)$$

Let's take a look at **Figure 11.30** one more time. The ideal scenario would be that the current and voltage values are equal to I_{sc} and V_{oc} which would yield the theoretical maximum value. In real life, this is not achievable. However, the product of these values sets a reference that the actual maximum power can be compared to. This ratio is called the *fill factor*, which is defined as:

$$FF = \frac{I_{mp} \, V_{mp}}{I_{sc} \, V_{oc}} \qquad (11.47)$$

Temperature is an important parameter that affects the performance of a PV cell. Increasing temperature results in a slight increase in current, but a noticeable decrease in voltage. Hence a lower power output is achieved. The effect of temperature variations on a PV module can be calculated based on the temperature coefficients declared by the manufacturer. These coefficients are obtained experimentally for current and voltage, as well as directly for power. In datasheets of different manufacturers, these coefficients are provided along with electrical data for the modules. The values can be given as changes in percentage or actual value with respect to one unit degree of difference from the nominal operating cell temperature (NOCT). Convective cooling of PV modules employing coolants such as water or air can enhance the performance of panels by minimizing the temperature difference between the actual and nominal operating temperatures.

Actual electrical values of a module at a given operating temperature can be determined by using temperature coefficients for power (β_P), current (β_I), and voltage (β_V):

$$P = P_{max} \left[1 + \beta_P \Delta T \right] \qquad (11.48)$$

$$I = I_{mp} + I_{sc} (\beta_I \Delta T) \qquad (11.49)$$

$$V = V_{mp} + V_{oc} (\beta_V \Delta T) \qquad (11.50)$$

and,

$$\Delta T = T_c - NOCT \qquad (11.51)$$

Note that β_P and β_V have negative values, hence the actual power and voltage values are less than those obtained at the reference temperature when the operating temperature is higher.

Example 11.7 Determining PV Module Output at a Given Cell Temperature

A client is considering purchasing CS1U-405MS photovoltaic modules having the electrical and temperature characteristics provided below.

a. Calculate the fill factor for these panels.
b. Find the actual power output under 800 W/m² solar irradiance if the cell temperature has reached 58 °C. First obtain your result using the temperature coefficient for power only, and then compare the result to the power output calculated by using current and voltage corrections.

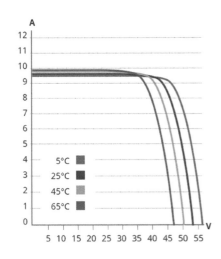

ELECTRICAL DATA | NMOT*

CS1U	400MS	405MS	410MS	415MS	420MS
Nominal Max. Power (Pmax)	296 W	300 W	304 W	307 W	311 W
Opt. Operating Voltage (Vmp)	40.8 V	41.0 V	41.2 V	41.4 V	41.5 V
Opt. Operating Current (Imp)	7.26 A	7.32 A	7.37 A	7.43 A	7.48 A
Open Circuit Voltage (Voc)	49.9 V	50.0 V	50.1 V	50.2 V	50.3 V
Short Circuit Current (Isc)	7.75 A	7.79 A	7.83 A	7.87 A	7.91 A

* Under Nominal Module Operating Temperature (NMOT), irradiance of 800 W/m² spectrum AM 1.5, ambient temperature 20°C, wind speed 1 m/s.

TEMPERATURE CHARACTERISTICS

Specification	Data
Temperature Coefficient (Pmax)	-0.37 % / °C
Temperature Coefficient (Voc)	-0.29 % / °C
Temperature Coefficient (Isc)	0.05 % / °C
Nominal Module Operating Temperature	43±3 °C

Electrical and temperature characteristics of a CS1U-405MS module. Source: Canadian Solar

Solution

a. Fill factor can be calculated using **Equation 11.47**:

$$FF = \frac{I_{mp} \, V_{mp}}{I_{sc} \, V_{oc}} = \frac{(7.32 \text{ A})(41.0 \text{ V})}{(7.79 \text{ A})(50.0 \text{ V})} = 0.77$$

b. First, let's calculate the actual power output based on the power correction as given in **Equation 11.48**:

$$P = P_o[1 + \beta_P \Delta T]$$

From the datasheet, β_P is obtained as $-0.37\%/°C$. Temperature difference is:

$$\Delta T = T_c - NOCT = 58 - 43 = 15\,°C$$

Then,

$$P = 300\,W\left[1 + \left(\frac{-0.37}{100}\right)(15\,°C)\right] = 283.4\,W$$

Now, we calculate the power as a product of corrected current and voltage:

$$I = I_{mp} + I_{sc}(\beta_I \Delta T) = (7.32\,A) + \left[(7.79\,A)\left(\frac{0.05}{100}\right)(15\,°C)\right] = 7.378\,A$$

and,

$$V = V_{mp} + V_{oc}(\beta_V \Delta T) = (41.0\,V) + \left[(50.0\,V)\left(\frac{-0.29}{100}\right)(15\,°C)\right] = 38.825\,V$$

Applying **Equation 11.46**:

$$P = IV = (7.378\,A)(38.825\,V) = 286.5\,W$$

Comparison of this corrected calculated power output to the one we obtained using **Equation 11.48** yields a difference of about 1%, which could come from rounding errors, as well as the errors within the temperature coefficient of each quantity.

11.3 Applications and Case Study

In this section, two real-world applications and a case study are covered. The applications include a solar-powered greenhouse application in Malta, and a floating solar farm in Singapore. The case study is on Noor Solar Power Station in Morocco, which has been the world's largest concentrated power plant.

Application 1 Solar-Powered Greenhouses for Sustainable Agriculture (Malta)

Solar-Powered Greenhouses
Location: Fiddien, Malta
GHI: ~ 1800 kWh/m^2/year [5]
Panel Type: Monocrystalline
Solar tracking: —
www.d-maps.com

Mediterranean Sea

Solar energy can also be used in farming and agriculture. Combining agriculture with photovoltaic systems, which is referred as *agrivoltaics*, provides benefits for both the crops and the solar panels. This symbiotic relationship provides benefits to

FIGURE 11.37 A sustainable agriculture application in Fiddien, Malta, using PV panels to power greenhouses. Source: Felix Cesare/Moment via Getty Images.

the *water–food–energy nexus*, which is pivotal for sustainable development. Greenhouses are multifunctional spaces for growing crops, collecting water, and generating electricity. The largest solar greenhouse in Malta is in Fiddien (**Figure 11.37**). It covers an area of $11,500 \, \text{m}^2$, which is equivalent to the size of one and a half soccer fields. While utilizing integrated PV systems can benefit farming and agriculture businesses in terms of lowering or eliminating utility costs and providing alternative energy sources in rural areas, the shading effects of these panels on plants should also be accounted for to avoid restraining photosynthesis conditions.

Application 2 Tengeh Floating Solar Farm (Singapore)

Tengeh Reservoir Floating Solar Farm
Location: Tuas, Singapore
GHI: ~1790 kWh/m^2/year [6]
Owner: Sembcorp Industries
Size: 122,000 panels
Capacity: 60 MWp
www.d-maps.com

Singapore Strait

Floating solar panels come with the benefit of not occupying land space. Therefore, they can be utilized in both rural and urban areas. Also, the cooling effect of the water provides lower operating temperatures for the modules, which results in enhanced efficiency. In certain applications where evaporation loss is a concern, such as dams or reservoirs, there is a mutual benefit to floating systems and the water as the panels lessen the evaporative losses due to shading, while the water enhances panel efficiency due to the convective cooling it provides. Sembcorp Tengeh Floating Solar Farm in Tuas, Singapore, is one of the world's largest floating PV applications

FIGURE 11.38 Solar construction crew assembling the PV panels onshore before attaching them to the floating arrays. Source: Roslan Rahman/AFP via Getty Images.

(Figure 11.38). It covers an area of 450,000 m^2 with 122,000 panels, offering a maximum capacity of 60 MWp. An Environmental Impact Assessment (EIA) was conducted for the project to evaluate its impact on biodiversity and water quality. The study showed that there were no significant findings of change in water quality or impact on surrounding wildlife [7]. The floating solar farm contributes towards Singapore's efforts to reduce GHG emissions.

Case Study Noor Concentrated Solar Power Plant (Morocco)

Morocco
Population: 36,561,813
   (2021 est.)
GDP: $259.42 billion
   (2020 est.)
GDP per capita: $6900
   (2020 est.)
Renewable energy world
   rank: 59
Data: CIA Factbook [8]
www.d-maps.com

INTRODUCTION Noor Concentrated Solar Power Plant, with 510 MW capacity, is the largest CSP plant in the world. The solar complex was commissioned in 2016. It covers an area of 2500 hectares, which is equivalent to the size of 3500 soccer fields. It has a parabolic mirror array, a salt-thermal storage system, and a water-cooled steam circuit. The plant is also equipped with an auxiliary diesel fuel system to maintain minimum required temperatures for the working fluid during times when the sunlight is not sufficient. The plant is located about 200 km southeast of the city of Marrakesh, near the town of Ouarzazate, which has been a spot for the film industry, including movies *The Mummy*, *Gladiator*, and *Kingdom of Heaven*.

OBJECTIVE The objective of this case study is to learn about the impacts of a high-capacity CSP plant on energy security, economy, and sustainable development in Morocco.

METHOD This case study was conducted through researching articles and reports.

RESULTS During the 21st session of the COP in Paris, Morocco announced their renewable share goal as 52% by 2030 with a mix of 20% solar, 20% wind, and 12% hydropower. The implementation of the Noor Solar Complex is a significant step towards achieving these goals and has already provided benefits at both regional and national scales. In terms of regional welfare, the plant has contributed to the development of regional infrastructure, availability of regional services, access to reliable energy, improved transportation intensity, employment opportunities, and increased local business activities. Technology improvement and increased energy security are two major benefits of the project at a national scale [9].

DISCUSSION For countries, increasing the share of total installed power capacity from renewable energy resources offers crucial benefits in terms of their energy security, economy, and sustainable development. In such investments, it is also important not to put all the eggs in one basket. Diversifying the renewable sources such as solar, wind, and hydro provides a more effective solution to the energy demand problem. In fact, an energy mix can be achieved even with the same energy source, as is the case with the power plant discussed in this case study. The plant houses both CSP and PV sites, generating electricity from both heat and light.

RECOMMENDATIONS Renewable energy initiatives, especially large-scale projects, offer energy independence and autarky, but they do entail significant budgets. Financial load is one of the challenges. Another major challenge is public acceptance. To address this issue, effective communication with the community on climate change awareness, risks and benefits of the project, transparency in the decision-making process, and the regional impacts is needed. Social acceptance of such projects paves the way for amplified benefits.

CLASSROOM DEBATE

Considering the amount of clean energy produced by solar panels and the potential fact that all panels completing their lives will go to landfill, debate the benefits and potential recycling impacts of solar panels for a rapidly growing solar energy market.

11.4 Economics

The total cost of a solar energy project, whether it is passive solar thermal, active solar thermal, CSP, or PV, is the sum of initial costs and O&M costs. For certain applications, there are incentives which are deducted from the initial costs. Both initial and O&M costs differ significantly from smaller scale projects (i.e., residential passive solar, active solar, or PV) to larger scale sites (i.e., CSP plants, land-based or floating solar farms). For passive solar applications, the major costs are for the glazing system and the thermal mass. Solar collectors are the main budget items in active solar thermal applications. On the PV side, module prices have been decreasing in an encouraging manner over the past few decades. **Figure 11.39** illustrates the average monthly module prices by different technologies and manufacturing countries between 2010 and 2020. Global weighted-average total installed costs, capacity factors, and LCOE for PV and CSP applications can be seen in **Figures 11.40** and **11.41**, respectively. For both technologies, total installed costs and LCOE values have decreased drastically in a decade. During the same time period, an increase in capacity factors is another uplifting development for the solar industry. Capacity factor is the ratio of energy generated over a time period to the installed capacity. Hence, getting more use out of the solar applications reduces the payback period as well.

FIGURE 11.39 Average monthly solar PV module prices by technology and manufacturing country sold in Europe, 2010–2020. Source: IRENA, Renewable Power Generation Costs.

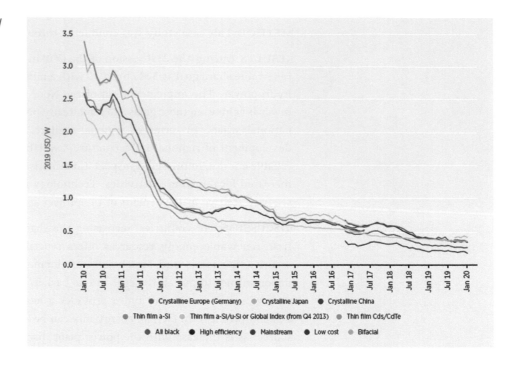

FIGURE 11.40 Global weighted-average total installed costs, capacity factors, and LCOE for PV applications, 2010–2019. Source: IRENA, Renewable Power Generation Costs.

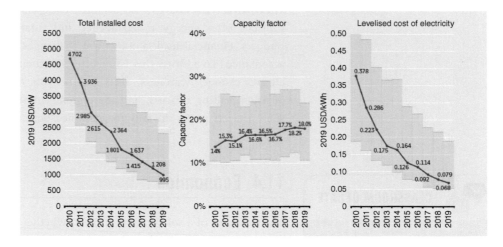

FIGURE 11.41 Global weighted-average total installed costs, capacity factors, and LCOE for CSP applications, 2010–2019. Source: IRENA, Renewable Power Generation Costs.

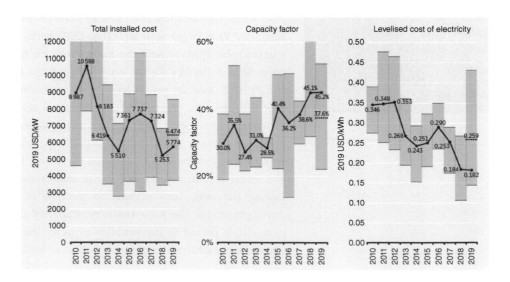

Two important measures in assessing the economics of solar energy systems are levelized cost of electricity (LCOE) and payback period. The LCOE is a helpful tool for assessing the cost competitiveness of a solar energy system. It can be calculated by:

$$\text{LCOE} = \frac{I_A + \sum_{t=1}^{n} \dfrac{OM_t}{(1+r)^t}}{\sum_{t=1}^{n} \dfrac{E_t}{(1+r)^t}(1-\delta)^t} \tag{11.52}$$

where

I_A: annualized investment costs

OM_t: operating and maintenance costs in year t

E_t: electrical energy generated in year t

r: annual effective discount rate (amount of interest paid or earned as a percentage of the balance at the end of the annual period)

n: estimated lifespan of the solar energy system

δ: system degradation rate

In Equation 11.52, the numerator gives the total cost at the present time and in the future, as its *net present value*, NPV. The denominator gives the amount of energy produced each year. As the system will be aging, a *degradation factor*, δ, is introduced into the equation to avoid underestimation of the LCOE. Annualized investment cost can be calculated with known values of the total investment cost (I_T), annual discount rate (r), and anticipated lifespan (n) of the PV system, as given in **Equation 11.53**:

$$I_A = I_T \frac{r(1+r)^n}{(1+r)^n - 1} \tag{11.53}$$

Payback period is another measure for having an understanding of the economic feasibility of a solar energy application. For PV systems, payback period can be calculated based on different approaches. Simple payback is one of the most utilized tools and is expressed as follows:

$$\text{Payback} = \frac{\text{Initial cost (\$)}}{\left[\text{Annual production}\left(\dfrac{\text{kWh}}{\text{year}}\right) \times \text{Unit price}\left(\dfrac{\$}{\text{kWh}}\right)\right] - \text{O\&M}\left(\dfrac{\$}{\text{year}}\right)} \tag{11.54}$$

While this is a straightforward method, it has some disadvantages. The method does not take into account the time value of money, changes in energy unit prices, and variable rate tariffs.

Example 11.8 Calculating Payback Period for a PV Application

A 4 kW grid-connected residential PV application has a net cost of $8648 after federal and state incentives. The cost includes the modules, inverter, array mounting parts, electrical components, labor, shipping, and taxes. Average consumption for the unit is 880 kWh/month. Unit price for electricity is 0.123 $/kWh. Annual O&M cost includes two regular maintenances which adds up to 2.5 person-hours at a rate of 75 $/h. Calculate the payback period for this PV application.

Solution

Simple payback is calculated by employing **Equation 11.54**:

$$\text{Payback} = \frac{\text{Initial cost(\$)}}{\left[\text{Annual production}\left(\frac{\text{kWh}}{\text{year}}\right) \times \text{Unit price}\left(\frac{\$}{\text{kWh}}\right)\right] - \text{O\&M}\left(\frac{\$}{\text{year}}\right)}$$

Plugging in the values given, we get:

$$= \frac{8648\ \$}{\left[(880 \times 12)\,\text{kWh} \times 0.123\,\frac{\$}{\text{kWh}}\right] - (2.5\ \text{person-h})\left(75\,\frac{\$}{\text{person-h}}\right)} = 7.8\ \text{years}$$

According to the simple payback calculation method, the break-even point for the application is approximately 7.8 years. It should be noted that the calculation does not consider the net present value and hence provides a more optimistic result.

Example 11.9 Retrieving Solar Energy Data and Plotting Graph Using Python

Using EIA data,[1] plot the monthly net energy generation from solar thermal and photovoltaic energy sources between January 2019 and October 2021 for California, Florida, Nevada, and the United States.

Solution

First, we load the libraries that will be used.

```
import matplotlib.pyplot as plt
import pandas as pd
import numpy as np
```

In the next step, we download the spreadsheet from the EIA website provided in the exercise.

```
xls = pd.ExcelFile("https://www.eia.gov/electricity/data/state/
generation_monthly.xlsx")
```

The EIA has organized each year's data in a separate worksheet. We first list these sheets in the order of years they belong to, and then we will be combining the years of interest using the **append** command.

```
tmp=0
for index in xls.sheet_names:
  print(tmp," - ", index)
  tmp=tmp+1
```

```
0  -  2001_2002_FINAL
1  -  2003_2004_FINAL
2  -  2005-2007_FINAL
```

---

[1] https://www.eia.gov/electricity/data/state/generation_monthly.xlsx

```
 3  -  2008-2009_FINAL
 4  -  2010-2011_FINAL
 5  -  2012_Final
 6  -  2013_Final
 7  -  2014_Final
 8  -  2015_Final
 9  -  2016_Final
10  -  2017_Final
11  -  2018_Final
12  -  2019_Final
13  -  2020_Final
14  -  2021_Preliminary
15  -  EnergySource_Notes
```

To plot the data between 2019 and 2021, we need to combine the worksheets starting from sheet #12. The last worksheet (#15) will not be included as it does not include data.

```
starting_sheet=12
df=pd.read_excel(xls,starting_sheet,skiprows=4)
for sheet_name in range(starting_sheet+1,len(xls.sheet_names)):
  tmp=pd.read_excel(xls,sheet_name,skiprows=4)
  df=df.append(tmp, ignore_index=True)
```

Now, let's generate a DATE column in the form of YEAR.MONTH.

```
df['DATE'] = pd.to_datetime(df[['YEAR', 'MONTH']].assign(DAY=1)).
dt.strftime('%Y.%m')
```

We also check the unique states and energy sources listed in the EIA datasheet.

```
df["STATE"].unique()
```

```
array(['AK', 'AL', 'AR', 'AZ', 'CA', 'CO', 'CT', 'DC', 'DE', 'FL', 'GA',
       'HI', 'IA', 'ID', 'IL', 'IN', 'KS', 'KY', 'LA', 'MA', 'MD', 'ME',
       'MI', 'MN', 'MO', 'MS', 'MT', 'NC', 'ND', 'NE', 'NH', 'NJ', 'NM',
       'NV', 'NY', 'OH', 'OK', 'OR', 'PA', 'RI', 'SC', 'SD', 'TN', 'TX',
       'US-Total', 'UT', 'VA', 'VT', 'WA', 'WI', 'WV', 'WY'],
      dtype=object)
```

```
df["ENERGY SOURCE"].unique()
```

```
array(['Total', 'Coal', 'Hydroelectric Conventional', 'Natural Gas',
       'Other', 'Petroleum', 'Other Biomass', 'Wind',
       'Wood and Wood Derived Fuels', 'Nuclear', 'Other Gases',
       'Solar Thermal and Photovoltaic', 'Pumped Storage', 'Geothermal'],
      dtype=object)
```

Then, we filter only the "Solar Thermal and Photovoltaic" energy source from the list.

```
df_solar=df[df["ENERGY SOURCE"]=='Solar Thermal and Photovoltaic']
```

We can now form a pivot table with index of date, columns as states and values as the sums of "Type of producers".

```
df_solar_states=pd.pivot_table(df_solar, values='GENERATION\n
(Megawatthours)', index=['DATE'], columns=['STATE'], aggfunc=np.sum)
```

Finally, we plot the monthly solar thermal and photovoltaic energy generation data for CA, FL, NV, and US total.

```
plt.figure(figsize=(13,7))
plt.xticks(rotation=90)
plt.title("Net Electricity Generation from Solar")
plt.xlabel("Date")
plt.ylabel("GWh")
plt.grid()
plt.plot(df_solar_states['NV']/1000, label="Nevada")
plt.plot(df_solar_states['CA']/1000, label="California")
plt.plot(df_solar_states['FL']/1000, label="Florida")
plt.plot(df_solar_states['US-Total']/1000, label="US Total")
plt.legend()
```

```
<matplotlib.legend.Legend at 0x7fd9ae14be90>
```

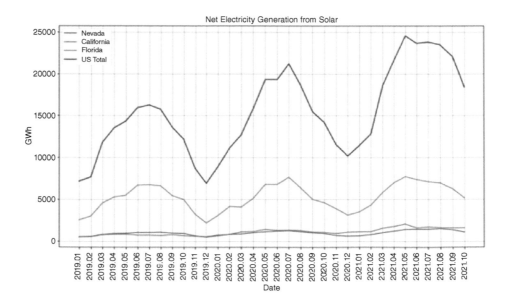

Although, the steps may seem long, this is a straightforward method for retrieving specific data from a large database with multiple worksheets. As such, the data shared by the EIA (or by any other sources) can be processed in a similar manner to yield an extensive collection of plots which can be useful for drawing conclusions for the energy industry.

11.5 Summary

Solar energy can either be harvested directly for space heating, hot water, daylighting, or power generation, or it can be utilized for electricity generation by use of the photovoltaic effect. In this chapter, a wide spectrum of solar energy systems including passive solar thermal applications, solar collectors, parabolic troughs, parabolic dishes, concentrated solar power plants, heliostats, and photovoltaic systems (land-based and floating) have been discussed. Theoretical analysis was made on the fundamentals and applications, including the Sun's motion, solar radiation, solar collectors in active solar thermal applications, solar heat gain with passive solar thermal applications, and photovoltaic systems. In the economic analysis, both LCOE and payback periods were covered. Engineers and energy professionals working on solar energy projects require clear understanding of how solar applications differ and which parameters they rely on. Utilizing the equations with acceptable assumptions is important for obtaining realistic estimates. In addition to capacity calculations, economic analysis is also important as budget is a key deciding factor in renewable energy projects for both small- and large-scale applications.

REFERENCES

[1]   REN21, "Renewables 2021 Global Status Report," 2021. Accessed: Aug. 11, 2021 [Online]. Available: www.ren21.net/wp-content/uploads/2019/05/GSR2021_Full_Report.pdf.

[2]   D. Williams, "Sun Fact Sheet," NASA, 2018. https://nssdc.gsfc.nasa.gov/planetary/factsheet/sunfact.html (accessed Aug. 11, 2021).

[3]   Oak Ridge National Laboratory, "Heating Data," ORNL, Jun. 2005. https://web.ornl.gov/sci/buildings/tools/heating-data/ (accessed Sep. 6, 2021).

[4]   F. M. Vanek, L. D. Albright, and L. T. Angenent, *Energy Systems Engineering: Evaluation and Implementation*. New York: McGraw Hill Education, 2012.

[5]   Solargis, "Solar Resource Maps of Malta." https://solargis.com/maps-and-gis-data/download/malta (accessed Sep. 9, 2021).

[6]   Solargis, "Solar Resource Maps of East Asia and Pacific." https://solargis.com/maps-and-gis-data/download/east-asia-and-pacific (accessed Sep. 9, 2021).

[7]   Sembcorp Industries, "Sembcorp and PUB Officially Open the Sembcorp Tengeh Floating Solar Farm," Jul. 14, 2021. www.sembcorp.com/en/media/media-releases/energy/2021/july/sembcorp-and-pub-officially-open-the-sembcorp-tengeh-floating-solar-farm (accessed Oct. 1, 2021).

[8]   Central Intelligence Agency, "Morocco: The World Factbook," Oct. 25, 2021. www.cia.gov/the-world-factbook/countries/morocco/ (accessed Oct. 14, 2021).

[9]   A. Laaroussi, A. Bouayad, Z. Lissaneddine, and L. A. Alaoui, "Impact study of NOOR 1 project on the Moroccan territorial economic development," *Renewable Energy and Environmental Sustainability*, vol. 6, p. 8, 2021, doi: 10.1051/rees/2021008.

[10]   International Renewable Energy Agency, "Power Generation Costs in 2019," IRENA, Abu Dhabi, 2019. Accessed: Oct. 16, 2021 [Online]. Available: www.irena.org/-/media/Files/IRENA/Agency/Publication/2020/Jun/IRENA_Power_Generation_Costs_2019.pdf.

CHAPTER 11 EXERCISES

11.1 Determine the standard daylight savings time on May 2, 2026, in San Antonio, Texas, when the local solar time is 9:29 AM.

11.2 Calculate the hour angles for the given times below:
   a. 2:35 PM
   b. 11:10 AM
   c. 4:18 PM
   d. 7:36 AM

11.3 Find the solar altitude (β) and azimuth (φ) angles at 2:30 PM local solar time on June 21 in Punta Cana, Dominican Republic.

11.4 A laundromat utilizes a flat-plate solar collector on its roof for hot water use. Both the absorptivity and transmissivity values for the collector are 0.90. The plot by the collector manufacturer depicts efficiency and temperature difference per irradiance relation as given below. Determine:
   a. the heat removal factor, F_R (–)
   b. the overall heat transfer coefficient for the collector, U (W/m²K)
   c. the efficiency if the water inlet temperature is 28 °C and the outside ambient temperature is 16 °C, with an irradiance value of 550 W/m²
   d. the exit temperature of water, T_e (°C), if the mass flow rate of water is 0.11 kg/s, and the solar collector area, A_{sc}, is 8 m² ($c_{p,water} \approx 4.18$ kJ/kg K)

Flat-plate solar collector performance plot.

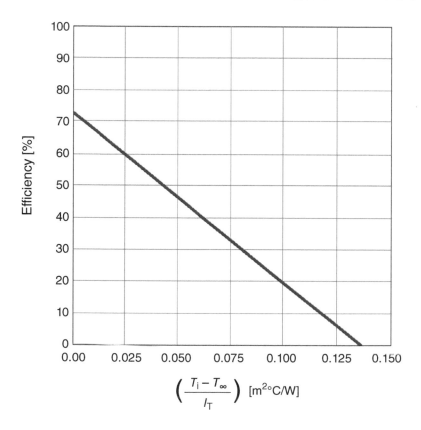

11.5 For the solar collector in Exercise 11.4, under what conditions would the highest solar collector temperature ($T_{sc,max}$) be obtained and what would that temperature be?

11.6 Using the *f*-chart for liquids and air in Figure 11.36, compare the total energy absorbed by the collector for both fluids for the *f*-value of 0.8 and normalized heat loss value of 12.

11.7 Using HDD data from Table 11.2, (a) calculate the total heating energy required by a house in Calgary, Alberta, if the *UA*-value for the house is 675 W/°C. (b) What would be the total heating load of the house if the same construction was built in Memphis, Tennessee.

11.8 A house in Charleston, South Carolina, benefits from indirect heat gain utilizing a vented Trombe wall. Net building load coefficient is 11,500 kJ/°C day. Determine:

    a. the total heating load (kWh)

    b. SSF if Q_{supply} = 1764 kWh

11.9 A thin-film solar panel has amorphous silicon (a-Si) as the semiconductor material, which has a band gap energy of approximately 1.75 eV. Determine the minimum amount of energy required to excite an electron from the valence band of the semiconductor and find the corresponding wavelength.

11.10 Consider the PV module list and the datasheet in Example 11.7. A building owner is considering having CS1U-415MS photovoltaic modules installed on their roof. Determine:

    a. the fill factor for these panels

    b. the actual power output under 800 W/m^2 solar irradiance if the cell temperature has reached 62 °C. First obtain your result using the temperature coefficient for power only, and then compare the result to the power output calculated by using current and voltage corrections.

11.11 Using NREL's PVWatts® calculator, compare the annual average solar radiation (kWh/m^2/day), total AC energy (kWh), and savings value ($) for a residential building in Tucson, Arizona, for the below orientation and tracking cases. Use default parameters. Summarize your results in a table and interpret them.

PV panel application types based on orientation and tracking. Source: EIA

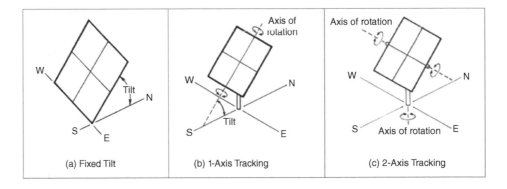

(a) Fixed Tilt   (b) 1-Axis Tracking   (c) 2-Axis Tracking

11.12 Streetlights in Everglades City, Florida, have an LED solar lighting system. The components of the system include a solar panel, a 12 V/30 Ah lithium-ion battery, and a 60 W LED lighting fixture. If the battery is fully charged and can be discharged by up to 85%, how many hours of light can this system provide in the absence of sunlight?

Solar-powered streetlights in Everglades City, Florida. Photo: Serdar Celik.

11.13 A PV power plant with an installed capacity of 70 MW has an initial investment of $80 M. Annual O&M costs are recorded as $90,000. The power station operates at 17.7% capacity factor and has a degradation of 0.05%. Assume a 7% annual effective discount rate and 25-year lifespan for the plant. Calculate the LCOE for this PV plant in $/MWh.

11.14 A 6.5 kW commercial PV application for a grocery store has a net cost of $11,054 after the incentives. Average consumption for the building is 1370 kWh/month. Unit price for electricity is 0.112 $/kWh. Annual O&M cost for the PV system is $240. Calculate the payback period for this PV application.

CHAPTER 12
Wind Energy

12.1 Introduction

Wind energy is one of the most promising and mature renewable energy technologies. Similar to hydropower, wind energy also is a result of solar energy which causes the formation of air currents in the atmosphere. Wind is caused by non-uniform heating of the Earth's surface due to different bodies making up the surface. An example of this is coastal winds. As the Sun rises, it heats the land faster than it heats the sea. Air above the land becomes less dense by heating up and rises. This results in a pressure gradient, pulling the air from over the sea towards the land. This is the *sea breeze*. After sunset, the land cools off faster than the water. Hence, the air over land becomes denser and tends to lower while the air over the sea rises. This causes an air movement from land towards the sea in the evening which is the *land breeze*. Formation of both breezes can be considered analogous to the free convection cycle of air in front of a heated or cooled vertical plate.

Energy from the wind can be extracted in different ways for various purposes. The two main applications are wind-driven locomotion (i.e., sailboats, kitesurfs) and wind-powered systems (i.e., windmills, wind turbines). Wind was used to move boats along the River Nile dating back to 5000 BC. By 200 BC, water pumps driven by wind power were utilized in China and windmills for grinding grain were in use in the Middle East. Over the following centuries, wind pump and windmill experience increased significantly in the Middle East. The technology was later brought to Europe, and then spread to the Western Hemisphere. Traditional Dutch windmills in the Netherlands, and American multi-blade turbines are some examples of this influence from both sides of the Atlantic [1].

12.1.1 Global Overview

Wind power global capacity has experienced a steady increase in the first two decades of the new millennium. A record capacity addition of 93 GW took place in 2020, mainly because of the national record additions in China and the United States. In both countries, policy changes seemed to be a key factor fueling the ascent. Global installed capacity reached 743 GW, with China, the United States, and Germany being the top countries [2]. Global wind power capacity and annual additions between 2010 and 2020 are illustrated in **Figure 12.1**. On a country basis, the top 10 nations in terms of capacity are listed in **Figure 12.2** along with the annual additions recorded. Some countries, although not in the top 10 ranking due to their comparatively lower energy production and consumption, are heavily invested in wind energy. In terms of electricity generation, wind has a share of over 58% in Denmark, 40.4% in Uruguay, and 38% in Ireland [2].

In the United States, the top states in terms of installed capacity were Texas, Iowa, Oklahoma, Kansas, and California in 2022 [3]. Texas, despite its fame with oil and natural gas, was the top wind energy producer as of 2022. Wind energy investments are

FIGURE 12.1 Global wind power capacity and annual additions. Source: REN21 Renewables 2021 Global Status Report (Paris: REN21 Secretariat).

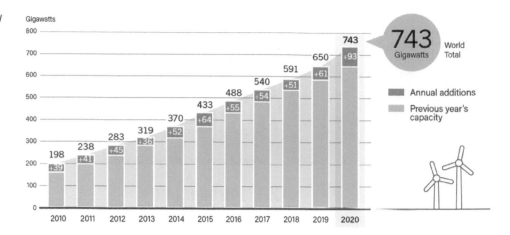

FIGURE 12.2 Top 10 countries in wind power capacity and annual additions. Source: REN21 Renewables 2021 Global Status Report (Paris: REN21 Secretariat).

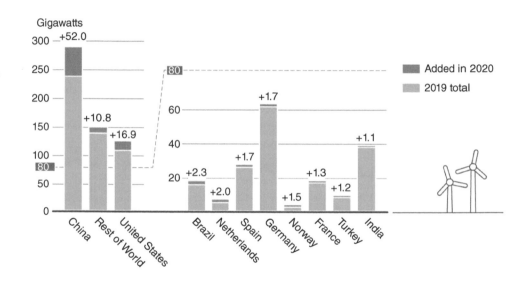

FURTHER LEARNING

WindExchange: Wind Energy Maps and Data
https://windexchange.energy.gov/
maps-data?category=land-based

FURTHER LEARNING

Wind Europe
https://windeurope.org

stimulated by the geographic and wind conditions, as well as policies in each state. The capacities for planned wind farms can be predicted based on average wind speeds over a time period. As wind speed changes with height due to the no-slip condition, hub height is another parameter that is considered in these prediction analyses. The NREL's Wind Integration National Dataset (WIND) Toolkit provides useful information on designing wind energy projects. The information includes instantaneous meteorological conditions, wind speeds, and power curves to estimate power produced at different geographic locations with varying hub heights. **Figure 12.3** shows the wind speed distribution over North and Central America, at a hub height of 100 m.

It is crucial for the wind energy industry to have some idea about the general pattern of wind speed variations during their feasibility studies. This helps investors and industry to have a better understanding of expected power generation throughout the year. Wind speed distribution can be anticipated by mathematical models. The probability of a wind speed being observed can be determined using the Weibull density function:

$$f(v) = \left(\frac{k}{c}\right)\left(\frac{v}{c}\right)^{k-1}\exp\left[-\left(\frac{v}{c}\right)^k\right]$$

(12.1)

FIGURE 12.3 Wind speed map of North and Central America generated by NREL's WIND toolkit. Source: NREL, www.nrel.gov/grid/wind-toolkit.html.

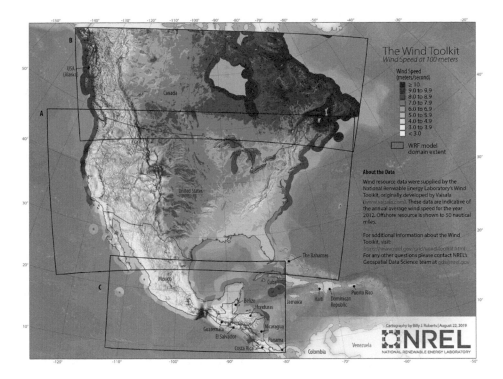

FIGURE 12.4 Weibull wind speed distribution for various shape parameters at $c = 20$ kph.

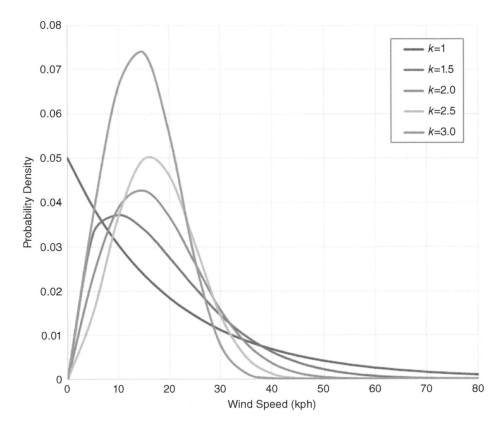

In this equation v is the wind speed for which the probability is sought, k is the dimensionless Weibull shape parameter, and c is the scale parameter. An example probability density distribution for a scale parameter of 20 kph and various shape parameters can be seen in **Figure 12.4**.

FIGURE 12.5 Projected turbine sizes and capacities for onshore and offshore wind energy applications. Source: Lawrence Berkeley National Laboratory (LBNL).

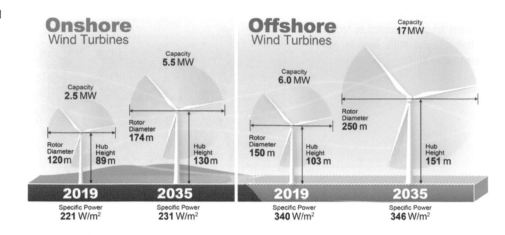

12.1.2 Onshore and Offshore Wind Energy

Wind energy applications can be built onshore or offshore. Onshore wind farms have turbines installed on land while offshore projects have wind turbines installed over the sea surface where wind flow is more powerful and more consistent. In **Figure 12.5**, a comparative scale drawing of onshore and offshore wind turbines is presented. The illustration includes the size and specific power comparisons for both technologies in the years 2019 and 2035. As can be seen from the figure, a significant increase in size and power generation capacity is anticipated for both the onshore and offshore turbines by 2035. Enhancement in offshore systems is projected to be greater according to a study led by Berkeley Lab [4].

Onshore wind energy systems have lower installation costs due to the land being easier to work on. Infrastructure costs are also lower as they can usually be connected to an existing grid. In terms of technical knowledge, there is more experience and trained personnel in this field. The cost of maintenance and repairs is also less compared to offshore applications. Onshore systems have some disadvantages as well. Space limitation can be a hurdle in certain areas where wind speeds are encouraging. Public acceptance can sometimes be an issue, especially for farms that are planned to be built close to residential areas. Even though the planning is done based on mandated regulations concerning public safety and noise problems, these projects can encounter *not in my backyard* (NIMBY) opposition from the local community. Another flaw is that, depending on the location, onshore wind turbines with lower hub heights can experience *wind shadow*, which is disturbed airflow downstream of obstacles such as buildings, trees, or hills.

Offshore projects have far lower space limitations or legislative planning issues compared to onshore applications. The winds are more predictable, consistent, and powerful. This results in higher power generation. There are no public safety, aesthetic, and noise complaints associated with these systems. They may, however, require regulations for marine safety. Offshore wind turbines can be built significantly larger than their onshore peers. This not only increases the power generation capacity, but also contributes to the cost-effectiveness of each turbine. Offshore turbines also have the advantage of eliminated wind shadows away from the coast. In terms of disadvantages of offshore wind applications, one of the drawbacks is the cost of installation and energy transmission lines. Installing these comparatively much bigger turbines over the sea requires a significant amount of planning and higher budget. Connecting the offshore wind farm to the grid also

FIGURE 12.6 Onshore wind farm atop hills in Manawatu, New Zealand. Source: New Zealand Transition/Moment Open via Getty Images.

FIGURE 12.7 Offshore wind turbines at Middelgrunden, east of Copenhagen, Denmark. Source: PARETO/E+ via Getty Images.

adds to the project costs. One other *Achilles' heel* of offshore systems is that restoring the system back can cost more and may take longer than in the case for onshore applications, following a potential system failure due to technical reasons or severe weather conditions. The interaction of offshore turbines with marine life should also be studied carefully as the impact of turbines on marine mammals and seabirds can be concerning.

12.1.3 Wind Turbine Technologies

Two basic types of wind turbines exist based on the orientation of the air flow around the axis of rotation. These types are vertical-axis and horizontal-axis wind turbines and they are discussed in the following sections.

12.1.3.1 VERTICAL-AXIS WIND TURBINES

Vertical-axis wind turbines (VAWTs) have blades that are attached to a vertical rotor. These turbines can receive air at any angle; therefore, they do not need an additional system to rotate the blades such that they face the wind. Savonius and Darrieus designs are the most common configurations of VAWTs. The Savonius design, which takes its name from the Finnish engineer *Sigurd Johannes Savonius*, consists of scoops that force the rotor to turn under the effect of wind drag. The scoops can be straight or twisted in a helical form. In **Figure 12.8**, a helical Savonius turbine can be seen. The Darrieus design is another VAWT configuration, named after the French engineer *Georges Jean Marie Darrieus*. Instead of a scooped design, Darrieus turbines consist of curved airfoil blades which generate the shaft torque. Some common types of Darrieus turbines include helical, D-type, H-type, and φ-type designs. A D-type Darrieus configuration is illustrated in **Figure 12.9**. Savonius turbines are considered to be drag systems as the rotation of the rotor occurs due to the drag force of wind into the vertical pockets. Darrieus turbines on the other hand are more of a lift system, as the rotor mainly receives the torque from the lift force of the wind. There are also hybrid designs which incorporate both Savonius and Darrieus applications in one unit. Both drag and lift forces can be significant for these systems.

12.1.3.2 HORIZONTAL-AXIS WIND TURBINES

Horizontal-axis wind turbines (HAWTs) have their axis of rotation horizontal to the ground. They can be found in many configurations. Some of these configurations include single-bladed, double-bladed, three-bladed, or multi-bladed designs. Most wind energy applications utilize HAWT designs. These systems, however, require a yaw mechanism, which can control the turbine blades so that they always face the wind. HAWTs operate primarily on the lift principle. Wind flow through the turbine blades yields a lift force which rotates the shaft. Windmills and American multi-blade turbines are also considered as HAWTs, although they may not necessarily be used for electricity generation. Multi-blade turbines also have a mechanism called a *tail vane* which turns the rotor into the wind. A group of American multi-blade turbines can be seen in **Figure 12.10**. Three-bladed rotors are the most common type in electricity generation in the wind energy industry because of their much higher efficiencies. In fact, two-bladed turbines also provide high efficiencies, although not as high as their three-bladed competitors. Two-bladed designs come with the advantage of reduced weight. This weight reduction is approximately 30%, with one of the three blades

FIGURE 12.8 A helical Savonius wind turbine in Granville, France. Source: Popolon, Wikimedia Commons author, https:// commons.wikimedia.org/wiki/File:Granville .twisted_Savonius.jpg.

FIGURE 12.9 A Darrieus wind turbine in Martigny, Switzerland. Source: Lysippos, Wikimedia Commons author, https:// commons.wikimedia.org/wiki/File:Darrieus_ rotor_martigny_ch.jpg.

FIGURE 12.10 American multi-blade turbines in Oklahoma. Source: Walter Bibikow/DigitalVision via Getty Images.

FIGURE 12.11 Maintenance of a three-blade wind turbine. Source: Billy Hustace/ Corbis Documentary via Getty Images.

being eliminated. Another advantage of two-bladed designs is that they can be transported preassembled as the blades can be rested horizontally along with the tower. The cost of constructing these turbines is also less due to reduced materials. Adding blades to turbines enhances efficiency while also adding to the weight and cost of the turbine. The reason for three-bladed turbines being more common is actually a choice of efficiency, cost, and weight balance. A three-bladed wind turbine at a wind farm undergoing maintenance is illustrated in **Figure 12.11**.

These wind turbines mainly consist of a tower, blades, and a *nacelle*. Towers are used to support the turbine structure. Increasing the tower height enables increased power generation capacity as the wind speed increases with height. Also, the turbine blade lengths can be taller with higher towers. Another advantage of higher towers is the reduced effects of turbulence. Irregular terrain, trees, buildings, or other obstacles cause eddies which affect turbine efficiency. As for offshore wind turbines, towers can be fixed to the seabed with different techniques including monopoles, tripods, semi-submersible platforms, tension legs, floating spars, or barges. Hub height, which is the distance from the ground to the center of the turbine rotor, can reach over 90 m for onshore turbines. Tower heights for offshore systems are much higher and are projected to get to 150 m by 2035 [5].

Turbine blades are the critical parts of the electricity generation process. They receive the energy from the air and transmit it to the generator. Most blades are made of fiberglass-reinforced polyester or epoxy. Alternative materials such as carbon fiber are promising since they weigh much less than fiberglass. This enables the turbine designs to have longer blades, which means capturing more energy from the wind. Increased blade length gives higher rotor diameters, which results in larger swept area, hence more power output. Some other alternative materials include *aramid* (aromatic polyamide) fibers, basalt fibers, hybrid composites, and wood.

The nacelle is a streamlined container that houses the mechanical and electrical components of the turbine. These components are the brake, low-speed and high-speed shafts, the gearbox, generator, and the controller. A yaw mechanism beneath the nacelle rotates the system so that the turbine always faces the wind as the wind direction changes. A yaw motor powers the yaw mechanism for the rotation. At the back of the nacelle, a wind vane is mounted to determine the wind direction and signal the yaw mechanism accordingly. To the back of the nacelle, there is also an anemometer placed to measure the wind speeds and transmit the measured values to the controller. The components of a wind turbine are illustrated in **Figure 12.12**.

FIGURE 12.12 Components of a three-bladed HAWT. Source: U.S. Department of Energy.

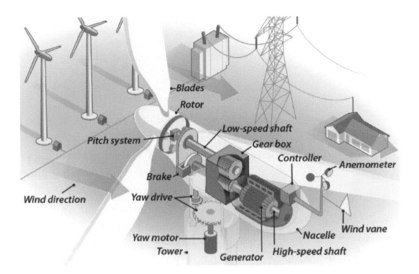

FIGURE 12.13 Phases of turbine speed control.

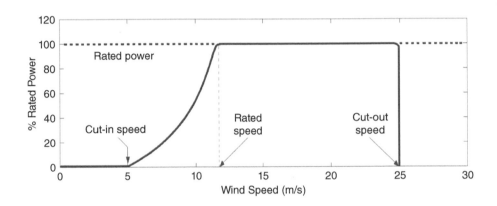

Between the rotor and the nacelle there is another critical unit, the pitch system, which controls the rotor speed. The pitch system controls how much energy the blades can harvest from the wind flow by adjusting the angle of the blades. When the wind speeds are too high, the pitch system "feathers" the blades along their longitudinal axis, changing the blades' angle of attack so that the force exerted on the blades is minimized. This action protects the blades and the rotor from being damaged due to wind speeds over the safe limits. Turbine speed control is performed at both a lower and an upper threshold. Below a certain wind speed, the turbine blades do not rotate as it is inefficient. When the wind speed reaches the *cut-in speed*, the generator is turned on and the turbine starts producing power. As the wind flow reaches the *rated speed*, the turbine generates its *rated power*. The pitch system feathers the blades at the *cut-out speed* to protect the system from damage. Phases of turbine speed control can be seen in **Figure 12.13**.

FURTHER LEARNING

Wind Energy Virtual Lab
www.youngscientistlab.com/sites/default/files/interactives/wind-energy/

12.2 Theory

The power available from wind strongly depends on the wind speed and is given by the expression:

$$P_{\text{avail}} = \frac{1}{2}\rho A V^3 \qquad (12.2)$$

where ρ is the density of air, A is the area swept by the turbine blades, and V is the wind speed.

Density of air can be calculated using the ideal gas law:

$$\rho = \frac{p}{RT} \tag{12.3}$$

where p is the absolute pressure in Pa, R is the gas constant for air in J/kg K, and T is the absolute temperature in K. Then, it becomes obvious that the power available from the wind is a function of air temperature due to its effect on the density.

Wind turbines operate by harvesting energy from the wind by slowing down the wind. For a turbine to extract all the energy from the wind, the rotor would need to be a solid disk. However, in such a case the disk would not rotate, and no energy conversion would be observed. On the other hand, if the turbine had a single blade, then most of the wind would by-pass the blade; hence a lesser amount of energy would be extracted from the wind. The *Betz limit* is the theoretical maximum efficiency for a wind turbine, which is equal to 59.3% of the power available from the wind. This ratio can be obtained using an optimizer and an actuator disk model. Hence, the maximum power that can be harvested from the available wind power is:

$$P_{\text{max}} = 0.593\, P_{\text{avail}} \tag{12.4}$$

In real-world applications, wind turbine efficiencies cannot reach the Betz limit. The efficiency values are typically within the 35–48% range. The operational power generation capacity of these wind turbines is determined by the actual power coefficient of the turbine:

$$P_{\text{actual}} = c_p\, P_{\text{avail}} \tag{12.5}$$

Power coefficient ranges differ for different wind turbine configurations. **Figure 12.14** illustrates a graph of power coefficients for various turbine designs based on their tip

FIGURE 12.14 Power coefficient vs tip speed ratio for different turbine designs.

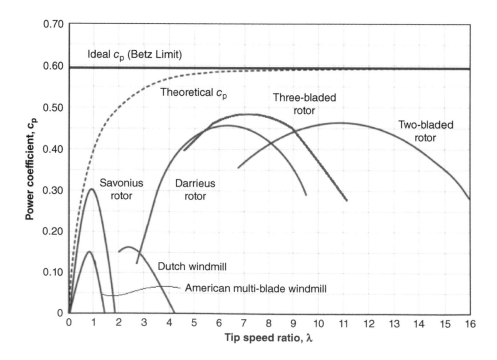

FIGURE 12.15 R&D engineering of a wind turbine involves both structural and performance analyses. Enhancing the efficiency of the wind turbine requires overall improvements across all major components of the turbine including the blades, rotor, gearbox, and the generator. Source: Andrew Merry/Moment via Getty Images.

 **FURTHER LEARNING**

OpenFAST: Wind Turbine Simulation Tool
www.nrel.gov/wind/nwtc/openfast
.html

speed ratio, λ, which is the ratio of the rotational speed of the tip of the blade to the actual wind velocity. Tip speed ratio is obtained by:

$$\lambda = \frac{r\omega}{V} \tag{12.6}$$

where r is the radius of the rotor, ω is the angular velocity of the blade (rad/s), and V is the wind velocity.

In wind energy engineering, there is another useful parameter, called *turbine specific power*, for designing wind farms. Specific power of a wind turbine is the ratio of its power rating to its rotor-swept area. This ratio basically provides information on how much power can be attained per unit area swept by the blades. Turbine specific power is given by:

$$P_{sp} = \frac{P_{cap}}{A} \tag{12.7}$$

This ratio is analogous to how much horsepower can be obtained per cc (cubic centimeter) of cylinder-swept volume in a car engine. The more efficient the engine becomes, the more horsepower can be achieved from the same amount of engine volume. This is similar for wind turbines. The more efficient the blades, rotor, and other mechanical and electrical components of the turbine are, the higher the power output that can be obtained from the same rotor diameter.

Example 12.1 Calculating Wind Turbine Power Output

A three-bladed turbine at a wind farm in Katete, Zambia, has a rotor diameter of 80 m. Air pressure and temperature are recorded as 100.9 kPa and 26 °C, respectively. For a wind speed of 38 km/h and a tip speed ratio of 7, determine:

a. the available power, P_{avail} [kW]
b. the maximum power, P_{max} [kW]
c. the actual power, P_{actual} [kW]
d. the angular velocity of the blade, ω [rpm]

Solution

a. To find the available power from the wind, first the density of air should be calculated utilizing **Equation 12.3**:

$$\rho = \frac{p}{RT} = \frac{100.9 \text{ kPa}}{(287.05 \text{ J/kg K})(26 + 273.15)\text{K}} \left| \frac{10^3 \text{ J/m}^3}{1 \text{ kPa}} \right| = 1.175 \text{ kg/m}^3$$

Then, available power can be determined employing **Equation 12.2**:

$$P_{avail} = \frac{1}{2}\rho A V^3$$

where

$$A = \frac{\pi(80 \text{ m})^2}{4} = 5026.6 \text{ m}^2$$

$$V = 38 \text{ km/h} = 10.556 \text{ m/s}$$

Then,

$$P_{avail} = \frac{1}{2}\left(1.175\ \frac{kg}{m^3}\right)(5026.6\ m^2)\left(10.556\ \frac{m}{s}\right)^3 \left|\frac{1\ kW}{10^3\ kg\frac{m^2}{s^3}}\right| = 3473.6\ kW$$

b. Maximum achievable power is determined by the Betz limit. Using **Equation 12.4**:

$$P_{max} = 0.593\ P_{avail} = (0.593)(3473.6\ kW) = 2059.8\ kW$$

c. To calculate the actual power, the c_p value needs to be obtained from Figure 12.14. For a tip speed ratio of 7, the power coefficient for a three-bladed turbine can be approximated to be 0.48. Then:

$$P_{actual} = (0.48)(3473.6\ kW) = 1667.3\ kW$$

d. Angular velocity of the blade is determined using **Equation 12.6**:

$$\lambda = \frac{r\omega}{V}$$

Rearranging the equation, we get:

$$\omega = \frac{\lambda\ V}{r} = \frac{7\left(10.556\ \frac{m}{s}\right)}{40\ m} = 1.847\ s^{-1}\left|\frac{60\ rpm}{1\ rps}\right| = 110.8\ rpm$$

Example 12.2 Turbine Specific Power Calculation

The diameter of the *London Eye* by the River Thames is 135 m. Determine the turbine specific power for a wind turbine that has the same diameter as this observation wheel and a nameplate capacity of 3.25 MW.

Solution

Turbine specific power is calculated using **Equation 12.7**:

$$P_{sp} = \frac{P_{cap}}{A} = \frac{3.25\ MW}{\frac{\pi D^2}{4}} = \frac{3.25 \times 10^6\ W}{\frac{\pi\ (135\ m)^2}{4}} = 227\ W/m^2$$

Example 12.3 Determining Wind Power at Varying Air Temperatures Using Python

At 1 atmosphere pressure (101.325 kPa), plot the available, maximum, and actual power values for a wind turbine for an air temperature range of 5–35 °C. The turbine has a rotor diameter of 40 m and is exposed to a wind speed of 42 km/h. Power coefficient for the turbine is given as 38%.

Solution

```
import matplotlib.pyplot as plt
import numpy as np

rotor_diam=40
wind_speed_kmh=42
power_coeff=38/100
p=101325

Area= np.pi * (rotor_diam**2)/4
Area
1256.6370614359173

R=287
T=1
density_air=p/(R*T)
wind_speed_ms=wind_speed_kmh*1000/(60*60)

wind_speed_ms
11.666666666666666

T=np.arange(5,35)+273.15
density_air=p/(R*T)
P_avail=(1/2)*density_air*Area*(wind_speed_ms**3)
P_max=0.593*P_avail
P_actual=0.38*P_avail

plt.grid()
plt.title("Air Temperature vs Power (MW)")
plt.xlabel("Air Temperature (C)")
plt.ylabel ("Power(MW)")
plt.plot(T-273.15,P_avail/1e6,label="P available")
plt.plot(T-273.15,P_max/1e6,label="P max")
plt.plot(T-273.15,P_actual/1e6,label="P actual")
plt.legend()

<matplotlib.legend.Legend at 0x7fccc71c8450>
```

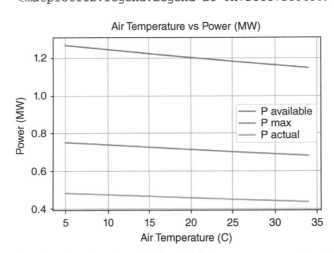

Note: *We can also plot a 3D parametric surface with colormap for power corresponding to a range of both temperature and wind speed.*

- Temperature (`T_array`) ranges from $1\,°C$ to $40\,°C$ (temperature values will be converted to kelvin to be used in the ideal gas law equation)
- Wind speed (`V_array`) ranges from 1 m/s to 20 m/s

We will plot the 3D graph using the `plot_surface` command. Note that temperature values are converted back from K to $°C$ using `T_array-273.15` in this command so that the graph shows temperature values in degrees Celsius.

```
from matplotlib import cm

T_array = np.arange(1, 40, 0.5)+273.15
V_array = np.arange(1, 20, 0.5)
V_array, T_array = np.meshgrid(V_array, T_array)

density_air=p/(R*T_array)
P_avail=(1/2)*density_air*Area*(V_array**3)

fig = plt.figure(figsize=(8,5))
ax = fig.add_subplot(projection='3d')

#ax.set_ylim(0, 80)
plt.title("Power available vs Temperature and Wind Speed")

surf=ax.plot_surface(T_array-273.15,V_array,  P_avail/1e6,cmap=cm.
coolwarm)
fig.colorbar(surf, shrink=0.5, aspect=5, label="MW")

ax.set_zlabel('Power(MW)')
plt.xlabel("Air Temperature(C)")
plt.ylabel("Wind speed(m/s)")

Text(0.5, 0, 'Wind speed(m/s)')
```

Power Available vs Temperature and Wind Speed

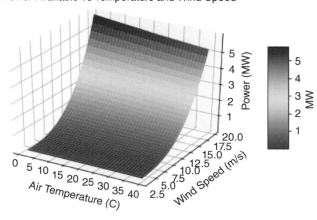

12.3 Applications and Case Study

In this section, two applications and a case study are discussed as real-world examples. The applications include Cavar onshore wind farm in Spain, and Donghai Bridge offshore wind farm in China. The case study discusses Horns Rev III wind energy site on the North Sea in Denmark.

Application 1 Cavar Onshore Wind Farm (Spain)

Cavar Wind Farm
Location: Navarre, Spain
Type: Onshore
Hub height: 101.5 m
Rotor diameter: 132 m
Number of units: 32
Nameplate capacity: 111 MW
www.d-maps.com

Wind energy has been one of the growing sectors in Spain, which was amongst the top five countries worldwide in terms of installed capacity by the end of 2021. Renewable sources accounted for 47% of its total electricity generation, with about half of it coming from wind [6]. Navarre is one of the regions in Spain that has substantial wind energy investment. Cavar wind farm complex is an example onshore wind energy project in this region with a nameplate capacity of 111 MW. The site houses 32 Siemens SG3.4-132 turbines, with each of them having a power rating of 3.4 MW. Hub height for the turbines is 101.5 m, and the rotors have a diameter of 132 m. The blade length in this application is 64.5 m, and the turbines are designed to operate at medium and high winds.

Application 2 Donghai Bridge Offshore Wind Farm (China)

Donghai Bridge Wind Farm
Location: Shanghai, China
Type: Offshore
Hub height: 91 m
Rotor diameter: 90 m
Number of units: 102
Nameplate Capacity: 102 MW
www.d-maps.com

China is the leading country in wind energy capacity by a large margin as of 2022. Having a large land area and long coastline provides significant advantages for the country. Following a 52 GW addition in 2020, it has added another record amount of 101 GW to its wind energy capacity, opening the gap with the other countries in the top list. Of this addition, 17 GW was offshore wind. By the end of 2021, China had reached a total offshore capacity of 26 GW, which is almost the half the global offshore sum of 54 GW [7]. Donghai Bridge wind farm is one of the offshore sites just outside Shanghai on the East China Sea. The farm has a nameplate capacity of 102 MW. The turbines have a hub height of 91 m, and a rotor diameter of 90 m. The complex is the first commercial offshore wind farm in China.

Case Study Horns Rev III Wind Farm (Denmark)

Horns Rev III
Location: North Sea, Denmark
Type: Offshore
Hub height: 105 m
Rotor diameter: 164 m
Number of units: 49
Nameplate Capacity: 406.7 MW
www.d-maps.com

INTRODUCTION Denmark is the leading country in terms of wind electricity per capita. Almost half of its electricity comes from wind energy. The country has invested in both onshore and offshore wind projects. Horns Rev III is one of the offshore wind farms in Denmark. It is built on the North Sea, spreading 25–40 km off the Jutland coast on the western side of the country. The site houses 49 turbines, providing a capacity of 406.7 MW. The turbines are MHI Vestas V164–8.3, with a rotor diameter of 164 m, capacity of 8.3 MW, and a hub height of 105 m [8].

OBJECTIVE The objective of this case study is to learn about an offshore wind energy application, its benefits to the environment, and its potential impacts on marine life.

METHOD This case study was conducted through researching articles and reports.

RESULTS The construction of Horns Rev III started in 2016, and its inauguration was in 2019. It is the second largest offshore wind farm in Denmark, preceded by Kriegers Flak wind farm which has a capacity of 604.8 MW. Annual production from Horns Rev III is expected to be 1,700,000 MWh which could power 425,000 homes [8]. The project was an addition to the earlier Horns Rev I and Horns Rev II farms, contributing to the offshore electricity generation of Denmark on the west coast. It has increased Danish wind energy production by 12% and serves as an important step for the 100% renewable energy goal of Denmark by 2030.

DISCUSSION The third phase of Horns Rev offshore wind farm has clearly contributed to the renewable energy mix of Denmark. Research on other wind energy sites in Denmark would show that there is a diversification in selection of wind turbines in different wind farm applications. This might be government policy or might be based solely on the selection of the companies winning the tender. This kind of a diversity can be considered an advantage as it provides enhanced experience with different turbine models.

RECOMMENDATIONS From an energy performance point of view, offshore wind energy technologies already seem to be mature. Developments in materials technology can improve the efficiency of these systems while reducing their weight and cost. Another important topic to be considered is the impact of these wind farms on marine life. Populations of native fish should be closely monitored as the implementation of turbines can interfere with certain species' habitats or their reproduction. Such studies require multidisciplinary teams involving engineers, marine biologists, and oceanographers.

CLASSROOM DEBATE

Consider a coastal region with adequate wind speeds both onshore and offshore. Would you invest on an onshore or on an offshore farm with a certain budget allocated for such a project?

12.4 Economics

Similar to other renewable energy technologies, the economics of wind power generation has two main components which are total installed costs and O&M costs. Total installed costs include feasibility, planning, design, infrastructure, grid connection, and equipment. The equipment includes the tower, blades, rotor, generator, gear box, pitch system, control units, and other parts discussed earlier in the chapter. Operational costs include utilities, wages and salaries, workers' compensation, medical insurance, and taxes. Maintenance costs are made up of routine maintenance, troubleshooting, and replacement of equipment or parts.

Global weighted-average total installed costs, along with capacity factors and levelized cost of electricity (LCOE) values for onshore and offshore wind projects between 2010 and 2020, are illustrated in **Figures 12.16** and **12.17**, respectively. In 2020, the global average for total installed cost was $1355 per kW of capacity, which is about 31% less than the costs a decade ago. A similar encouraging cost reduction is observed

FIGURE 12.16 Global weighted-average total installed costs, capacity factors, and LCOE values for onshore wind projects between 2010 and 2020. Source: IRENA Power Generation Costs in 2020.

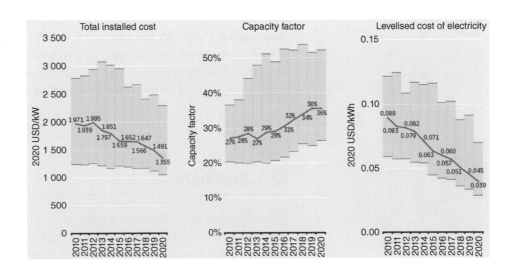

FIGURE 12.17 Global weighted-average total installed costs, capacity factors, and LCOE values for offshore wind projects between 2010 and 2020. Source: IRENA Power Generation Costs in 2020.

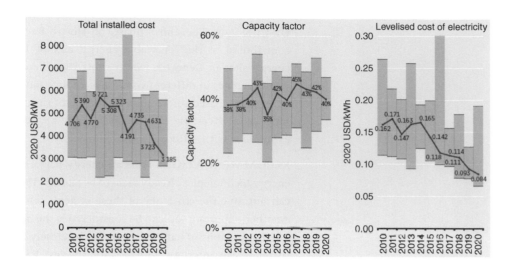

with offshore systems. About 32% reduction in total installed costs per kW of power generation was recorded within the 10-year time frame. In 2020, LCOE values for onshore and offshore wind were noted as 0.039 $/kWh and 0.084 $/kWh, respectively. These numbers are much lower than the LCOE values a decade ago [9].

The LCOE of a wind energy application is determined by:

$$\text{LCOE} = \frac{I_A + \sum_{t=1}^{n} \dfrac{\text{OM}_t}{(1+r)^t}}{\sum_{t=1}^{n} \dfrac{E_t}{(1+r)^t} (1-\delta)^t} \tag{12.8}$$

where

I_A: annualized investment costs

OM_t: operating and maintenance costs in year t

E_t: electrical energy generated in year t

r: annual effective discount rate (amount of interest paid or earned as a percentage of the balance at the end of the annual period)

n: estimated lifespan of the wind farm

δ: system degradation rate

In **Equation 12.8**, the numerator is the total cost at the present time and in the future, as its *net present value*, NPV. The denominator represents the amount of energy produced each year. As the wind farm will be aging over time, a *degradation factor*, δ, is also introduced into the equation. Annualized investment cost can be calculated using the total investment cost (I_T), annual discount rate (r), and anticipated lifespan (n) of the wind turbines, as provided in **Equation 12.9**:

$$I_A = I_T \frac{r(1+r)^n}{(1+r)^n - 1} \tag{12.9}$$

12.5 Summary

Wind energy is one of the most mature renewable energy applications and there is still ongoing improvement in wind turbine technologies. The applications can be onshore or offshore. Onshore wind has a lower LCOE than offshore projects, but they also come with lower capacity factors and power capacities. Offshore wind farms have the

FIGURE 12.18 Wind turbines blended with solar panels, forming a "mechanical garden" at the Arc Centre in Hull, United Kingdom. Source: abzee/E+ via Getty Images.

advantage of more powerful and consistent winds. There are two main types of turbine configurations: vertical-axis wind turbines (VAWTs) and horizontal-axis wind turbines (HAWTs). Darrieus and Savonius are some common VAWT designs. Single-bladed, two-bladed, three-bladed, and American multi-bladed turbines are some examples of the HAWT configuration. In this chapter, the theory of available, maximum, and actual power for wind turbines was covered. Theoretical analysis involving power coefficient, tip speed ratio, and turbine specific power was discussed, supported with in-chapter examples. Three global examples were shared as applications and a case study.

As for the future of wind energy, smart rotor technologies and innovative blade manufacturing techniques may contribute more to enhanced system performance and reduced costs. On the application side, offshore wind projects are expected to grow in number. Hybrid renewable technologies such as wind and solar energy are predicted to find more applications in buildings, industry, and the utility sector (**Figure 12.18**). Remote control and supervision of wind farms via tools such as SCADA (Supervisory Control and Data Acquisition) are essential and expected to find wider use.

REFERENCES

[1] U.S. Energy Information Administration, "History of Wind Power," EIA, 2016. www.eia.gov/energyexplained/wind/history-of-wind-power.php (accessed Jan. 25, 2022).

[2] REN21, "Renewables 2021 Global Status Report," Paris, 2021. Accessed: Jan. 25, 2022 [Online]. Available: www.ren21.net/wp-content/uploads/2019/05/GSR2021_Full_Report.pdf.

[3] U.S. Department of Energy, "WINDExchange: Wind Energy State Information," https://windexchange.energy.gov/states (accessed Jan. 26, 2022).

[4] J. Chao, "Experts' Predictions for Future Wind Energy Costs Drop Significantly," News Center, Lawrence Berkeley National Laboratory, Apr. 15, 2021. https://newscenter.lbl.gov/2021/04/15/experts-predictions-for-future-wind-energy-costs-drop-significantly/ (accessed Jan. 27, 2022).

[5] U.S. Department of Energy, "Wind Turbines: the Bigger, the Better," Aug. 30, 2021. www.energy.gov/eere/articles/wind-turbines-bigger-better (accessed Jan. 27, 2022).

[6] Yale Environment 360, "Wind Became Spain's Biggest Power Source in 2021," Yale E360, Dec. 16, 2021. https://e360.yale.edu/digest/wind-became-spains-biggest-power-source-in-2021 (accessed Jan. 30, 2022).

[7] D. Vetter, "China Built More Offshore Wind in 2021 Than Every Other Country Built in 5 Years," Forbes, Jan. 26, 2022. www.forbes.com/sites/davidrvetter/2022/01/26/china-built-more-offshore-wind-in-2021-than-every-other-country-built-in-5-years/?sh=6f5aa1ff4634 (accessed Jan. 30, 2022).

[8] Vattenfall, "Power Plants: Horns Rev 3." https://powerplants.vattenfall.com/horns-rev-3 (accessed Jan. 30, 2022).

[9] International Renewable Energy Agency, "Renewable Power Generation Costs in 2020," IRENA, Abu Dhabi, 2021. Accessed: Jan. 30, 2022 [Online]. Available: www.irena.org/publications/2021/Jun/Renewable-Power-Costs-in-2020.

CHAPTER 12 EXERCISES

12.1 Generate a table listing advantages and disadvantages of VAWT and HAWT systems against each other. Which type is more common? Explain.

12.2 Using **Equation 12.1**, plot the Weibull distribution for a scale parameter of 10 kph and shape parameters of $k = 1.5$, 2.0, and 2.5. Obtain the plot for a wind speed range of 0–75 kph.

12.3 Research and explain the relation between the Betz limit, actuator disk model, and Bernoulli's law.

12.4 Consider the turbine speed control plot in Figure 12.13. A project site has a wind velocity distribution that is approximated to be sinusoidal with a minimum of 0 m/s and a maximum of 30 m/s over a cycle of 24 hours. For a cut-in speed of 5 m/s, and a cut-out speed of 25 m/s, for how many hours of the day would the turbine blades not be rotating?

12.5 The log law is one approach in estimating wind speeds over different landscape types. Wind speed V at a height of z can be approximated by:

$$V(z) = V_{\text{ref}} \frac{\ln\left(\dfrac{z}{z_0}\right)}{\ln\left(\dfrac{z_{\text{ref}}}{z_0}\right)}$$

where V_{ref} is a known velocity value at a reference height of z_{ref}, z is the height at which the velocity is being estimated, and z_0 is the roughness length which is determined by the surface properties of the area the wind is blowing over. Plot the velocity values at hub heights of 60 m, 80 m, and 100 m, for roughness lengths of 0.0002 m (over water) and 0.0024 m (open terrain with a smooth surface) if the reference velocity is 4.5 m/s at 30 m.

12.6 Engineers are conducting a feasibility analysis for a wind energy project. Hub height for the planned turbines is 60 m. Wind speed measured at 20 m is 5.5 m/s. If the roughness length of the projected wind farm is 0.005 m, determine the wind speed at the hub height.

12.7 Determine the turbine specific power values for all four wind turbines shown in Figure 12.5.

12.8 Obtain the power coefficient values for a Darrieus turbine with a rotor diameter of 4 m, rotating at 60 rpm at wind speeds of 1.33 m/s, 1.6 m/s, and 2 m/s.

12.9 A two-bladed wind turbine with a rotor diameter of 30 m is exposed to 55 km/h wind. Air pressure and temperature are given as 1.011 bar and 23 °C, respectively. If the tip speed ratio for the turbine is 8, determine:

a. the available power, P_{avail} [kW]

b. the maximum power, P_{max} [kW]

c. the actual power, P_{actual} [kW]

d. the angular velocity of the blade, ω [rpm]

12.10 Consider Exercise 12.9. Obtain the actual power plot for the same turbine specifications and air pressure for an air temperature range of 5–35 °C.

12.11 The aerodynamic torque of a turbine extracted from the wind is given by

$$\tau = \frac{P_{\text{actual}}}{\omega}$$

Determine the torque for a wind turbine that has an available power of 180 kW and a power coefficient of 0.40 if the angular velocity of the blades is 80 rpm.

12.12 SCADA (Supervisory Control and Data Acquisition) systems are used broadly in energy applications. Research and explain how SCADA can be utilized in a wind farm and what benefits it can provide.

CHAPTER 13
Energy Storage

13.1 Introduction

Achieving sustainable development goals (SDGs) that are associated with energy relies on improvements in energy efficiency, clean power generation through conventional means, and renewable energy technologies. Yet, in achieving these goals there is another significant player in the game. The increasing contribution of renewables in total energy production and a global trend towards smart grids comes at the cost of the necessity to be able to store energy in a cost-effective manner with minimal losses possible. The need for such energy storage systems is greater than ever with the increase in energy consumption, and energy generation using both non-renewable and renewable sources. The increase in the capacity of electricity that renewable energy systems supply into the grid comes with the risk of increased uncertainty in managing the grid due to the intermittent and incalculable nature of these renewable sources. Energy storage systems become very useful in mitigating these risks. These systems can be utilized as an asset at the generation, transmission, distribution, or end-use phases. They offer a variety of benefits on the grid including higher integration capacity for renewable energy technologies, grid flexibility, reliability and resilience, and optimized utilization of the resources. A schematic of a future electrical grid and the benefits that energy storage systems can provide on the grid can be seen in **Figure 13.1**.

Energy storage systems can be categorized as mechanical, thermal, chemical, and electrical based on the physics and technology of the application. Different energy storage technologies under these categories are listed in **Figure 13.2**. Some applications can be hybrid, pertaining to different fields of physics such as electrochemical (electrical and chemical) batteries.

The whole network of electricity generation, delivery, storage, and utilization is called an *electrical grid* (or *power grid*). Depending on the area they encompass, grids can be *microgrids* or *macrogrids*. Microgrids can work in a synchronous manner with the macrogrid or operate independently when needed. Some examples of microgrids include remote areas, industrial facilities, military bases, university campuses, and islands.

Engagement of newer technologies into managing electrical grids has resulted in more energy-efficient and environmentally friendly networks called **smart grids**. Smart grids consist of computer networks, control systems, automation, communication infrastructure, and additional cutting-edge technologies and approaches to improve the electrical grid. This improvement includes benefits such as more efficient transmission and distribution of energy, reduced operational and maintenance costs for utilities, which results in lower energy costs for end-users, enhanced integration of renewable technologies, and more effective implementation of energy storage along with appropriate peak demand adjustment.

Managing energy consumption demand peaks is essential for securing grid stability. Energy demand management, also known as **demand-side management** (DSM) can be

FIGURE 13.1 Benefits energy storage systems can provide on future electrical grids. Source: U.S. Department of Energy, Office of Technology Transitions. www.energy.gov/sites/prod/files/2019/07/f64/2018-OTT-Energy-Storage-Spotlight.pdf.

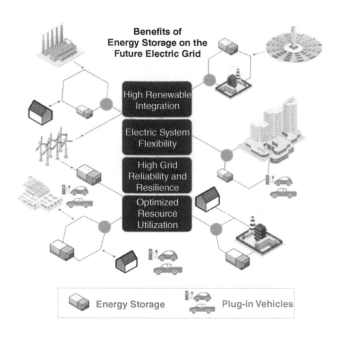

FIGURE 13.2 Categorization of energy storage systems.

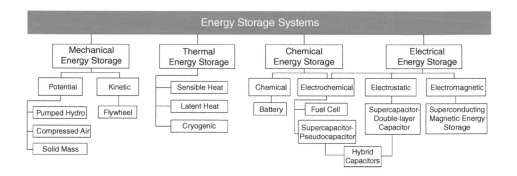

achieved using different methods such as energy efficiency or demand response. Energy efficiency is mostly an appliance-associated characteristic, while demand response involves utility companies and consumers. **Peak shaving** and **load leveling** are two common practices of demand response. In the peak shaving process, overall energy demand is reduced by means of reductions in energy consumption or use of energy stored when the load is lower. This process helps "shave" the peaks of the load and is also referred as **load shedding**. Another process of demand response is load leveling. In this process, there is no compromise in energy use. Overall energy consumption remains the same; however, the load is flattened by means of shifting it to other times of the day when it is not as high. This method requires more energy storage. From a business point of view, there is another action called *arbitrage*, which is buying energy during off-peak hours when the price is lower and selling it later when the price increases.

A successful energy storage technology needs to focus on five major criteria: (a) cost, (b) performance, (c) operational characteristics, (d) safety, and (e) environmental friendliness. Quantitative analysis of energy storage systems (ESSs) that addresses cost, performance, and operational characteristics can be performed based on the following parameters:

Table 13.1

Characteristics of selected energy storage technologies [1]						
Technology	Energy density (Wh/L)	Power rating (MW)	Storage duration	Lifetime (years)	Discharge time	Cycling times (cycles)
PHES	0.5–2	30–5000	h–month	40–60	1–24 h+	10,000–30,000
Flywheel	20–80	0.1–20	sec–min	15–20	sec–15 min	20,000
CAES	2–6	≥300	h–month	20–40	1–24 h+	8000–12,000
Capacitor	2–6	0–0.05	sec–h	1–10	millisec–1 h	50,000+
SMES	0.2–6	0.1–10	millisec–h	20–30	≥30 min	10,000+
TES	80–500	0.1–300	min–days	5–30	1–24 h+	–
Solar fuel	500–10,000	0–10	h–month	–	1–24 h+	–
H$_2$ fuel cell	500–3000	0–50	h–month	5–20	sec–24 h+	1000+
Li-ion	150–500	0–100	min–days	5–15	min–h	1000–10,000
Lead–acid	50–90	0–40	min–days	5–15	sec–h	500–10,000

Cost: cost of storing unit amount of energy ($/kWh)

Energy capacity: amount of energy that can be stored in the system

Power capacity (or *power rating*): rate of energy transfer during charging and discharging processes

Energy density: stored energy per unit volume or per unit mass

Round-trip efficiency: ratio of useful energy output during discharge to the amount of energy supplied for charging

Storage duration: energy storage time

Lifetime: lifespan of the ESS

Cycling times: number of cycles that the ESS can execute over its lifetime

Characteristics of some ESSs are listed in **Table 13.1** [1]. Average energy storage durations and system power capacities of different ESSs are illustrated in **Figure 13.3**.

The cost of energy storage systems can be quantified in terms of installed cost or levelized cost of storage. Installed costs account for all equipment and engineering, procurement, and construction (EPC) costs. Similar to levelized cost of electricity (LCOE), levelized cost of storage (LCOS) provides a fair comparison between different types of applications. The LCOS is the ratio of total cost of energy storage divided by the projected energy discharged over the entire life of the ESS. The LCOS can be calculated in a similar manner to LCOE with two differences. While the input and output are fuel cost and electricity generated in LCOE, for LCOS calculation input and output values are taken as the charging cost and discharged electricity, respectively.

$$\text{LCOS} = \frac{\sum_{t=1}^{n} \dfrac{I_t + M_t + C_t}{(1+r)^t}}{\sum_{t=1}^{n} \dfrac{E_t}{(1+r)^t}} \tag{13.1}$$

FIGURE 13.3 Average energy storage durations and system power capacities of different ESSs (pumped hydro not included due to its much larger global capacity compared to other systems). Source: Fu Ran, T. Remo, and R. Margolis. 2018. 2018 U.S. Utility-Scale Photovoltaics-Plus-Energy Storage System Costs Benchmark. Golden, CO: National Renewable Energy Laboratory. NREL/TP-6A20–71714. www.nrel.gov/docs/fy19osti/71714.pdf.

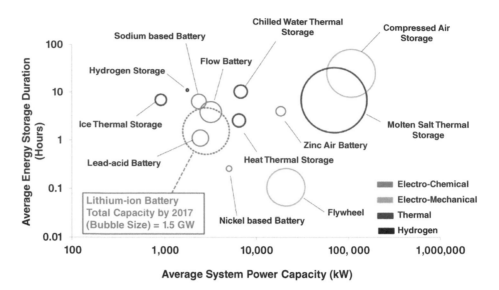

where

I_t: investment expenditure in year t

M_t: operating and maintenance expenditure in year t

C_t: charging costs in year t

E_t: electrical energy discharged in year t

r: annual effective discount rate (amount of interest paid or earned as a percentage of the balance at the end of the annual period)

n: estimated lifespan of the ESS

13.2 Mechanical Energy Storage

13.2.1 Pumped Hydroelectric Energy Storage

Pumped hydroelectric energy storage (PHES), also known as *pumped-storage hydroelectricity* (PSH), is a mechanical energy storage method based on charging water with potential energy. It is the most common energy storage technique with a big margin over the alternatives. It accounts for over 96% of global energy stored in grid-scale applications. Pumped storage has a total storage installed capacity of 158 GW, with China (30.3 GW), Japan (27.6 GW), and the United States (22.9 GW) making up half of the global capacity [2].

Pumped storage works as a "hydro battery." During times of low demand or when there is additional renewable energy generation, surplus is used to pump water up to a higher-elevation reservoir, hence granting it added potential. When the demand increases, water flows towards the lower reservoir, generating electricity while flowing through the turbine. The working principle of PHES can be seen in **Figure 13.4**. The energy storage system can be an open-loop type or a closed-loop type depending on the reservoirs' connection to other water sources. Open-loop systems have one or both reservoirs connected to a naturally flowing water source, whereas in closed-loop systems both reservoirs are independent of a free-flowing water body.

Besides the reservoirs, a PHES system consists of a penstock and a powerhouse. Penstocks include the channels carrying water and the gates controlling the water flow. The powerhouse is the enclosure that accommodates the turbine/pump assembly. Reversible turbomachines that can act as a pump during charging and as a turbine

FIGURE 13.4 Open- and closed-loop pumped hydroelectric energy storage systems. Source: U.S. Department of Energy.

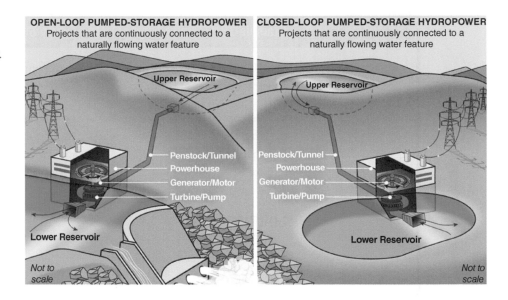

during the discharge process enable PHES systems to be more compact. This configuration is called a *binary set* with one pump/turbine and one motor/generator. *Pump as turbine* (PAT) is an example of a reversible turbomachine which is a reaction water turbine that can be used in a combined pump and motor/generator unit.

Energy stored in the upper reservoir is the potential energy of the water body stored at that elevation:

$$E_{\text{stored}} = mgh \tag{13.2}$$

where m is the mass of water, g is gravitational acceleration, and h is the height of the reservoir with respect to the turbine at a lower elevation. Power generation capacity is determined by the flow rate during the discharging process:

$$P = \frac{dE}{dt} = \frac{d}{dt}(mgh) = gh\frac{dm}{dt} = \rho gh\frac{dV}{dt} \tag{13.3}$$

where dV/dt is the volume flow rate of discharged water.

The cycle efficiency of a PHES system takes losses during both charging (pumping) and discharging (electricity generation) processes into account. Sources of inefficiency include losses from the motor, pump, generator, turbine, and transformer, and the head losses through the penstock both ways. Cycle efficiency, alternatively called *round-trip efficiency*, of a PHES system is:

$$\eta_{\text{cycle}} = \frac{E_{\text{out}}}{E_{\text{in}}} = \eta_p \eta_g \tag{13.4}$$

FIGURE 13.5 A PHES basin in Germany. Source: Ollo/E+ via Getty Images.

where η_p and η_g are the pumping and electricity generation efficiencies, respectively. These one-way efficiencies are determined as given in **Equations 13.5a** and **13.5b**.

$$\eta_p = \frac{E_{\text{stored}}}{E_{\text{in}}} = \eta_{\text{transformer}}\eta_{\text{motor}}\eta_{\text{pump}}\eta_{\text{penstock}} \tag{13.5a}$$

$$\eta_g = \frac{E_{\text{out}}}{E_{\text{stored}}} = \eta_{\text{penstock}}\eta_{\text{turbine}}\eta_{\text{generator}}\eta_{\text{transformer}} \tag{13.5b}$$

FURTHER LEARNING

Pumped Storage
www.hydro.org/waterpower/pumped-storage

Individual efficiency terms on the right-hand sides of these equations are listed in the order they come into play during the pumping and electricity generation processes. Round-trip efficiencies of PHES systems range between 70% and 85%.

Example 13.1 Pumped Hydroelectric Energy Storage

A PHES reservoir is situated 65 m above the powerhouse where the turbine and generator are housed. The reservoir holds 4 million cubic meters of water.

a. Determine the energy capacity of the reservoir (MWh).
b. Find the transformer efficiency ($\eta_{transformer}$) if the amount of energy used during the charging process was 1225 MWh, and the motor, pump, and penstock efficiencies are 92%, 88%, and 85%, respectively (%).

Solution

a. The amount of stored energy is:

$$E_{stored} = mgh = (\rho V)gh$$

$$= \left(1000 \; \frac{kg}{m^3}\right)(4 \times 10^6 \; m^3)\left(9.81 \; \frac{m}{s^2}\right)(65 \; m)\left|\frac{1 \; J}{1 \; \frac{kg \cdot m^2}{s^2}}\right|\left|\frac{1 \; MWh}{3.6 \times 10^9 \; J}\right|$$

$$= 709 \; MWh$$

b. Transformer efficiency can be determined using **Equation 13.5a**:

$$\eta_p = \frac{E_{stored}}{E_{in}} = \eta_{transformer} \, \eta_{motor} \, \eta_{pump} \, \eta_{penstock}$$

$$\eta_{transformer} = \frac{E_{stored}}{E_{in}} \left(\eta_{motor} \, \eta_{pump} \eta_{penstock}\right)^{-1}$$

$$\eta_{transformer} = \frac{709 \; MWh}{1225 \; MWh} \; \frac{1}{(0.92 \times 0.88 \times 0.85)} = 0.84$$

Hence, the efficiency of the transformer is 84%.

13.2.2 Compressed Air Energy Storage

Compressed air energy storage (CAES) is another type of mechanical energy storage system that utilizes the stored potential energy of a substance. In pumped hydro systems, energy storage is achieved with water, which is incompressible. In CAES systems, the energy storage substance is a compressible fluid, air. In fact, it is the compressibility of air that allows the charging. The working principle of CAES is also straightforward. The storage is charged with compressed air using compressors and discharged when power is needed. Electricity is generated by expansion of the air through a turbine. This principle is illustrated in **Figure 13.6**.

The storage can be constant volume or constant pressure. A *constant volume* storage could be a geological underground formation such as a salt cavern (**Figure 13.7**). The advantage of these natural vessels is that they can offer a large volume with no cost for the container itself. The disadvantage of such chambers is that the pressure keeps changing during charging (compression) and discharging (expansion) with the volume being constant. This poses a challenge to the compressors and turbines during operation. A *constant pressure* storage on the other hand eliminates the pressure fluctuation issue by maintaining the vessel pressure constant, owing to the vessel being a variable-volume system. The disadvantage of this

FIGURE 13.6 Compressed air energy storage (CAES) principle. Source: J. Wang, K. Lu, L. Ma, J. Wang, M. Dooner, S. Miao, J. Li, and D. Wang, "Overview of compressed air energy storage and technology development," *Energies*, vol. 10, no. 7, p. 991, 2017.

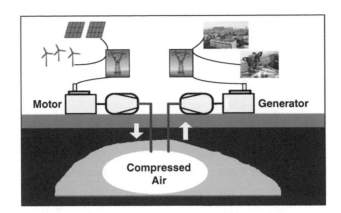

storage is the cost due to the need for keeping them underwater at great depths and the cost of the container itself.

CAES systems can be classified into three categories based on how heat is managed during compression and discharge (expansion) processes. These categories are:

- Diabatic systems (D-CAES)
- Adiabatic systems (A-CAES)
- Isothermal systems (I-CAES)

Diabatic compressed air energy storage (D-CAES) systems require heat addition to the compressed air during the discharge process when air is expanded. This is done to prevent condensation and frosting through the piping and machinery that the air passes through. Adiabatic designs (A-CAES) do not waste the compression heat. Heat generated during compression is stored in a thermal energy storage system. Stored heat is then utilized during the discharge process. As the heat addition during the discharge process does not require additional fuel combustion as in the D-CAES systems, adiabatic designs offer a higher efficiency potential. There is ongoing research on advanced designs of adiabatic compressed air energy storage systems, also known as AA-CAES. These designs can involve different forms of thermal energy storage such as packed bed heat exchangers or phase change materials (PCM). Isothermal compressed air storage (I-CAES) is an emerging technology which aims to maintain operating temperature by means of continuous heat exchange with the surroundings during both compression and expansion processes. Both processes being isothermal offers higher cycle efficiencies. The challenge for these systems is the need for a very large surface area to make up for the same heat transfer rate when the temperature change is very small.

The amount of energy that the compressed air possesses can be estimated using **Equation 13.6**, assuming isothermal storage and ideal gas behavior:

$$E_{\text{stored}} = \int_{V_f}^{V_i} p \, dV = \int_{V_f}^{V_i} \frac{(mRT)}{V} \, dV = mRT \ln \frac{V_i}{V_f} \tag{13.6}$$

where m is the mass of air, R is its gas constant, and T is the absolute temperature; V_i and V_f are the initial and final volumes of air before and after compression, respectively.

As compressed air is assumed to obey the ideal gas law:

$$pV = p_i V_i = p_f V_f \tag{13.7}$$

And,

$$\frac{V_i}{V_f} = \frac{p_f}{p_i} \tag{13.8}$$

FIGURE 13.7 Salt caverns can be utilized as storage sites for various energy storage media such as hydrogen, compressed air, crude oil, or natural gas . Source: KBB Underground Technologies.

Then, stored energy in terms of the initial pressure of air and its final pressure inside the cavern is:

$$E_{\text{stored}} = mRT \ln \frac{P_{\text{f}}}{P_{\text{i}}} \tag{13.9}$$

Cycle efficiencies of A-CAES and I-CAES plants can be calculated by:

$$\eta_{\text{cycle}} = \frac{E_{\text{net}}}{E_{\text{in}}} = \frac{\text{Power}_{\text{out}}}{\text{Power}_{\text{in}}} \tag{13.10}$$

where Power_{in} is the electrical energy supplied to the compressors during the storage process, and $\text{Power}_{\text{out}}$ is the electrical energy produced through the generators during the discharge process. Cycle efficiency of D-CAES plants includes an additional heat addition term as input:

$$\eta_{\text{cycle}} = \frac{E_{\text{out}}}{\sum E_{\text{in}}} = \frac{\text{Power}_{\text{out}}}{\text{Power}_{\text{in}} + Q_{\text{in}}} \tag{13.11}$$

It should be noted that the denominator values are not added to air simultaneously. Electrical energy, Power_{in}, is supplied during the storage process, while heat addition takes place during the discharge process to prevent dew and frost formation in the expansion units.

13.2.3 Solid Mass Gravitational Energy Storage

Solid mass gravitational energy storage charges solid materials such as concrete blocks (**Figure 13.8**) by lifting them up, hence storing potential energy in them. Different designs based on this gravitational principle exist. During the charging process, solid masses are transported to a higher elevation using mechanisms such as electric motors, winches, or electric locomotives. When electricity is needed, the blocks descend to lower elevations resulting in released energy to be converted into electricity. Solid mass energy storage systems can be above ground or below ground. Above-ground examples include solid block towers, rail cars, or high-altitude solar-powered balloon platforms. Below-ground applications can be performed in vertical mine shafts. Both above- and below-ground designs have siting limitations. Above-ground towers need to be in remote areas due to safety, aesthetic, and land use concerns. Below-ground systems on the other hand rely on the availability of old mines or require vertical shafts built for storage purposes which increases the capital cost.

There is also a hybrid design where solid mass is accompanied by water. Although the purpose is different, the principle is similar to the *Panama Canal* locks. These systems work based on the principle of hydraulic lifting. Excess energy to be stored powers pumps to fill a large cylinder in which a large mass of solid block acts as a free piston. As the water is pumped into the cylinder the mass is lifted and energy is stored. When energy is needed, the piston (solid mass) moves down and pressurizes water to run through a turbine to generate electricity.

13.2.4 Flywheel

Pumped hydroelectric, compressed air, and solid mass are mechanical techniques used to store potential energy. Flywheels are also mechanical energy storage systems;

FURTHER LEARNING

Feasibility Study: CAES
https://caes.pnnl.gov/

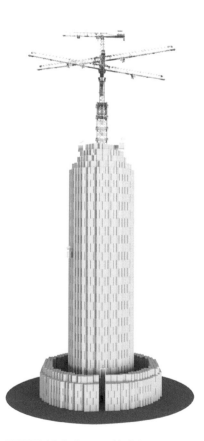

FIGURE 13.8 Concrete block tower acting as a gravity-fed battery. Source: https://energyvault.com.

FIGURE 13.9 Schematic and components of a flywheel. Source: M. Amiryar and K. Pullen, K., "A review of flywheel energy storage system technologies and their applications," *Applied Sciences*, vol. 7, no. 3, p. 286, 2017.

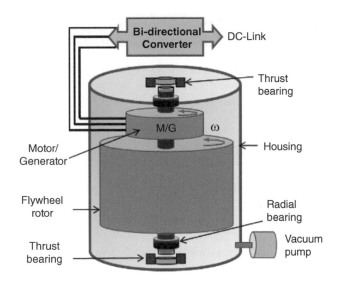

however, they differ from the previously discussed techniques as they store energy in the form of kinetic energy. Flywheel energy storage (FES) uses a rotating mass that stores energy. The charging phase for this "rotational battery" takes place by accelerating the rotor. When electricity is needed, the discharge phase is executed by extracting the energy from the rotating mass. This results in deceleration of the mass as per the conservation of energy.

The main components of a flywheel design include the rotor, motor/generator, bearings, vacuum pump, and the housing, as depicted in **Figure 13.9**. The motor/generator is used for charging and discharging the FES. Bearings support the rotation of the rotor and the shaft. They can be ball bearings or active magnetic type. Ball bearings cause more friction and are more prone to wear, as opposed to active magnetic bearings which do not come into contact with the solid surface and hence reduce frictional losses. A vacuum pump is used to provide a vacuum enclosure to reduce frictional losses and increase the amount of energy that can be harvested. Vacuum designs, however, have disadvantages regarding heat transfer and tribology. Thermal management of the rotor can be a problem as only radiative heat transfer is taking place within the vacuum. As for tribology, lubrication of the system is more difficult in vacuum.

Stored kinetic energy, which for a FES system is the rotational energy, can be expressed as:

$$E = \frac{1}{2}I\omega^2 \tag{13.12}$$

where I is the moment of inertia and ω is the angular velocity. Moment of inertia is a function of mass and the shape factor of the rotor. The maximum speed that the rotor can handle depends on the strength of the rotor material, which is quantified by maximum tensile strength, σ_{max}. Hence, capacity improvements for flywheels can benefit significantly from R&D in the materials science field, in addition to ongoing studies on geometric and configurational enhancements. Advanced flywheel designs utilizing such contemporary materials exist, including rotors made of carbon-fiber composites or aerospace steel which offer high strength and hence higher maximum speeds, resulting in increased energy storage capacity.

Example 13.2 Stored Energy in a Flywheel

A small flywheel energy storage unit employs a rotating disk with a diameter of 0.8 m and a mass of 1200 kg. The disk is rotating with a frequency of 900 rpm. Determine its kinetic energy.

Solution

Stored kinetic energy for a FES system is $E = \dfrac{1}{2} I \omega^2$ (1)

Moment of inertia for a solid disk is $I = \dfrac{1}{2} m r^2$ (2)

Angular velocity in terms of frequency of rotation is $\omega = 2\pi f$ (3)

where frequency in terms of revolutions per second is:

$$f = 900 \ \frac{\text{rev}}{\text{min}} = 15 \ \frac{\text{rev}}{\text{s}}$$

Plugging Equations (2) and (3) into (1), we get:

$$E = \frac{1}{2} I \omega^2 = \frac{1}{2} \left(\frac{1}{2} m r^2 \right) (2\pi f)^2$$

$$= \frac{1}{4} (1200 \ \text{kg})(0.4 \ \text{m})^2 [(2\pi)(15 \ \text{s}^{-1})]^2 \left| \frac{1 \ \text{kJ}}{10^3 \ \text{kg} \frac{\text{m}^2}{\text{s}^2}} \right| = 426.37 \ \text{kJ}$$

13.3 Thermal Energy Storage

13.3.1 Sensible Heat Thermal Energy Storage

Thermal energy can be stored in the form of sensible heat or latent heat. Sensible heat thermal energy storage (SHTES) systems utilize the heat capacity of substances and the temperature changes these substances experience during charging (heat addition) and discharging (heat extraction) processes. These substances act as the storage medium and can be in liquid or solid form. Some examples of SHTES media are water, oil, sand, rocks, and molten salt.

The amount of energy that can be stored in a material is:

$$Q = \int_{T_i}^{T_f} m \, c_p \, dT \tag{13.13}$$

Assuming constant specific heats:

$$Q = m \, c_p \, (T_f - T_i) = \rho V c_p (T_f - T_i) \tag{13.14}$$

where m is the mass, c_p is the specific heat, T_i and T_f are the initial and final temperatures, ρ is the density, and V is the volume of the storage medium, respectively. As can be seen in the equation, substances with higher density and specific heat values

Table 13.2

Properties of selected storage media for SHTES [3]				
Medium	Fluid type	Temperature range (°C)	Density (kg/m³)	Specific heat (J/kg K)
Sand	—	20	1555	800
Rock	—	20	2560	879
Brick	—	20	1600	840
Concrete	—	20	2240	880
Granite	—	20	2640	820
Aluminum	—	20	2707	896
Cast iron	—	20	7900	837
Water	—	0–100	1000	4190
Calorie HT43	Oil	12–260	867	2200
Engine oil	Oil	≤160	888	1880
Ethanol	Organic liquid	≤78	790	2400
Propane	Organic liquid	≤97	800	2500
Butane	Organic liquid	≤118	809	2400
Isopentanol	Organic liquid	≤148	831	2200
Octane	Organic liquid	≤126	704	2400

can store more thermal energy. It should be noted that there are other factors that affect the system performance such as operating temperature, thermal conductivity, thermal diffusivity, and the thermal harmony between the storage medium and the container. Properties of some common materials used in sensible heat storage are listed in **Table 13.2** [3].

SHTES can be performed below ground or above ground. Some underground applications include *aquifer thermal energy storage* (ATES), cavern storage, pit storage, and underground solid thermal batteries utilizing soil, sand, or rocks. *Borehole thermal energy storage* (BTES) is another underground application where fluids such as water or air are also involved, in a similar manner to geothermal heat pump systems. These underground TES systems can enable storage of not only excess energy from the Sun or wind, but also excess process heat from industry for later use in domestic hot water (DHW) and space heating at a residential scale, and for district heating at a larger scale.

Water tank storage is the most common above-ground SHTES application. It can be used for DHW, space heating, or a combination of both via solar combisystems. Heat pump water heaters (HPWHs) can also be used as thermal batteries. Large hot-water storage systems can be used for seasonal solar thermal storage accompanied by district heating systems. Newer designs can be integrated into *4th Generation District Heating* (4GDH) systems, operating as part of a smart thermal grid.

Heat can also be stored in a TES system to produce electricity during the discharge process. These systems act as *solid-state thermal batteries*. Some common substances used as storage media in these thermoelectric plants include molten salts, synthetic oils, or liquid metals. There are also *thermophotovoltaic* (TPV) systems, which convert thermal energy into electrical energy with a similar principle to that of conventional PV systems. The difference is that the source of light in TPVs is the heated solid (also called the *emitter* or *TPV radiator* in this process) instead of the Sun in the case of traditional PV systems. During the charging process, solid blocks are heated to high temperatures with power from the grid. During peak demand hours when electricity is needed, the high-temperature blocks are exposed to TPV panels, generating electricity.

13.3.2 Latent Heat Thermal Energy Storage

Latent heat thermal energy storage (LHTES) employs the phase change of a substance using *phase change materials* (PCMs). While this is mostly done through solid-to-liquid phase change, in some applications liquid-to-gas phase change can also be utilized. For a solid-to-liquid phase change of a PCM, the charging and discharging processes for energy storage and utilization are the melting and solidification of the PCM, respectively. A heat transfer fluid (charging fluid) heats the PCM during the charging process resulting in melting of the material. When this stored energy is needed, a discharge fluid will absorb the stored heat from the PCM, hence cooling and solidifying it. This method is useful in reducing cooling loads during air-conditioning seasons. During the day, PCMs absorb energy from the space to be cooled which results in lower cooling loads on AC units.

An efficient LHTES system requires an effective PCM and optimized enclosure geometry. Selection of an effective PCM for a project depends on a variety of parameters including melting point of the substance, its corrosivity and toxicity, and stability after completing a cycle. Geometric properties of the enclosure play a significant role as heat exchange between the substance and the surroundings is directly related to these properties. Certain thermophysical properties of PCMs can be altered by mixing other substances into them. Nano-enhanced PCMs are an example of such practice. Mixing nanoparticles into the PCM can enhance thermal conductivity; however, there might be a compromise on the heat capacity of the PCM mix. Some applications involve natural PCMs such as *beeswax*, which is a sustainable material that is low cost and non-toxic.

PCMs can be classified based on the nature of their contents, melting points, or phase transitions. Classification of different PCMs based on the nature of their contents is presented in **Figure 13.10**. In addition to organic and inorganic PCMS, there is also a *eutectic* PCM category, which defines a homogeneous mixture of organic and/or inorganic PCMs at a specific mixing ratio that yields the lowest possible melting point. The achieved melting point is lower than the melting point of each individual constituent.

LHTES capacity can be calculated as suggested by Lane (1983) as cited in Sharma *et al.* [4]:

$$Q = \int_{T_i}^{T_m} m\, c_p\, dT + m a_m \Delta h_m + \int_{T_m}^{T_f} m c_p\, dT \tag{13.15}$$

For constant mass and constant average specific heats for solid and liquid phases below and above the melting temperature, the integral equation can be expressed as:

$$Q = m\left[c_{p,s}(T_m - T_i) + a_m \Delta h_m + c_{p,l}(T_f - T_m)\right] \tag{13.16}$$

FIGURE 13.10 Classification of PCMs.

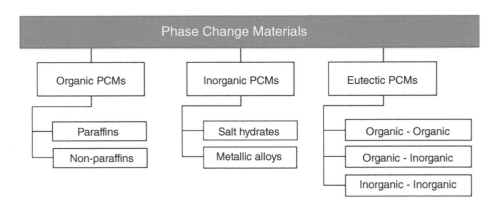

where

a_m: fraction melted (–)

c_p: specific heat (J/kg K)

$c_{p,s}$: specific heat of solid phase, between T_i and T_m (J/kg K)

$c_{p,l}$: specific heat of liquid phase, between T_m and T_f (J/kg K)

Δh_m: latent heat of fusion per unit mass (J/kg)

m: mass of PCM (kg)

Q: LHTES capacity, amount of energy gained by the PCM (J)

T_i: initial temperature (°C)

T_m: melting temperature (°C)

T_f: final temperature (°C)

Time rate of heat transfer from the charging fluid to the PCM is:

$$\dot{Q}_{ch} = \dot{m}\, c_{p,ch}\left(T_{ch,i} - T_{ch,o}\right) \tag{13.17}$$

where

$c_{p,ch}$: specific heat of the charging fluid (J/kg K)

$\dot{m}$: mass flow rate of the charging fluid (kg/s)

$\dot{Q}_{ch}$: heat transfer rate from the charging fluid (W)

$T_{ch,i}$: inlet temperature of the charging fluid (°C)

$T_{ch,o}$: outlet temperature of the charging fluid (°C)

There is ongoing research in different fields of PCM use including space heating and cooling, domestic water heating, waste heat recovery, and thermal energy storage systems. This research mainly pertains to materials and systems. Development of more effective, and preferably natural and sustainable PCMs, along with improved systems with enhanced heat transfer and energy storage abilities will increase the use of PCMs in the fields mentioned.

13.3.3 Cryogenic Energy Storage

The root *cryo-* comes from the Greek word *krýos* which means "icy cold." Cryogenics deals with the study and production of substances at very low temperatures. Cryogenic energy storage (CES) utilizes these low-temperature substances such as air, helium, or nitrogen in liquefied form. When air is used as the energy storage medium, the method is also referred to as **liquid air energy storage** (LAES). Unlike compressed air energy storage (CAES) which is a mechanical method, LAES is a thermal energy storage technique. One advantage of LAES over CAES is that in liquid air applications there

FURTHER LEARNING

Thermal Energy Storage in Europe
https://celsiuscity.eu/thermal-energy-storage/

FURTHER LEARNING

Phase Change Materials for Energy-Efficient Buildings
www.seas.ucla.edu/~pilon/PCMIntro.html

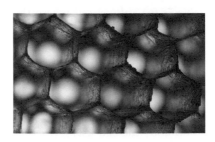

FIGURE 13.11 *Beeswax* is an organic non-paraffin PCM which is used as an alternative material for thermal energy storage applications. Source: Gulfu Photography/Moment via Getty Images.

is no need for underground geological formations such as salt caverns, which makes it feasible to use LAES applications almost anywhere.

There are three processes in CES: charging, energy storage, and discharging. The charging process in CES systems involves cleaning of the intake air and liquefaction of it by means of cooling it to very low temperatures using excess amount of energy from the grid during off-peak times. After being liquefied, air becomes about 710 times denser compared to the gas phase. In other words, 1 liter of liquid air is obtained from over 700 liters of air. The storage process requires well-insulated tanks that are maintained at low pressures. The discharging process takes place when energy is needed again. Liquid air goes through evaporation and expansion, which results in a significant increase in pressure as the volume of the rigid tank is constant. High-pressure air is then directed to a turbine to generate electricity.

Cooling units that are used to cool substances to cryogenic temperatures are called *cryocoolers*. Some cryocooler technologies besides the conventional systems include *Stirling refrigerators* (i.e., free-piston Stirling cooler), *pulse-tube refrigerators*, *Joule–Thomson coolers*, and *Gifford–McMahon coolers*.

13.4 Chemical Energy Storage

13.4.1 Battery

Batteries have a significant advantage over their peer energy storage technologies in terms of how much they can offer per unit volume. Both energy density (Wh/L) and power density (W/L) values of batteries are much higher than those for PHES or CAES. Battery energy storage systems (BESS), however, have a limited number of cycles compared to both mechanical energy storage methods. The cost of BESS can be expressed in terms of installed cost or LCOS. Installed cost would include cost of the batteries, balance of system (BOS), and engineering, procurement, and construction (EPC) costs. The BOS costs come from the auxiliary units including the containers, climate control systems, safety equipment, and monitoring and control systems. The LCOS can be calculated using **Equation 13.1**. Battery selection criteria are determined by the nature of the application. Some of the important factors in determining battery type are illustrated in **Figure 13.12**.

Batteries can be recharged by passing electric current through them. They behave as *electrolytic cells* during the charging process in which the anode is positive and cathode is negative. When there is need for electricity, stored energy is discharged. During the discharging process, the battery system acts as a *galvanic* (or *voltaic*) *cell* where the anode is negative and cathode is positive.

> *Charging:* Electrolytic cell (Electrical energy → Chemical energy)
> *Discharging:* Galvanic cell (Chemical energy → Electrical energy)

Battery applications can be categorized based on the orientation of the battery with respect to the meters that keep track of energy consumption by the end-users, whether they are residential, commercial, or industrial consumers. The two categories of battery applications are *front of the meter* (FTM) and *behind the meter* (BTM). FTM batteries are connected either to a power generation facility, or to transmission and distribution networks. These batteries can be functional in providing grid load relief. BTM

FIGURE 13.12 Parameters considered in battery selection.

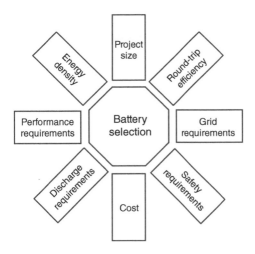

Table 13.3

Characteristics of different battery technologies				
	Li-ion	**Lead–acid**	**NiCd**	**NaS**
Anode	Graphite, silicon/carbon	Pb	Cd	Na (molten salt)
Cathode	Li-NMC or NMC	PbO_2	NiO (OH)	S
Electrolyte	Lithium salts (liquid), lithium metal oxides (solid)	H_2SO_4	KOH	Beta-alumina solid electrolyte
emf (V)	3.6–3.7	~2.1	1.2	~2.0
Round-trip efficiency (%)	95	60–70	60–80	~90

batteries, on the other hand, are connected to systems on the end-user side for different purposes, such as storing energy during off-peak hours and consuming it later, and hence reducing electricity bills.

Batteries can also be classified based on their technology. Common types include lithium-ion (Li-ion), lead–acid, nickel–cadmium (NiCd), and sodium–sulfur (NaS) batteries. Some of the characteristics of these batteries are listed in **Table 13.3**. These characteristics can differ in different battery applications. There are a number of anode, cathode, and electrolyte options for each battery technology. Commonly used materials have been listed.

Lithium-ion batteries are commonly used in personal electronics such as cell phones and laptops or used in EVs. They can also be utilized in larger scale projects such as energy storage for residential, commercial, or grid use. They are different from lithium batteries as lithium batteries are designed for single use and are not meant to be recharged while Li-ion batteries are rechargeable. They have one of the highest energy densities among all battery technologies and can yield up to 3.7 V, which is much higher than other types such as lead–acid and NiCd batteries. Besides their advantages, lithium-ion cells also have some drawbacks. These cells are flammable and can cause problems. Incidents in the past have been reported on devices such as cell phones,

FIGURE 13.13 Operating principle of a lithium-ion battery. Source: Argonne National Laboratory.

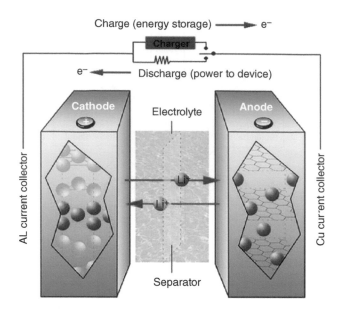

drones, domestic robots, and hoverboards. The operating principle of a Li-ion battery is shown in **Figure 13.13**.

Lead–acid batteries are employed in a wide range of applications including automobile starting, UPS (uninterruptible power supply), emergency power, PV systems, vehicles such as electric cars, golf carts, forklifts, and wheelchairs. Batteries having gel electrolytes instead of a liquid solution provide more mobility and flexibility in positioning. Absorbent glass mat batteries house fiberglass mats immersed in the electrolyte. Lead–acid batteries have a specific configuration with a one-way release valve to prevent pressure build-up within the casing. This configuration is known as a *valve-regulated lead acid* (VRLA) battery. Lead–acid batteries are low cost and easy to manufacture. They are capable of high discharge currents and perform well in a wide temperature range under different operating conditions. They require low maintenance which is another advantage of these battery types. Major disadvantages of lead–acid batteries are their adverse environmental impacts due to their lead content, limited cycle life, and low specific energy.

Nickel–cadmium batteries are rechargeable energy storage units utilized in small battery-operated devices such as portable electronics and toys. They have higher tolerance both to being fast-charged and to deep discharge. This is an advantage of NiCd batteries over Li-ion batteries, which can experience severe damage if discharged below a threshold voltage. They are easy to transport and store. They have a very low cost per cycle. These batteries can perform well under rough environmental conditions, which makes them a good choice for portable devices. Their major shortcomings are the toxicity due to cadmium and the low cell voltage they provide. A 1.2 V emf entails use of multiple cells for achieving higher voltages.

Sodium–sulfur batteries utilize molten salt and sulfur in the electrodes and a solid electrolyte such as beta-alumina solid electrolyte (BASE). The materials being low cost provides economic manufacturing opportunities for these types of batteries. They can be used in larger capacity applications such as grid-scale energy storage. They have high energy efficiency and high cycle life. A disadvantage of NaS batteries is that they require operating temperatures above 300 °C. This limits the use of these batteries due to logistic and safety limitations. In addition to high operating temperature ranges,

FIGURE 13.14 Schematic of a redox flow battery. Source: A. Clemente, G. A. Ramos, and R. Costa-Castelló, "Voltage H∞ control of a vanadium redox flow battery," *Electronics*, vol. **9**, p. 1567, 2020.

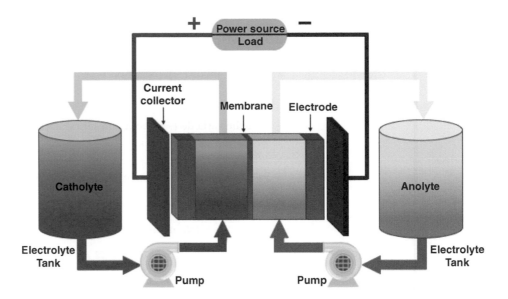

corrosive contents add to the safety concerns. Therefore, these batteries are preferred for non-portable and larger capacity applications. They also come with an extra cost of the containment built to secure the structure and surroundings, and to prevent potential leakages.

Flow Batteries

As an alternative to the conventional batteries discussed, there is another group of batteries called *flow batteries* which store energy in an electrolyte contained in tanks, unlike conventional batteries in which electricity is stored in the electrodes. These batteries utilize tanks of electrolyte and a membrane to control the flow of electrons. Energy storage capacity is proportional to the volume of electrolyte, and the power capacity is related to the surface area of the electrodes. Flow batteries have different categories such as redox, hybrid, membraneless, single liquid, and nano-flow batteries. A schematic of a *redox flow battery* can be seen in **Figure 13.14**. Some application examples include vanadium, zinc–bromine, zinc–iron, and nanofluid flow batteries.

Flow batteries have several advantages including long lifespan and unlimited cycles, a versatile configuration due to the energy generation (electrode) and energy storage (electrolyte) sections being split, and higher safety due to the materials being non-flammable. Flow batteries have three major disadvantages. These are: low energy density, higher charging/discharging duration, and complexity of the system due to added components with moving parts such as circulation pumps.

Example 13.3 An Electric Surfboard with a Li-Ion Battery

An electric hydrofoil surfboard (**Figure 13.15**) operates with a Li-ion battery which has a nominal voltage of 50 V, and an electrical storage capacity of 32 Ah. The battery weighs 9.5 kg.

a. Write the anode and cathode half-reactions and the full reaction for the battery.
b. Determine the capacity of the battery (kWh).
c. Find the energy density of the battery (Wh/kg).

Solution

a. At the anode $(-)$, oxidation takes place. The half-reaction is:

$$LiC_6 \rightarrow C_6 + Li^+ + e^-$$

At the cathode $(+)$, reduction takes place. The half-reaction is:

$$CoO_2 + Li^+ + e^- \rightarrow LiCoO_2$$

Full reaction (left-to-right: discharging, right-to-left: charging) is:

$$LiC_6 + CoO_2 \leftrightarrow C_6 + LiCoO_2$$

b. Capacity = (Nominal voltage) (Energy storage capacity)

$$= (50\ V)(32\ Ah) = 1600\ Wh \quad = 1.6\ kWh$$

c. Energy density = Capacity/Weight

$$= 1600\ Wh/9.5\ kg = 168.4\ Wh/kg$$

FURTHER LEARNING

Software Tools for Battery Design
https://www.nrel.gov/transportation/
caebat-modeling-design.html

FURTHER LEARNING

**Battery Initiative: Battery 2030+/
bold>**
https://battery2030.eu/

FIGURE 13.15 Electric surfboards can glide over water without the need for waves. They are powered by lithium-ion batteries. Source: Sergei Proschenko/500Px Plus via Getty Images.

13.4.2 Fuel Cell

A fuel cell is an electrochemical device that converts the chemical energy in fuels into electrical energy through redox chemical reactions, where a reducing agent (reductant) loses electrons to an electron recipient (oxidizing agent or oxidant). Fuel cells are different from conventional thermal power generation plants as there is no need for producing heat to achieve mechanical work. Therefore, they do not have a theoretical limit such as the Carnot efficiency for heat engines. They also do not release harmful gases to the surrounding environment like thermal power plants do, as there is no combustion process taking place. They are also different from batteries. Batteries release electrical energy that is converted from chemical energy which comes from the metals used. In fuel cells, there needs to be a continuous flow of reductant and oxidant to yield electrical energy. Batteries run out of energy when the chemical reactants are consumed, as opposed to fuel cells which can provide continuous energy if fuel and oxygen are supplied. A schematic of a fuel cell is illustrated in **Figure 13.16**. A fuel cell consists of an anode, an electrolyte, and a cathode. The electrolyte allows positively charged fuel (i.e., hydrogen) ions to migrate between the two electrodes. As hydrogen is neutral, each atom having one electron and one proton, it cannot pass through the electrolyte by itself to combine with oxygen. A catalyst is used on the anode side to split hydrogen molecules into electrons and protons. The positively charged particles (protons) can pass through the electrolyte; however, the electrons cannot penetrate through this medium. On the cathode side of the cell, oxygen molecules cannot combine with only the hydrogen protons. The electrons should also be present for hydrogen molecules to form again and react with oxygen to produce water. Hence the remaining electrons are forced through an alternative path which makes up the circuit, yielding electrical energy.

Fuel Cell Types

Fuel cells are mostly similar based on their structuring and operating principle. Different types of fuel cells exist due to the type of electrolyte utilized in the unit.

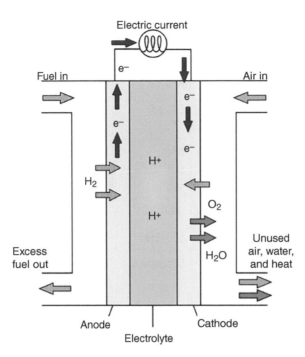

FIGURE 13.16 Operating principle of a proton-exchange membrane fuel cell (PEMFC). Source: Mattuci via Wikimedia Commons.

Six major types of fuel cells in the order of their operating temperature and power rating are:

- Direct methanol fuel cell (DCFC)
- Proton-exchange membrane fuel cell (PEMFC)
- Alkaline fuel cell (AFC)
- Phosphoric acid fuel cell (PAFC)
- Molten carbonate fuel cell (MCFC)
- Solid oxide fuel cell (SOFC)

Reactions within various fuel cell types can be seen in **Figure 13.17**. This is a superimposed illustration on the same electrode–electrolyte assembly showing reactions occurring in each fuel cell type. Characteristics and applications of the fuel cells discussed are listed in **Table 13.4** [5].

Direct Methanol Fuel Cell (DMFC)

Description: The DMFC is a type of proton-exchange membrane fuel cell (PEMFC). Most fuel cell types utilize hydrogen as the fuel. Hydrogen can be provided directly to the system or it can be produced within the fuel cell via a reforming process. DMFCs, however, do not require hydrogen reforming, whether it is done externally or internally. They utilize pure methanol (CH_3OH) as the fuel. The platinum/ruthenium catalyst on the anode side extracts the hydrogen from methanol, eliminating the need for the reforming process, which results in reduced costs. Also, utilization of methanol provides fuel flexibility in terms of both transportation safety and cost. DMFCs are mostly used for powering portable electronics.

Anode reaction: $CH_3OH + H_2O \rightarrow 6H^+ + 6e^- + CO_2$

Cathode reaction: $\dfrac{3}{2}O_2 + 6H^+ + 6e^- \rightarrow 3H_2O$

Efficiency: 30–40%

Table 13.4

Characteristics and applications of different fuel cell types [5]						
	DMFC	**PEMFC**	**AFC**	**PAFC**	**MCFC**	**SOFC**
Operating temperature (°C)	20–90	30–100	50–200	~220	~650	500–1000
Electrolyte type	Polymeric ion exchange membrane	Polymeric ion exchange membrane	Alkaline salt solution	Liquid phosphoric acid	Liquid molten carbonate	Solid oxide or ceramic
Charge carrier	H^+	H^+	OH^-	H^+	CO_3^{2-}	O^{2-}
Power range (W)	1–100	1–100k	500–10k	10k–1M	100k–10M+	1k–10M+
Applications	Portable electronics	Cars, boats, buses, portable electronics	Spaceships	Buses, stationary power generators	Buses, CHPs, distributed power generation	Auxiliary power units, CHPs, stationary power generation

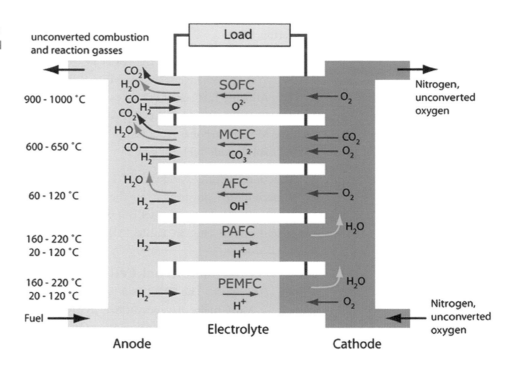

FIGURE 13.17 Superimposed illustration of reactions taking place in various fuel cell types depicted on the same electrode–electrolyte assembly. Source: University of Cambridge, DoITPoMS, www.doitpoms.ac.uk/tlplib/fuel-cells/types.php.

Advantages: Cost-effective fuel production, safer and easier fuel transport, low operating temperature, less wear on system components

Disadvantages: High cost, lower power density, lower efficiency, membrane sensitivity to abnormal moisture content, risk of methanol crossover to the cathode side

Proton-Exchange Membrane Fuel Cell (PEMFC)

Description: The PEMFC is also known as a *polymer electrolyte membrane fuel cell*. Similar to DMFCs, these fuel cells also use a polymer-exchange membrane electrolyte. PEMFCs, however, are fed with hydrogen which is externally reformed. Besides the electrolyte type, another similarity of PEMFCs with DMFCs is that they both operate at comparatively low temperatures. They use a solid polymer as the electrolyte, porous electrodes, and platinum or platinum alloy catalysts on the electrodes. Polymer electrolyte membrane design should be able to filter out the electrons and only allow protons to pass through, as the passage of electrons would result in short-circuiting of the fuel cell. Besides this selective filtering, the membrane prevents the gases on either side of it passing to the other side, which is an undesirable case known as *gas crossover*. The PEMFC is the most broadly used vehicular fuel cell technology. Because of their quick start-up and high power density, they are suitable for transportation applications such as cars, boats, and buses.

Anode reaction: $H_2 \rightarrow 2H^+ + 2e^-$

Cathode reaction: $\frac{1}{2}O_2 + 2H^+ + 2e^- \rightarrow H_2O$

Efficiency: 50–60%

Advantages: High power density, higher efficiency, low operating temperature, quick start-up, less wear on components, reduced corrosion and electrolyte management concerns due to solid electrolyte

Disadvantages: Requires a noble-metal catalyst (i.e., platinum) resulting in increased cost, platinum membrane is sensitive to CO poisoning, membrane is also sensitive to over-wetting or dryness which requires well-controlled evaporation of water

Alkaline fuel cell (AFC)

Description: Among different fuel cell types, AFCs are one of the most mature technologies. They were used by NASA in the Apollo and space shuttle programs to produce energy for on-board systems and drinking water for the crew, hence *"feeding two birds with one scone."* These fuel cells use an aqueous alkaline solution such as potassium hydroxide (KOH) as electrolyte. They can utilize inexpensive catalysts at both electrodes which reduces overall cost. In recent years, polymer membrane electrolytes have been developed for use in AFCs, making them alike with PEMFCs. The difference is that they employ alkaline membranes instead of acid membranes. As aqueous alkaline solutions are not CO_2-phobic, the fuel carries the risk of being poisoned. Therefore, pure oxygen or purified air is used as oxidant. A liquid electrolyte helps mitigate carbonate (CO_3^{2-}) formation; however, recirculation of the aqueous substance can cause problems associated with current shunts. AFCs provide quick start-up with relatively low operating temperatures. Besides their cost-effectiveness and maturity in technology, they are one of the most efficient fuel types.

Anode reaction: $H_2 + 2OH^- \rightarrow 2H_2O + 2e^-$

Cathode reaction: $O_2 + 2H_2O + 4e^- \rightarrow 4OH^-$

Efficiency: 60–70%

Advantages: Low operating temperature, quick start-up, higher efficiency over acidic fuel cell types, cost-effective components, safe for critical applications such as space shuttles

Disadvantages: High sensitivity to CO_2 (even a small amount of CO_2 within the intake air can negatively affect the cell performance), difficult to manage liquid electrolyte which can cause corrosion and pressure gradients within the cell, lower ionic conductivity of polymer electrolytes

Phosphoric acid fuel cell (PAFC)

Description: PAFCs represent another mature fuel cell technology. They were the first type to be commercially available. The electrolyte is liquid phosphoric acid (H_3PO_4) which is contained in a silicon carbide (SiC) matrix. The acid dissociates into phosphate and hydrogen ions (H^+), where the hydrogen ions act as the charge carrier. The advantage of phosphoric acid is that it is a chemically stable substance and transporting it is convenient. PAFCs have operating temperatures of around 220 °C. At this temperature phosphoric acid has a low vapor pressure making it difficult for it to evaporate and mix into the exhaust gases, which otherwise would mean loss of electrolyte solution and a decrease in fuel cell efficiency over time. The electrodes of these fuel cells are porous carbon coated with platinum film as the catalyst. PAFCs operate more efficiently when utilized in cogeneration. They do not perform as well when used for producing electricity directly. PAFCs are tolerant to CO_2 and to some extent to CO. This provides flexibility in fuel selection and supply. They can be expensive, mainly due to the cost of the catalyst applied. They have lower power density hence they require plenty of space to deliver a sufficient amount of power. Therefore, PAFCs are used in larger size applications such as stationary power generators. They can also be used in land and marine transportation such as buses or submarines.

Anode reaction: $2H_2 \rightarrow 4H^+ + 4e^-$

Cathode reaction: $O_2 + 4H^+ + 4e^- \rightarrow 2H_2O$

Efficiency: 40–45%

Advantages: Higher tolerance to fuel impurity (less vulnerability to CO poisoning), highly efficient for cogeneration (CHP) plants

Disadvantages: Lower power density, longer start-up time, higher sulfur sensitivity (if gasoline is used), expensive catalyst

Molten carbonate fuel cell (MCFC)

Description: MCFCs are considered as high-temperature fuel cells due to their operating temperature being significantly higher than those for PEM, alkaline, and phosphoric acid fuel cells. They employ a non-aqueous electrolyte composed of molten carbonate salt mixture suspended in a separator matrix consisting of lithium aluminum oxide ($LiAlO_2$). The electrodes do not need to be platinum which is expensive. They are nickel-based and inexpensive. The operating temperature of MCFCs is about 650 °C. This comparatively higher temperature provides some advantages. Different from low-temperature fuel cells, MCFCs do not require expensive metals as the catalyst, which reduces their cost. Another advantage of high-temperature operation is the elimination of the need for reforming. Instead of executing an external reforming process to convert natural gas or biogas to hydrogen, internal reforming takes place within the fuel cell. This also reduces overall cost as an external process and equipment are avoided. One other advantage of higher operating temperature is the increased ionic conductivity. This allows the electrodes to be thicker as there will be less concern for voltage drops and yet increasing mechanical strength of the fuel cell. Higher operating temperatures introduce some disadvantages as well. The major drawback is the shorter cell life due to wear and breakdown associated with high temperatures and corrosion risk due to the corrosive nature of the electrolyte. MCFCs can be used in transit buses and power plants. Excess heat can be used in industry, such as a CHP application, which comes with increased efficiency. They are not suitable for domestic applications due to operating conditions and safety requirements.

Anode reaction: $H_2 + CO_3^{2-} \rightarrow H_2O + CO_2 + 2e^-$

Cathode reaction: $\frac{1}{2}O_2 + CO_2 + 2e^- \rightarrow CO_3^{2-}$

Efficiency: 50–60%

Advantages: High efficiency (especially if used in CHP systems), fuel flexibility (no risk of CO or CO_2 poisoning), no need for an external reformer to convert natural gas or biogas to hydrogen, reforming process can be done internally resulting in reduced costs

Disadvantages: Low durability and cell life (corrosion, wear, and breakdown risks due to high operating temperatures), lower power density, long start-up time

Solid oxide fuel cell (SOFC)

Description: SOFCs have permeable electrodes and an impermeable electrolyte which is solid. In most cases, the anode is a cermet, which is a composite material consisting of ceramic and metal. The most common cermet used in these fuel cells is Ni–YSZ (yttria-stabilized zirconia). Nickel is low cost and it has two major benefits: it acts as the catalyst for the oxidation, and it has good electrical conductivity. As for cathodes, conductive perovskites are preferred. Lanthanum manganite ($LaMnO_3$) is a widely used cathode due to its high electrical conductivity and compatibility with YSZ electrolytes. The electrolyte can be solid oxide or ceramic. Like MCFCs, the high operating temperatures of SOFCs provide similar benefits including the ability to house an interior reforming process and the flexibility of choosing inexpensive materials. SOFCs can use a variety of fuels including methane, butane, and propane, which can be processed via internal reforming due to being light hydrocarbons; or they can utilize heavier hydrocarbons such as gasoline or biofuels after an external reforming process. In SOFCs, the charge carrier is the oxide ion (O^{2-}) which moves from the cathode to the anode side. They release their electrons and combine with hydrogen to form water, which results in electric current flowing from the anode to the cathode side, where the electrons reduce incoming oxygen to oxide ions to repeat the process. SOFCs have higher efficiency in converting fuel to electricity compared to most of the other fuel cell types, being about 60%. Fuel use efficiency values can go up to 85% if these fuel cells are used in cogeneration power plants.

Anode reaction: $H_2 + O^{2-} \rightarrow H_2O + 2e^-$

Cathode reaction: $\frac{1}{2}O_2 + 2e^- \rightarrow O^{2-}$

Efficiency: ~60%

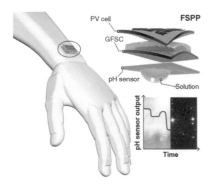

Advantages: High efficiency, fuel flexibility (not poisoned by CO and the most sulfur-resistant fuel cell type), reforming can be done externally or internally, no fluidic problems due to solid electrolyte

Disadvantages: Low durability and cell life (corrosion, wear, and breakdown risks due to high operating temperatures), long start-up time (some enhanced surface area cells can offer shorter start up durations)

13.4.3 Supercapacitor (Pseudocapacitor)

FIGURE 13.18 A graphene foam-based supercapacitor (GFSC) used as an energy storage system for a wearable nanorod-based chemi-resistive pH sensor. This supercapacitor is powered by a flexible PV cell which makes it a fully self-charging wearable technology. Source: L. Manjakkal, C. G. Núñez, W. Dang, and R. Dahiya, "Flexible self-charging supercapacitor based on graphene-Ag-3D graphene foam electrodes," *Nano Energy*, vol. 51, pp. 604–612, 2018. https://doi.org/10.1016/j.nanoen.2018.06.072.

Supercapacitors are charge storage devices like capacitors or batteries. They can store more energy than conventional capacitors. The difference from batteries is that they have much lower energy density; however, they have much higher power density. They offer faster charging and have longer cycle life. They are environmentally friendly, and they do not explode like batteries can when they are overcharged. They have high reliability and low costs. Easy fabrication and light weight make supercapacitors a good fit for a wide spectrum of applications as energy storage systems. These applications include portable electronics, backup power supply, hybrid vehicles, and smart grids. There are also foldable and flexible designs. Supercapacitors can be used in wearable electronic devices such as sensors (**Figure 13.18**), flexible displays, and personal health monitors.

A *Ragone plot* of supercapacitors and some other energy storage systems including fuel cells, batteries, and capacitors can be seen in **Figure 13.19** [13]. The Ragone plot is a graphing technique performed on logarithmic-scale axes which is used for comparing

FIGURE 13.19 Ragone plot of different energy storage systems. Source: S. Hussain, M. U. Ali, G.-S. Park, S. H. Nengroo, M. A. Khan, and H.-J.Kim, "A real-time bi-adaptive controller-based energy management system for battery-supercapacitor hybrid electric vehicles," *Energies*, vol. 12, no. 24, p. 4662, 2019.

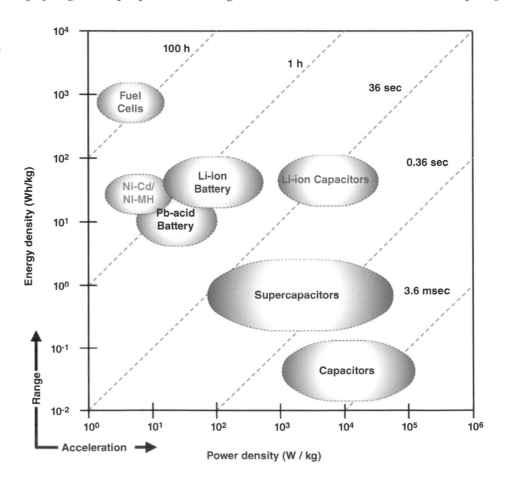

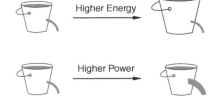

FIGURE 13.20 Energy–power comparison for supercapacitors and batteries. Batteries can store more energy than supercapacitors. However, supercapacitors can provide a much higher rate of energy discharge, which is power output. According to the analogy of water stored in a bucket representing energy, the upper-right bucket would be a battery, and the lower-right bucket would be a supercapacitor.

energy and power densities of various energy storage technologies. Dashed lines on the plot are isotherms of characteristic time constants.

Supercapacitors can be categorized into two main types: *pseudocapacitors* and *double-layer capacitors*. If the electrode allows intercalation of Li ions, then a Faradaic reaction will take place at its surface. The kinetics of the surface are diffusion-controlled. These are pseudocapacitors. If the reaction is non-Faradaic, then the charges can only be physically absorbed. The kinetics of the process are surface-controlled. These are double-layer capacitors. The process taking place in these supercapacitors is solely electrostatic. Therefore, the pseudocapacitor type is discussed as part of chemical energy storage, and the double-layer capacitor is discussed in the following section under electrical energy storage.

Pseudocapacitors are a type of supercapacitors that store energy in a Faradaic manner through reduction or oxidation (redox reaction) of a chemical substance. The prefix *pseudo-* denotes a close or deceiving resemblance to the word following it. A pseudocapacitor can be considered as a device that is somewhat in between a capacitor and a battery. It can store more energy than a double-layer capacitor, and it can charge or discharge faster than a battery (**Figure 13.20**).

13.5 Electrical Energy Storage

13.5.1 Supercapacitor (Double-Layer Capacitor)

Electrostatic double-layer capacitors (EDLCs) are another type of supercapacitors. As discussed in the previous section, their difference from pseudocapacitors is that they operate based on a non-Faradaic process. The working principle of these capacitors relies on the existence of an electrical double layer which forms on the solid surface of an object interacting with a fluid. The first layer is the surface charge. The second layer of ions is attracted to the first layer due to the Coulomb force, and it acts as electrical screening for the first layer. Electrical energy is stored by charge separation in the double layer based on electrostatic effects occurring between two carbon electrodes. These electrodes need to have high specific surface area per unit volume, such as *activated carbon* (also called *activated charcoal*). It is a processed form of carbon that provides increased surface area within a limited volume to enhance adsorption or chemical reactions. The idea is analogous to the extended surface heat exchangers used to enhance heat transfer in thermal systems. The electrodes are immersed in an electrolytic fluid in which a separator is placed. Charge is stored electrostatically.

There is also a hybrid version of the supercapacitor called a **hybrid capacitor**. This storage system is a combination of pseudocapacitors and double-layer capacitors. It utilizes two electrodes with different characteristics. One of these electrodes reveals electrochemical capacitance similar to pseudocapacitors, and the other electrode yields electrostatic capacitance like double-layer capacitors. This combination helps improve both energy and power densities of the capacitor. A common hybrid capacitor application is a **lithium-ion capacitor** (LIC). These capacitors have higher energy density than EDLCs and they exhibit much higher power densities than lithium-ion batteries. The charging and discharging principles of supercapacitor types and batteries can be seen in **Figure 13.21**.

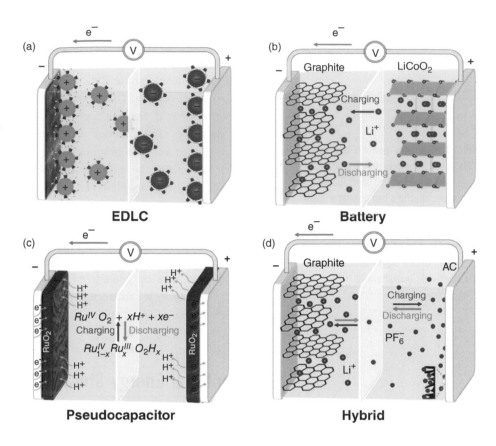

FIGURE 13.21 Schematic of (a) electric double-layer capacitor (EDLC), (b) battery, (c) pseudocapacitor, and (d) hybrid capacitor. Source: A. Noori, M. F. El-Kady, M. S. Rahmanifar, R. B. Kaner, and M. F. Mousavi, "Towards establishing standard performance metrics for batteries, supercapacitors and beyond," *Chemical Society Reviews*, vol. 48, no. 5, pp. 1272–1341, 2019.

FIGURE 13.22 Heike Kamerlingh Onnes (1853–1926) was a Dutch physicist who discovered superconductivity in 1911. The Nobel Prize in Physics was awarded to Onnes in 1913 for his investigations on the properties of matter at low temperatures. The famous Onnes effect was named after him. The Onnes effect refers to the ability of superfluid liquids to creep over substances at a higher level due to the capillary forces dominating both body and viscous forces. Source: ullstein bild. Dtl. / Contributor via Getty Images.

13.5.2 Superconducting Magnetic Energy Storage (SMES)

Superconductivity is the absence of electrical resistance of a material when its temperature is below a critical value, known as the critical temperature, T_c. The phenomenon of resistance of certain materials vanishing when their temperatures fall below a critical temperature was first found in 1911 by *Heike Kamerlingh Onnes* (**Figure 13.22**), a Dutch physicist, when he conducted an experiment on a mercury wire. When the material is cooled down below T_c, its resistance suddenly drops down to zero. Such materials are superconductors.

Superconducting magnetic energy storage (SMES) technology utilizes these superconducting materials carrying electric current at cryogenic temperatures at which the material has negligible losses. Therefore, SMES systems can be used in both small-scale and large-scale energy storage applications with the advantage of fast charging and discharging. A schematic of a SMES system is depicted in **Figure 13.23**.

SMES systems store energy in the magnetic field formed by the flow of current through a coil which is superconducting due to being cooled below its critical temperature. The cryogenic temperature is achieved by immersing the coil in liquid helium or nitrogen, resulting in electrical resistance being reduced to a negligible amount. During the charging process, electrical energy supplied from a source (e.g., a renewable source) is converted into magnetic energy and is maintained within a magnetic field. When energy is needed, the discharge process takes place in a reverse manner by converting the magnetic energy into electrical energy.

FIGURE 13.23 Schematic of a SMES system. Source: X. Liu, and K. Li, "Energy storage devices in electrified railway systems: A review," *Transportation Safety and Environment*, vol. 2, no. 3, pp. 183–201, 2020.

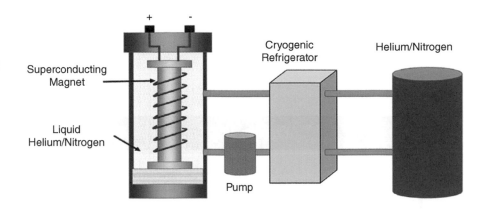

Charging: Electrical energy → Magnetic energy
Discharging: Magnetic energy → Electrical energy

Magnetic energy stored by the coil through which the current flows with no resistance can be calculated by:

$$E = \frac{1}{2}LI^2 \tag{13.18}$$

where E is the energy stored (J), L is the inductance of the coil in henry (H), and I is the current flowing through the coil (A). Inductance for a solenoid is given by:

$$L = \mu\frac{N^2 A}{l} \tag{13.19}$$

where μ is the magnetic constant (permeability), N is the number of turns, A is the cross-sectional area, and l is the length of the coil.

The operating temperature of the SMES system is also important in terms of the performance of energy storage. The coefficient of performance (COP) of the cooling system is:

$$COP = \frac{Q_c}{W_{net,in}} \tag{13.20}$$

where Q_c is the cooling capacity of the cryogenic cooler and $W_{net,in}$ is the net amount of work done by the components of the cooling system. The theoretical limit for the performance of the cooling system is defined by the Carnot COP:

$$COP_{Carnot} = \frac{T_C}{T_H - T_C} \tag{13.21}$$

The overall performance of SMES systems depends on all the processes taking place during charging and discharging the unit, including electrical, magnetic, and thermal losses in the coil or cooling system sections. Some sources of these losses are current leads, splices, magnetization, and heat leakage.

SMES systems have several strengths over some other energy storage applications. As the major components do not employ moving parts, they are highly reliable. They have very fast charging and discharging capabilities and they have a very high cycle life. Losses are minimal due to negligible electrical resistance. SMES systems have high power density, and they provide high round-trip efficiencies, exceeding 95%. Some of the disadvantages and challenges associated with these systems are their need for large space and the extremely high *Lorentz force* (electromagnetic force) they generate in grid-scale applications. This comes with the necessity for mechanical support which adds to the cost. Another challenge of SMES systems is that the coil material can lose its superconducting behavior when the magnetic field strength is above a threshold known as the *critical field*. This would introduce resistance to the electric current and can diminish the performance of the storage device significantly.

Example 13.4 Energy Stored in a SMES Unit

A superconducting magnetic energy storage unit has a solenoid that is 0.7 m in diameter and 1 m in length. The solenoid coil has an air core and houses 3500 turns of wire. Determine:

a. the inductance of the coil (H)
b. the energy that can be stored in the coil if the current flowing is 800 A (MJ)

Solution

a. Solenoid inductance can be calculated by using **Equation 13.19**:

$$L = \mu \frac{N^2 A}{l}$$

Permeability for air is ~1.2566×10^{-6} H/m (from tables).
Area of the solenoid is $\frac{\pi D^2}{4} = 0.385 \text{ m}^2$, then

$$L = \left(1.2566 \times 10^{-6} \frac{\text{H}}{\text{m}}\right) \frac{(3500)^2 (0.385 \text{ m}^2)}{1 \text{ m}} = 5.93 \text{ H}$$

b. Energy storage capacity of the coil is calculated using **Equation 13.18**:

$$E = \frac{1}{2} L I^2 = \frac{1}{2}(5.93 \text{ H})(800 \text{ A})^2 \left|\frac{1 \text{ MJ}}{10^6 \text{ H A}^2}\right| = 1.9 \text{ MJ}$$

13.6 Hydrogen

Hydrogen has been gaining popularity towards becoming one of the potential solutions to the need for net-zero emissions. In this section, the term hydrogen economy is discussed followed by hydrogen production, distribution, and storage. Hydrogen vehicles, including both hydrogen internal combustion engine vehicles (HICEVs) and fuel cell electric vehicles (FCEVs), are also covered in this section.

13.6.1 Hydrogen Economy

Hydrogen economy is a term that refers to the idea of including hydrogen in an economy that utilizes it as a low-carbon energy source to replace fossil fuels. This idea will become realistic only if major challenges associated with *safety*, *infrastructure*, and *cost* are overcome. If the production, distribution, and storage can be handled in a safe,

FIGURE 13.24 Jules Verne, in his novel *The Mysterious Island* (1874) wrote, "water will one day be employed as fuel, that hydrogen and oxygen which constitute it, used singly or together, will furnish an inexhaustible source of heat and light, of an intensity of which coal is not capable." Source: John Parrot/Stocktrek Images via Getty Images.

 CLASSROOM DEBATE

Debate on whether hydrogen can be the future solution to our energy problem or not. Both sides should point out the strengths of the side they are advocating for and the weaknesses of the opposite opinion.

reliable, environmentally friendly, and economic manner, hydrogen can become part of the solution for achieving net-zero emissions. While there is strong belief among engineers, energy professionals, policymakers, and industry officials that hydrogen is the clean fuel solution for the future, the challenges mentioned need to be addressed. Working towards these challenges requires extensive R&D, supporting policies and regulations, and major investments.

Hydrogen is the most abundant element in the universe. It is not found in pure form on Earth and exists naturally in compound form with other elements. It is the lightest element in the periodic table. Hydrogen is not a primary energy source; it is an energy carrier, and it can be used as an energy storage medium. It is a secondary energy source like electricity, heat, or biofuels, which needs to be produced from other substances such as fossil fuels, wastes, or water. It has high energy density by weight; however, its energy density by volume is low.

Energy density by weight → ~ three times that of gasoline
Energy density by volume → ~ a quarter that of gasoline

Hydrogen can be used to harvest electrical energy using fuel cells, or energy in the form of heat through combustion. **Equation 13.22** shows the anode reaction in a solid-oxide fuel cell. Burning of hydrogen with oxygen, releasing heat as output and water as exhaust is shown in **Equation 13.23**.

$$H_2 + O^{2-} \rightarrow H_2O + 2e^- \qquad (13.22)$$

$$2H_2 + O_2 \rightarrow 2H_2O + \text{Heat} \qquad (13.23)$$

Hydrogen can serve as an energy storage method that can be utilized by different sectors such as buildings (residential or commercial), industry, and transportation. The energy storage options it can provide to future smart grids are an important characteristic which could be instrumental in benefiting from electricity generation through intermittent renewable energy technologies. Despite the major challenges it carries, it can contribute to achieving SDGs to some extent.

13.6.2 Hydrogen Production, Distribution, and Storage

In this section, the production, distribution, and storage phases of hydrogen are explained.

13.6.2.1 PRODUCTION
Hydrogen can be produced through a variety of methods [6]:

- Natural gas reforming
- Renewable electrolysis
- Gasification
- Renewable liquid reforming
- Nuclear high-temperature electrolysis
- High-temperature thermochemical water splitting
- Biological production
- Photoelectrochemical (PEC) production

FIGURE 13.25 Colors and attributions for hydrogen produced from non-renewable and renewable energy sources.

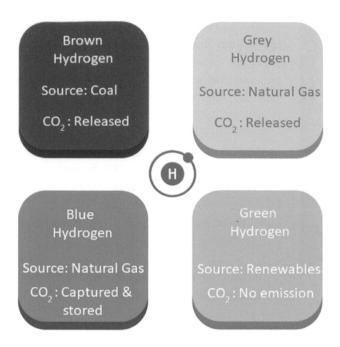

Brown Hydrogen

Source: Coal

CO_2: Released

Grey Hydrogen

Source: Natural Gas

CO_2: Released

Blue Hydrogen

Source: Natural Gas

CO_2: Captured & stored

Green Hydrogen

Source: Renewables

CO_2: No emission

H

Natural gas reforming uses high-temperature steam which reacts with natural gas to produce syngas. This process is called steam–methane reforming. This is the most economic and most common technique for producing hydrogen. **Renewable electrolysis** produces hydrogen based on the electrolysis principle, which uses electric current to split water into hydrogen and oxygen. Electrical energy needed for the process can come from non-renewable or renewable sources. If the source of electricity for the electrolysis process is renewable energy, then the process is called renewable electrolysis. **Gasification** is another method of obtaining syngas which reacts with steam to produce hydrogen. Coal or biomass can be used as the input fuel. With appropriate carbon capture and storage (CCS) methods, hydrogen production from these fuels can almost have near-zero greenhouse gas (GHG) emissions. **Renewable liquid reforming** utilizes renewable liquid fuels such as ethanol or bio-oil which yield hydrogen when reacted with high-temperature steam. **Aqueous-phase reforming** (APR) is a similar technology in producing hydrogen. **Nuclear high-temperature electrolysis** is a type of electrolysis technique where water is heated to high temperatures using nuclear energy. Increasing its temperature enables water to be split more easily, hence resulting in less electrical energy being required for the electrolysis process. **High-temperature thermochemical water splitting** is another method for splitting water. High temperatures achieved by solar concentrators or nuclear reactors drive a series of chemical reactions that split water to produce hydrogen. **Biological production** of hydrogen takes place by use of microbes, such as green algae. These microbes split water into hydrogen and oxygen when they are exposed to sunlight, as part of their metabolism. **Photoelectrochemical (PEC) production** has to do with photons, electrons (electricity), and a chemical reaction, as its name reveals. A special class of semiconductors absorb energy from sunlight and utilize this energy to split water into hydrogen and oxygen [6].

While all these methods described have their own challenges, it is an advantage of hydrogen that it can be produced from a variety of sources rather than being strictly reliant on a single source. There exists a color coding for hydrogen based on the sources it is produced from and the way emissions are handled. In **Figure 13.25**, four major color codes for produced hydrogen can be seen. The diagram depicts the source of hydrogen and the emission characteristic for each category.

The efficiency rate of hydrogen can be low considering all the processes from its production through the different methods discussed, to its compression, transportation, utilization in a fuel cell, and electrical energy output to run an electric motor. As there are losses in each step involved, about 70% of the energy can be lost. That is, if 100 units of energy are generated by an energy source (e.g., wind turbine), the amount of electrical energy obtained to operate an electric motor will be approximately 30 units. Therefore, there is significant room for R&D and improvement to reduce losses in each step and achieve higher electrical output when discharged.

Another challenge for engineers, or broadly scientists, is finding economic ways to increase the rate of hydrogen production. The rate at which hydrogen is produced can be increased by increasing the amount of input fuel or energy, or it can be improved by enhanced materials or processes. For example, the production rate of hydrogen through the electrolysis method can easily be increased by increasing the voltage supplied to the system. However, this would mean higher energy consumption. Another solution would be achieving higher electrical conductivity within the electrolyte and the electrodes. This example involves an interdisciplinary R&D involving mainly the fields of electrochemistry and materials science. This is where science and engineering can contribute to generating increased energy storage with the same amount of input energy.

13.6.2.2 DISTRIBUTION

Transportation and distribution of hydrogen can be conducted in the form of compressed gaseous or liquid hydrogen. The most common methods for transportation and distribution are [7]:

- Compressed gas tube trailers
- Liquefied hydrogen tankers
- Pipelines

Compressed gas tube trailers can transport compressed hydrogen in high-pressure tube trailers. These tube trailers can come in a variety of designs such as modular, jumbo, or composite tube trailers. **Liquefied hydrogen tankers** carry hydrogen in well-insulated cryogenic tanks. As the specific volume of liquid hydrogen is much lower than that of gaseous hydrogen, more hydrogen for a given volume of container can be transported. Even though the tankers are insulated, liquid hydrogen can heat up during long journeys or in warmer climates, which causes the pressure inside the tankers to increase. Besides trucks, hydrogen either as a compressed gas or as a cryogenic liquid can also be transported on railways or waterways in railcars, ships, or barges. Using **pipelines** for transporting hydrogen is the least expensive method. The initial investment outlay for this method is high; however, the payback time of these webs can become reasonably short with increasing demand for hydrogen.

13.6.2.3 STORAGE

Production, transportation, and distribution of hydrogen are key components in utilizing this energy source in different applications; however, they do not complete the picture for use of hydrogen to be realistic. It should also be storable in a safe and cost-effective manner. Hydrogen storage is another key element for sustaining and enhancing technologies that use hydrogen as fuel, such as fuel cells. Different hydrogen storage methods exist. Some of these methods are categorized in **Figure 13.26**.

FIGURE 13.26 Methods of hydrogen storage. Source: U.S. Department of Energy, Hydrogen Storage.

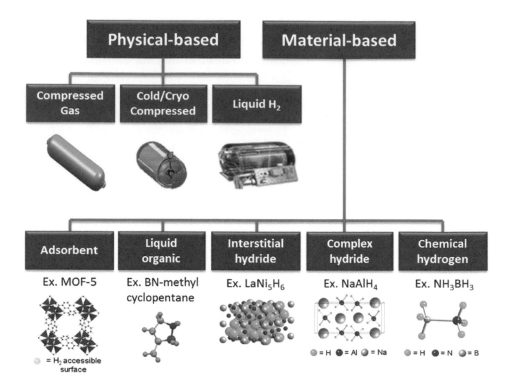

Physical-based hydrogen storage methods are similar to the common transportation and distribution methods. Hydrogen can be stored physically in containers in the form of compressed gas or cryogenic liquid. Material-based storage involves depositing of hydrogen in a variety of solids. Hydrogen can be stored on the surfaces of solids via *adsorption*, or it can be stored within the solids via *absorption*. One is a surface phenomenon, and the other is a bulk volumetric phenomenon. Let us think about this with an analogy. Imagine a handful of sewing pins. These can be stored on the surface of a magnetic pin holder or they can be stowed inside a pin box. The pins are hydrogen in this analogy. The magnetic pin holder is the hydrogen storage material experiencing adsorption, and the pin box represents the storage medium in the absorption process.

A challenge for hydrogen storage is its low density at the operating temperatures of the applications. Even though it has the highest energy per unit mass (gravimetric density, MJ/kg) as a fuel, it has a very low energy per unit volume (volumetric density, MJ/L) compared to its competitors. This results in a need for larger pressure vessels with higher pressures. Both requirements have limitations. Larger volume is a concern due to space limitations, and achieving higher pressures requires advanced, safe, and cost-effective materials for vessels. R&D studies focus on ways of attaining hydrogen storage technologies that are safe, economic, and offer higher energy densities.

13.6.3 Hydrogen Vehicles

Hydrogen vehicles are means of transportation using hydrogen as the fuel for traction. These vehicles can be automobiles, buses, tractors, motorcycles, trains, ships, airplanes, and spacecraft. Hydrogen vehicles can burn hydrogen in an internal combustion engine in a similar manner to traditional automobiles using fossil fuels, or they can utilize

FIGURE 13.27 *Energy Observer*, the first hydrogen vessel to travel around the world, passing through St. Petersburg, Russia. Source: Peter Kovalev/TASS via Getty Images.

hydrogen in fuel cells to generate electricity for powering the vehicles. These two types, hydrogen internal combustion engine vehicle (HICEV) and fuel cell electric vehicle (FCEV) are discussed in this section.

13.6.3.1 HYDROGEN INTERNAL COMBUSTION ENGINE VEHICLE (HICEV)

HICEVs employ an internal combustion engine like conventional vehicles burning fossil fuels such as gasoline or diesel, or renewable fuels such as bioethanol. The advantage of hydrogen over these fuels is that the hydrogen internal combustion has near-zero emissions. While bioethanol yields reduced emissions compared to non-renewable fossil fuels, it still generates CO_2 and NO_x emissions. Another advantage of HICEVs is that the additional cost of production and assembly lines is not as high due to the similarities with traditional automobile manufacturing techniques.

13.6.3.2 FUEL CELL ELECTRIC VEHICLE (FCEV)

Electric vehicles (EVs) can be all-electric or plug-in hybrid electric vehicles (PHEVs). All-electric (or pure-electric) vehicles include battery electric vehicles (BEVs), solar vehicles, or fuel cell electric vehicles (FCEVs). PHEVs are hybrid technologies employing both conventional internal combustion engines and electric motors.

FCEVs, also known as fuel cell vehicles (FCVs), utilize fuel cells to power a variety of vehicles such as automobiles, motorcycles, boats, and buses. Larger scale applications include submarines, trains, and airplanes. They use hydrogen as fuel like HICEVs, and they are electric vehicles like BEVs. Therefore, they can be compared with both technologies. FCEVs cost more than HICEVs due to the cost of the fuel cell itself. Most fuel cell vehicles have zero emissions. As an electric vehicle, FCEVs can provide higher energy capacities than BEVs, depending on the fuel cell design. Other advantages of FCEVs over BEVs include lower weight, elimination of charging time, and battery degradation. Their disadvantage against BEVs is the number of steps involved for providing traction. BEVs use electrical energy from the grid (which could have been generated by non-renewable or renewable means) to charge the battery and run on an electric motor. FCEVs need hydrogen to be produced, transported to a station, and stored within the vehicle. Stored hydrogen is then processed within the fuel cell to either directly power the motor or charge the batteries.

13.7 Role of Artificial Intelligence in Energy Storage

Artificial intelligence (AI) is a multidisciplinary branch of computer science with other fields including engineering, economics, education, health, and public safety. It is the intelligence of machines designed with the purpose of achieving systems that think and operate like humans. Some examples of AI that we are using currently in our daily lives are navigation systems that optimize the route for us depending on traffic congestion or other conditions affecting the arrival time to destination, online search and recommendation algorithms, chatbots, and facial detection and recognition for access to personal electronics, for surveillance, or for security such as passenger screening at airports.

Energy storage technologies driven by AI can collect and process data on renewable resources (e.g., weather forecasting for solar or wind), existing energy conversion efficiencies, and demand. This processed data then feeds a simulation to optimize energy storage and utilization options, as well as predicting possible deficiencies. Optimization of energy use and storage involves maximization of financial gains as well, accounting for the actual energy costs and incentive program parameters for the region of interest.

Three key steps in applications of AI for energy storage and its integration into the modern energy grid are:

- *Machine learning* (ML) for recognizing patterns to predict energy supply and demand
- Running the simulation that is built based on a model
- Utilization of the *Internet of Things* (IoT) to monitor and control the energy systems

FURTHER LEARNING

Machine Learning in Energy Storage
https://www.nrel.gov/transportation/
machine-learning-for-advanced-
batteries.html

Applying AI to energy storage systems and its implementation into future smart grids can have some challenges as well. These challenges are similar to those for *digitalization* in the energy sector. One of the challenges is the need for additional machines and computing power to be able to process the enormous amount of data to run the smart grid effectively. Another concern is *cyber-security*. While AI and digitalization offer many benefits, they can also make the energy systems more vulnerable to cyber-attacks.

Use of AI will keep increasing along with digital transformation. These technologies are expected to help the pieces of smart grids connect in a better way by making energy systems more intelligent and more efficient while ensuring sustainability, reliability, and resilience.

13.8 Summary

Energy storage systems play a significant role in many aspects of the energy sector from portable electronics and wearable technologies to recreational devices, health equipment, electric vehicles, and power generation by renewable or non-renewable sources. Energy storage enables mobility, grid reliability and resilience, improved demand management, increased renewable integration, and improved end-user quality of life. There is a diverse pool of energy storing technologies, which is an advantage for the energy sector. These technologies have their own strengths and weaknesses. Not every energy storage technology is applicable everywhere. Some of them have geographic or space limitations. Other parameters in determining optimum energy storage method include cost, energy and power capacities and densities, round-trip efficiency, storage duration, discharge duration, lifespan of the system, and the cycling times. For a given application, all of these generic parameters along with application-specific criteria should be evaluated to determine the best energy storage option. More research and investment into energy storage technologies is expected towards 2050 with the increasing share of renewable energy, growth of smart grids, expansion of the EV market, and development in portable electronics. This is mainly of interest to policymakers, engineers, the energy sector, and businesses.

REFERENCES

[1] J. Wang, *et al.*, "Overview of compressed air energy storage and technology development," *Energies*, vol. 10, no. 7, p. 991, Jul. 2017. doi: 10.3390/en10070991.

[2] International Hydropower Association, "2020 Hydropower Status Report: Sector Trends and Insights," IHA, 2020. Accessed: Nov. 17, 2020 [Online]. Available: www.hydropower.org/publications/2020-hydropower-status-report.

[3] I. Sarbu and C. Sebarchievici, "A comprehensive review of thermal energy storage," *Sustainability*, vol. 10, no. 2, p. 191, Jan. 2018. doi: 10.3390/su10010191.

[4] A. Sharma, V. V. Tyagi, C. R. Chen, and D. Buddhi, "Review on thermal energy storage with phase change materials and applications," *Renewable and Sustainable Energy Reviews*, vol. 13, no. 2, pp. 318–345, Feb. 2009. doi: 10.1016/j.rser.2007.10.005.

[5] University of Cambridge, DoITPoMS, "TLP Library Fuel Cells: Types of Fuel Cells." www.doitpoms.ac.uk/tlplib/fuel-cells/types.php (accessed Nov. 18, 2020).

[6] U.S. Department of Energy, Fuel Cell Technologies Office, "Hydrogen Production," Sep. 2014. Accessed: Nov. 21, 2020 [Online]. Available: www.energy.gov/sites/prod/files/2014/09/f18/fcto_hydrogen_production_fs_0.pdf.

[7] U.S. Department of Energy, Alternative Fuels Data Center, "Hydrogen Production and Distribution," 2019. https://afdc.energy.gov/fuels/hydrogen_production.html (accessed Nov. 21, 2020).

CHAPTER 13 EXERCISES

13.1 Describe the following terms and highlight their similarities and differences:
   a. Microgrid
   b. Macrogrid
   c. Supergrid
   d. Megagrid
   e. Smart grid

13.2 Some important terms associated with smart grids and AI are listed below. Research the definition of each term and determine their relevance to AI and energy storage.
   a. T&D: Transmission and Distribution
   b. AMI: Advanced Metering Infrastructure
   c. DER: Distributed Energy Resources
   d. DMS: Distribution Management System
   e. SAIDI: System Average Interruption Duration Index
   f. SCADA: Supervisory Control and Data Acquisition

13.3 **Case Study:** The *Smart Grid Gotland Project* is a microgrid example from a Swedish island on the Baltic Sea. Research this project and prepare a case study report that includes the following sections:
   - *Introduction:* Introduce the case with its background and previous work (if any)
   - *Objective:* Describe the objective of the project
   - *Methodology:* Explain how the project was planned, initiated, and carried out.
   - *Results:* Provide your findings through your analysis.

- *Discussion:* Explain the significance of this project and the conclusions that can be drawn from this particular project.
- *Recommendations:* Provide your recommendations for any key problems you might have identified during your analysis, or to improve the effectiveness of the existing project.

Smart Grid Gotland Project

Location: Gotland, Sweden
Size: Microgrid
Project objective: Demonstration, R&D
Sponsor: Swedish Energy Agency
www.d-maps.com

13.4 The upper reservoir of a pumped hydro energy storage (PHES) facility in Tajikistan is 85 m above the powerhouse. The reservoir has a surface area of 1 km^2 with an average depth of 3.5 m.

a. Find the energy capacity of the reservoir (MWh).

b. Determine the penstock efficiency (η_{penstock}) if the amount of energy used during the charging process is 1450 MWh, and the transformer, motor, and pump efficiencies are 86%, 85%, and 82%, respectively (%).

13.5 If the energy that can be retrieved from the upper reservoir in Exercise 13.4 during the discharging process is 750 MWh, determine the round-trip efficiency of the PHES system.

13.6 An adiabatic compressed air energy storage (A-CAES) facility in Manchester, United Kingdom, stores energy in a salt cavern 300 m beneath the surface. The volume of the cavern is estimated to be 280,000 m^3. Air at atmospheric pressure is compressed to a pressure of 3.87 MPa in the cavern. Determine:

    a. the amount of energy that is stored in the salt cavern (MJ)

    b. the amount of air that is compressed in the cavern if the average temperature in the cavern is 45 °C (kg)

13.7 A flywheel energy storage (FES) unit houses a rotating disk with a diameter of 1.2 m and a mass of 1800 kg. The disk is rotating with a frequency of 700 rpm. Determine the amount of energy stored in the unit (MJ).

13.8 A large well-insulated hot-water tank is used for sensible heat thermal energy storage (SHTES). The volume of the tank is 12 million liters. The water in the tank is heated by means of solar collectors placed atop the tank. Water enters the collectors at a temperature of 18 °C. The temperature of the water maintained in the tank is 42 °C. Determine the amount of energy stored in the hot-water tank (MJ).

13.9 In a latent heat thermal energy storage (LHTES) application for residential cooling, beeswax is used as a phase change material (PCM). During the daytime, beeswax is heated from 25 °C solid to 64 °C liquid. The amount of beeswax used is 8 kg. Determine the cooling capacity of it at nighttime if 80% of the PCM melts. Melting temperature of the beeswax used is 60.1 °C, and its latent heat is given as 145 kJ/kg. Specific heat values for solid and liquid phases are approximated as 3.4 kJ/kg K and 7.5 kJ/kg K, respectively.

13.10 Write the anode and cathode reactions and draw the schematics of the fuel cell types below:

    a. Proton-exchange membrane fuel cell (PEMFC)

    b. Phosphoric acid fuel cell (PAFC)

    c. Solid oxide fuel cell (SOFC)

13.11 An electric toy car employing a sealed lead–acid (SLA) battery has a nominal battery voltage of 12 V and an electrical storage capacity of 18 Ah. The battery weighs 3.2 kg.

    a. Write the anode and cathode half-reactions and the full reaction for the battery.

    b. Determine the capacity of the battery (kWh).

    c. Find the energy density of the battery (Wh/kg).

Source: Bluecinema/E+ via Getty Images.

13.12 A superconducting magnetic energy storage (SMES) unit has a solenoid that is 0.6 m in diameter and 1.2 m in length. The solenoid coil has an air core and houses 4500 turns of wire. Determine:

a. the inductance of the coil (H)

b. the energy that can be stored in the coil if the current flowing is 900 A (MJ)

CHAPTER 14
Energy Efficiency

14.1 Introduction

Energy use is increasing along with the growth in population and the changes in our energy consumption habits. This brings up the question: How can we better prepare for a future so that we use less energy? This question can be answered with two phrases: *energy conservation* or *energy efficiency*. The two terms may seem similar; however, there is a basic difference between the two, which results from quality engineering. Energy conservation is an action of using less energy, either through personal choice, or because of necessity. Energy efficiency, on the other hand, is the engineering outcome of technologies that consume less energy to conduct the same function without giving up the standards. Turning off the lights in the evening, when there is a need for light, is energy conservation. Energy is being conserved; however, the standard of living is sacrificed due to the decision. Instead of deciding to sit in the dark, one could choose to replace an incandescent light bulb with an LED (light emitting diode) bulb, which consumes much less energy for the same amount of lumens. This would be an energy-efficiency example. The more efficiency improvement in such systems, the more energy savings we would get for the same function.

Energy efficiency can either misleadingly be grouped together with renewable energy technologies or simply be disregarded. Energy efficiency is not a renewable

Table 14.1

Orders of magnitude of power ratings of systems and devices used in various applications		
Application	**System/device**	**Order of magnitude of power (W)**
Buildings (general)	LED light	10^0
Buildings (general)	Laptop	10^1
Health/medicine	Dental chair	10^2
Buildings (residential or small offices)	Air-conditioner	10^3
Agriculture	Tractor (100 hp)	10^4
Transportation (personal)	Electric vehicle (250 hp)	10^5
Power generation	Nuclear reactor coolant pump	10^6
Transportation (commercial)	Airliner jet engine (30,000 hp)	10^7

energy source. Renewable energy systems *generate* energy, while energy efficiency is a term associated with *consuming* energy in a way that the consumption is less to achieve the same amount of output from the system that is utilized. The second mistake, as mentioned above, would be disregarding energy efficiency. It is the *low-hanging fruit* towards sustainability and a cleaner ecosystem. Efforts to enhance the energy efficiency of buildings, vehicles, and electrical devices used in various sectors such as manufacturing, agriculture, health, and power generation will help in achieving sustainable development goals (SDGs). It only makes sense to consider renewable energy technologies and investing in these systems after ensuring that energy efficiency has been given priority. Imagine a family with limited income and yet spending all of it, sometimes on things which are not essential. Due to this, the members of the family decide to take overtime work or part time jobs to help with the bills. The more rational act would be to first identify some expenditure items that can be totally cut or reduced without compromising health and comfort standards. This would be the analogy of energy efficiency. Then, the family members can do additional work to increase the net income of the house. This would be the renewable energy.

Obviously, the ideal scenario would be to ensure the development and use of energy-efficient systems, supported with renewable energy generation. This would significantly help in achieving the SDGs defined by the United Nations. In 2012, the European Union released the Energy Efficiency Directive (EED) 2012/27/EU, which mandates energy-efficiency improvements in various sectors. Member states submitted their National Energy Efficiency Action Plans (NEEAP) by 2014, specifying the steps to be taken towards improving energy efficiency and reporting the energy savings expected to result from the measures taken.

According to IRENA, renewable energy and energy-efficiency measures together have the potential to achieve 90% of the carbon reductions that are required to limit the global temperature increase to a maximum of 2 °C above pre-industrial levels with a 66% probability, aligned with the *Paris Agreement* goals [1].

Therefore, it is important to understand, further study, and invest in energy efficiency. This would apply to efficient use of energy in all sectors including mainly buildings, industry, and transportation.

14.2 Buildings

14.2.1 Building Energy Consumption

Building energy consumption adds up to significant numbers globally. In the United States, buildings account for approximately 40% of the total energy consumption [2]. This indicates that any improvement in building technologies and enhancement in building energy efficiency would result in remarkable reductions in total global energy consumption and carbon emissions.

As can be seen in **Figure 14.1**, residential buildings consume about one-fifth of the total energy in the United States. This consumption comes mainly from heating, cooling, cooking, and lighting equipment.

According to the United Nations Economic Commission for Europe's (UNECE) Committee on Sustainable Energy report [3], buildings are central in combating sustainability challenges we are facing. High-performance buildings can help achieve a number of SDGs including:

- *SDGs 1 and 7:* combating poverty by making energy more affordable,
- *SDGs 9, 11, and 12:* promoting sustainable urban development by considering buildings as parts of energy networks at all levels, and

FIGURE 14.1 Shares of total energy consumption by residential and other end-use sectors in the United States (2019). Source: U.S. EPA.

Share of total U.S. energy consumption by end-use sectors, 2019

Total = 100.2 quadrillion British thermal units

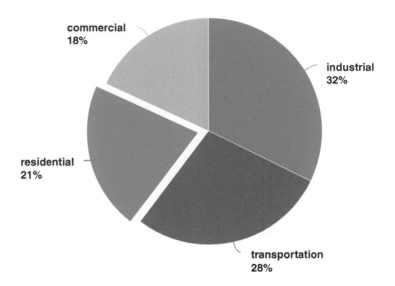

- *SDG 13:* supporting climate action by reducing the energy demand of buildings to a point that can be met by no or low-carbon energy sources.

There are numerous organizations dedicated to improving buildings for a more sustainable world. Among these, an example is the World Green Building Council (WorldGBC) which defines its goals as [4]:

- *Climate Action:* total decarbonization of the built environment.
- *Health and Wellbeing:* a built environment that delivers healthy, equitable, and resilient buildings, communities, and cities.
- *Resources and Circularity:* a built environment that supports the regeneration of resources and natural systems, providing socio-economic benefit through a thriving circular economy.

Many municipalities all around the world are trying to promote public awareness of building energy use. Some are implementing *smart city* programs with the purpose of making cities more citizen-friendly and sustainable.

Some of the commonly used phrases used for buildings that are environmentally friendly are:

- Green buildings
- Sustainable buildings
- Smart buildings
- Net-zero buildings

While these are not all exactly the same, the core value in all of them is that these buildings are aiming to reduce their carbon footprint. For such buildings, selection of materials for their construction is important. There are standards, codes, or regulations that can be specific to a city, state, or country which address the selection of construction materials in consideration of:

- *Thermal performance*
- *Mechanical properties*

- *Lifespan of materials*
- *Acoustic performance*
- *Indoor air quality*
- *Safety*
- *Cost*

Some factors that affect the energy performance of the building are:

- *Location*
- *Weather*
- *Building configuration*
- *Equipment and devices in the building*
- *Occupancy*

Some effective ways of improving building energy efficiency are:

- *Sealing air leaks and better insulation*
- *Installing energy-efficient windows*
- *Switching to high-performance lighting*
- *Utilizing passive solar designs for heating and cooling*
- *Installing green roofs*

Evaluation and calculation of energy and water use for buildings can be challenging, depending on geography, building complexity, and reliability of data including material properties, building details, and meteorological data. Software is available for conducting analysis on energy consumption, modeling, and efficiency. *eQuest* is a freely available building energy simulation tool that was funded mostly by the U.S. Department of Energy. Building performance modeling is done by a combination of three components: a building design wizard, an energy efficiency measure (EEM) wizard, and a graphical results display module. *EnergyPlus* is another building energy simulation program which is also free and open source. The software is funded by the U.S. Department of Energy's Building Technologies Office (BTO) and managed by the National Renewable Energy Laboratory (NREL). The program can be used to model both energy and water consumption in buildings. Both simulation tools, eQuest and EnergyPlus, have their own advantages and disadvantages. eQuest is easier to use and quick in generating the results. This can be an advantage during the design phase of a project in terms of time efficiency. EnergyPlus can deal with more complex building designs and yield more accurate results. This comes with a cost of more time invested in the analysis. *OpenStudio* is another freely available cross-platform collection of simulation tools to provide help in energy modeling for building professionals. It is a graphical energy-modeling tool that uses EnergyPlus. It includes visualization and editing of schedules, editing of materials and constructions, a user-defined interface to apply resources to the spaces and zones in the building, and a visual HVAC (heating, ventilating, and air-conditioning) design tool. OpenStudio also utilizes *Radiance*, which is a lighting simulation tool for conducting advanced daylight analysis. Radiance is used by architects and engineers to predict illumination inside a building, to design innovative spaces, and to evaluate new lighting and daylighting technologies. There are other programs and tools available to simulate building energy and water use. Some of these programs are commercial products. All this software helps architects, engineers, and energy professionals design energy-efficient buildings which promote sustainability. This enables smoother integration of new and existing buildings into sustainable and smart cities.

FURTHER LEARNING

eQuest: Quick Energy Simulation Tool
www.doe2.com/equest

FURTHER LEARNING

EnergyPlus
https://energyplus.net/

FURTHER LEARNING

Building Design Tools
www.wbdg.org/additional-resources/tools

Example 14.1 Building Energy Consumption Analysis Using eQuest Simulation Tool

Perform an energy consumption analysis of the campus building given below for four different global locations using eQuest.

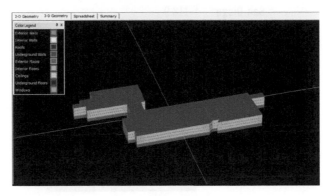

Perspective view of the engineering building.

The locations are:

- *Berlin, Germany*
- *Istanbul, Turkey*
- *San Diego, United States*
- *Shanghai, China*

Global locations (www.d-maps.com).

Results

Energy consumption analysis of the same building in different locations requires weather data for heating and cooling energy use. eQuest uses a weather database which can be obtained from *doe2.com*. Non-US locations can also be found in this database. Weather files from the EnergyPlus database can also be converted to be implemented into the eQuest simulation. A summary of the results for each location along with graphically displayed details can be seen in the table and following figures.

City	Space cooling electricity use (MWh)	Space heating fuel use (million Btu)
Berlin	71.9	772.5
Istanbul	146.1	200.8
San Diego	118.9	0.30
Shanghai	217.8	25.2

Berlin

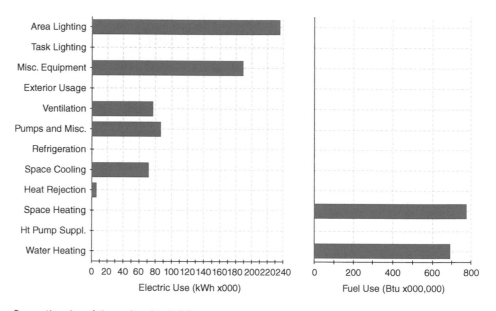

Perspective view of the engineering building.

Istanbul

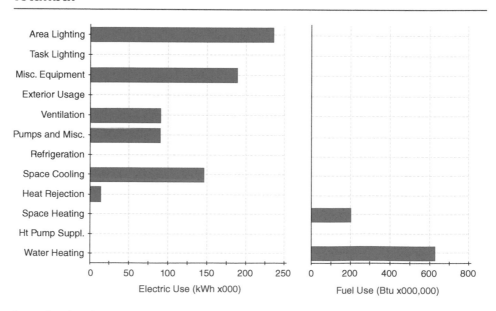

Perspective view of the engineering building.

San Diego

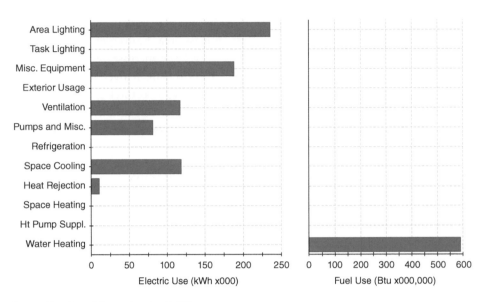

Perspective view of the engineering building.

Shanghai

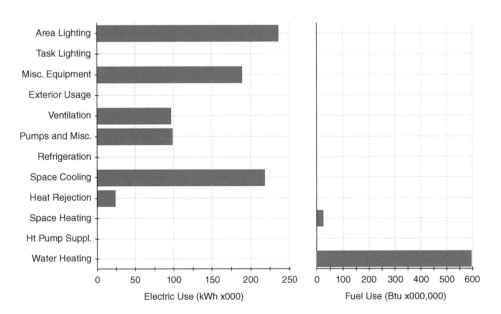

Perspective view of the engineering building.

Discussion

Cities such as Istanbul and Shanghai, which are hot and humid, require more energy consumption in both summer and winter as their winters are also tough. For these two cities, while cooling energy consumption does not increase significantly with an increase in glass area, heating energy consumption increases by 82.5% for Shanghai. This is a very large increase, which can be explained by the conduction losses in Shanghai being dominant over the solar radiation heat gain in winter. The notable results for San Diego portraying a mild climate hence lower energy consumption are observed for both cooling and heating seasons.

14.2.2 HVAC Systems

We have emphasized the importance of buildings in energy consumption. Looking at the shares of residential and commercial buildings in total energy consumption, it becomes obvious that energy efficiency in buildings is crucial in addressing a sustainable environment.

Figure 14.2 illustrates the shares of consumption items for both residential and commercial buildings. These EIA charts include both 2019 numbers and the projected consumption for 2050. If we zoom in on the breakdown of energy consumption items, we realize how significant a portion of it comes from heating and cooling. Hence, it is important to focus on reducing the HVAC energy consumption for buildings.

There are three key elements to achieving reduced HVAC energy consumption:

a. High-efficiency HVAC and refrigeration units: Consumes less energy for the same amount of heating or cooling required
b. Well-insulated building envelopes: Helps lower the heating and cooling loads of the building
c. Acquainted end-users: Ensures functional *and* economic use of HVAC systems

High-Efficiency HVAC and Refrigeration Units

Air-conditioners, furnaces, and refrigerators are life-saving systems whether we are at home or in our workplaces. These systems can come in different configurations depending on required heating and cooling capacities, space limitations, environmental

FIGURE 14.2 Residential and commercial purchased electricity intensities for 2019, and the projected values for 2050. Source: U.S. Energy Information Administration, Annual Energy Outlook 2020.

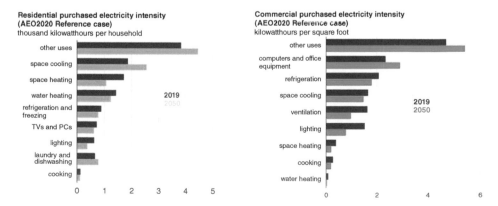

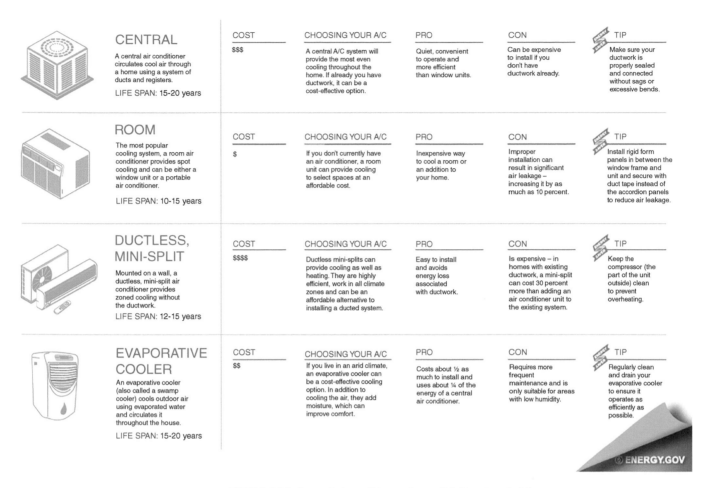

CENTRAL	COST	CHOOSING YOUR A/C	PRO	CON	TIP
A central air conditioner circulates cool air through a home using a system of ducts and registers. LIFE SPAN: 15-20 years	$$$	A central A/C system will provide the most even cooling throughout the home. If already you have ductwork, it can be a cost-effective option.	Quiet, convenient to operate and more efficient than window units.	Can be expensive to install if you don't have ductwork already.	Make sure your ductwork is properly sealed and connected without sags or excessive bends.
ROOM	COST	CHOOSING YOUR A/C	PRO	CON	TIP
The most popular cooling system, a room air conditioner provides spot cooling and can be either a window unit or a portable air conditioner. LIFE SPAN: 10-15 years	$	If you don't currently have an air conditioner, a room unit can provide cooling to select spaces at an affordable cost.	Inexpensive way to cool a room or an addition to your home.	Improper installation can result in significant air leakage – increasing it by as much as 10 percent.	Install rigid form panels in between the window frame and unit and secure with duct tape instead of the accordion panels to reduce air leakage.
DUCTLESS, MINI-SPLIT	COST	CHOOSING YOUR A/C	PRO	CON	TIP
Mounted on a wall, a ductless, mini-split air conditioner provides zoned cooling without the ductwork. LIFE SPAN: 12-15 years	$$$$	Ductless mini-splits can provide cooling as well as heating. They are highly efficient, work in all climate zones and can be an affordable alternative to installing a ducted system.	Easy to install and avoids energy loss associated with ductwork.	Is expensive – in homes with existing ductwork, a mini-split can cost 30 percent more than adding an air conditioner unit to the existing system.	Keep the compressor (the part of the unit outside) clean to prevent overheating.
EVAPORATIVE COOLER	COST	CHOOSING YOUR A/C	PRO	CON	TIP
An evaporative cooler (also called a swamp cooler) cools outdoor air using evaporated water and circulates it throughout the house. LIFE SPAN: 15-20 years	$$	If you live in an arid climate, an evaporative cooler can be a cost-effective cooling option. In addition to cooling the air, they add moisture, which can improve comfort.	Costs about ½ as much to install and uses about ¼ of the energy of a central air conditioner.	Requires more frequent maintenance and is only suitable for areas with low humidity.	Regularly clean and drain your evaporative cooler to ensure it operates as efficiently as possible.

ENERGY.GOV

FIGURE 14.3 Types of air-conditioners. Source: U.S. Department of Energy.

conditions, and consumer needs. Some common types of air-conditioners for both domestic and office use can be seen in **Figure 14.3**.

Window-type air-conditioners were widely in use in many of the developed and developing countries for a long time during the twentieth century. The problem with them was that the compressor unit was nested in the same frame that the evaporator was attached to. While the short distance for tubing and compact size was an advantage, the acoustic concern due to the compressor having direct access to the indoor environment for airborne sound transmission led consumers to split-type air-conditioners. The idea of these systems is that the compressor is kept away from the indoor environment, hence lowering the sound pressure level of the system by the receivers, the building occupants. The outdoor units include the condenser as well, for heat rejection to the outside environment. **Figure 14.4** depicts an air-conditioning system and the components of its indoor and outdoor units.

For larger cooling capacities, chiller units are used. These systems are utilized along with air-handling units (AHUs). **Figure 14.6** shows a refrigerant-based air-conditioner working with an AHU, air-cooled condensing unit, and a thermal-energy storage tank. Air-cooled units reject heat to the outside environment by means of forced convection with the help of fans. Water-cooled systems on the other hand, either have an open-loop system where the heat is dissipated into a large reservoir (heat sink) such as a river or a lake, or they operate along with a closed-loop system employing a cooling tower.

FURTHER LEARNING

History of Air-Conditioning
www.energy.gov/articles/history-air-conditioning

FIGURE 14.4 Indoor and outdoor units of an air-conditioner. Source: U.S. Department of Energy.

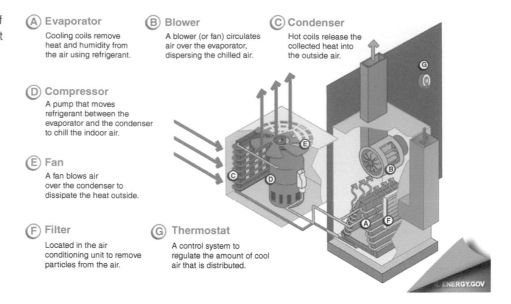

Ⓐ **Evaporator**
Cooling coils remove heat and humidity from the air using refrigerant.

Ⓑ **Blower**
A blower (or fan) circulates air over the evaporator, dispersing the chilled air.

Ⓒ **Condenser**
Hot coils release the collected heat into the outside air.

Ⓓ **Compressor**
A pump that moves refrigerant between the evaporator and the condenser to chill the indoor air.

Ⓔ **Fan**
A fan blows air over the condenser to dissipate the heat outside.

Ⓕ **Filter**
Located in the air conditioning unit to remove particles from the air.

Ⓖ **Thermostat**
A control system to regulate the amount of cool air that is distributed.

© ENERGY.GOV

As for heating systems, for domestic applications in typical US homes, a furnace is attached to ductwork for space heating, and a water heater for hot water needs. For larger applications such as commercial buildings, high-capacity boilers are required for both space heating and hot-water needs. Different types of heating systems can be seen in **Figure 14.7.**

Well-Insulated Building Envelopes

Insulation updates for buildings not only improve energy efficiency, but also durability and comfort level. Whether it is a residential, commercial, public, or industrial building, insulation of building envelopes helps reduce energy bills and outside noise (**Table 14.2**).

For residential buildings, insulation is used as both a thermal and an acoustic barrier for the structural components such as the walls, ceilings, roofs, and floors. In

FIGURE 14.5 Willis Carrier (1876–1950) was the inventor of the first modern air-conditioner, which he designed in 1902. He started his experiments on humidity control for a printing plant in Brooklyn, New York. He used existing refrigeration concepts and applied them, coming up with a system where chilled water ran through coils and air was sent to flow around the coils. This resulted in cooling and dehumidification of the air. Source: Wikimedia Commons, Carrier Corporation, https://en.wikipedia.org/wiki/Willis_Carrier#/media/File:Willis_Carrier_1915.jpg.

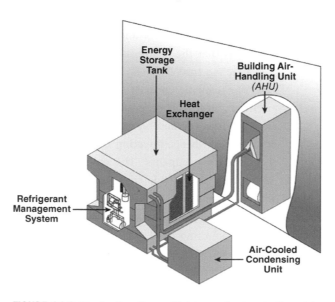

FIGURE 14.6 The *Ice Bear Storage Module* was developed with assistance from the U.S. Department of Energy Inventions and Innovation Program. Source: U.S. Department of Energy, Office of Energy Efficiency and Renewable Energy (EERE).

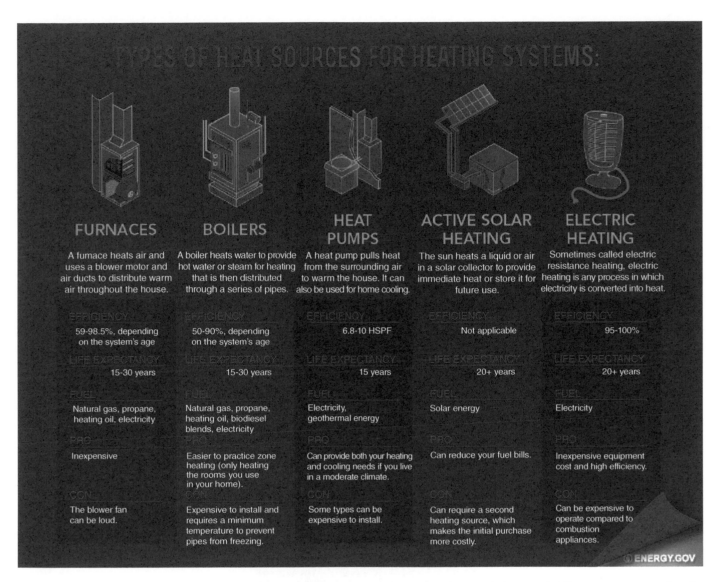

	FURNACES	BOILERS	HEAT PUMPS	ACTIVE SOLAR HEATING	ELECTRIC HEATING
	A furnace heats air and uses a blower motor and air ducts to distribute warm air throughout the house.	A boiler heats water to provide hot water or steam for heating that is then distributed through a series of pipes.	A heat pump pulls heat from the surrounding air to warm the house. It can also be used for home cooling.	The sun heats a liquid or air in a solar collector to provide immediate heat or store it for future use.	Sometimes called electric resistance heating, electric heating is any process in which electricity is converted into heat.
EFFICIENCY	59-98.5%, depending on the system's age	50-90%, depending on the system's age	6.8-10 HSPF	Not applicable	95-100%
LIFE EXPECTANCY	15-30 years	15-30 years	15 years	20+ years	20+ years
FUEL	Natural gas, propane, heating oil, electricity	Natural gas, propane, heating oil, biodiesel blends, electricity	Electricity, geothermal energy	Solar energy	Electricity
PRO	Inexpensive	Easier to practice zone heating (only heating the rooms you use in your home).	Can provide both your heating and cooling needs if you live in a moderate climate.	Can reduce your fuel bills.	Inexpensive equipment cost and high efficiency.
CON	The blower fan can be loud.	Expensive to install and requires a minimum temperature to prevent pipes from freezing.	Some types can be expensive to install.	Can require a second heating source, which makes the initial purchase more costly.	Can be expensive to operate compared to combustion appliances.

ENERGY.GOV

FIGURE 14.7 Different types of heating systems. Source: U.S. Department of Energy, www.energy.gov/sites/prod/files/2014/01/f6/homeHeating.pdf

FURTHER LEARNING

Insulation
www.energy.gov/energysaver/
insulation

FIGURE 14.8 A welder is working on a multi-tubular heat exchanger for the boiler industry. Source: Mediterranean/E+ via Getty Images.

commercial or public buildings, while thermal benefits of insulation are the priority, some buildings such as libraries can take advantage of the acoustic performance benefits of the insulation systems. In industrial facilities, insulation materials can be used to cover HVAC ductwork to prevent heat loss or heat gain. Insulation can also be applied to equipment such as turbines in power generation plants.

Green Roofs

Besides the insulation systems already listed, there is also an organic, live insulation method that has been gaining popularity in the new millennium. This application is called green roofs, and is also associated with other terms such as vegetable roofs or roof gardens. Plants can also be applied to walls, in which case such applications would be called green walls or vertical gardens.

Green roof systems are composed of layers, including a plant canopy layer, growth medium, and a drainage layer, as illustrated in **Figure 14.10**. While synthetic insulation materials are most common in the market to combat high energy costs and promote sustainability for buildings, this "live" method provides several benefits in addition to

FIGURE 14.9 A *genetic algorithm* is one of the tools used in optimal design approach for heat exchangers. The algorithm encodes possible solutions to a design problem on a chromosome-like data structure, which can be mutated and altered through an iterative process to yield optimum solutions. Source: Malte Mueller via Getty Images.

FURTHER LEARNING

Live Insulation Systems: Green Roofs
https://greenroofs.org/

reducing HVAC loads on the buildings, such as reducing storm water run-off, increasing the lifespan of roofing membranes, adding an aesthetic appeal to the buildings, reducing acoustic noise, decreasing the heat-island effect in large cities, and contributing to carbon sequestration.

Acquainted End-Users

In using and operating HVAC systems, the role of the building occupants, who could be family members for a residential unit or facility crew members for a commercial building, is important in ensuring that the system is running effectively. Effective operation of an HVAC system translates into utilizing these systems in an energy-efficient way, while ensuring that the health and comfort conditions for the occupants in the building are not compromised and the life of these systems is maximized.

For residential buildings, the householders should either be self-trained or be encouraged to learn the aspects of using the HVAC systems in the most effective way possible through the utility services, the city, or the government.

For commercial, public, or industrial buildings, the facilities team should be well trained to be able to operate, maintain, and when needed troubleshoot the HVAC systems for achieving the best performance from these systems in terms of energy efficiency, occupant comfort, and lifespan of the equipment.

14.2.3 Load Calculations

In HVAC projects, load calculation is one of the most critical steps. Theories of cooling load (heat gain) calculations for the air-conditioning season or applications, and heating load (heat loss) calculations for the heating season or applications are briefly described in this section.

FIGURE 14.10 Green roof layers. Source: U.S. Department of Interior, National Park Service.

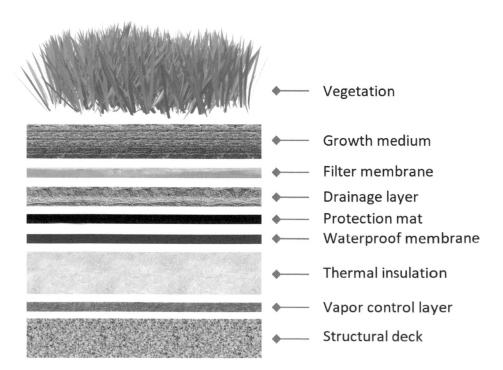

Vegetation

Growth medium

Filter membrane

Drainage layer

Protection mat

Waterproof membrane

Thermal insulation

Vapor control layer

Structural deck

Table 14.2

Insulation types [5]				
Type	**Material**	**Where applicable**	**Installation methods**	**Advantages**
Blanket: batts and rolls	Fiberglass Mineral (rock or slag) wool Plastic fibers Natural fibers	Unfinished walls, including foundation walls Floors and ceilings	Fitted between studs, joists, and beams	Do-it-yourself Suited for standard stud and joist spacing that is relatively free from obstructions. Relatively inexpensive
Concrete block insulation	Foam board, to be placed on outside of wall (usually new construction) or inside of wall (existing homes) Some manufacturers incorporate foam beads or air into the concrete mix to increase R-values	Unfinished walls, including foundation walls New construction or major renovations Walls (insulating concrete blocks)	Require specialized skills Insulating concrete blocks are sometimes stacked without mortar (dry-stacked) and surface bonded	Insulating core increases wall R-value Insulating outside of concrete block wall places mass inside conditioned space, which can moderate indoor temperatures Autoclaved aerated concrete and autoclaved cellular concrete masonry units have 10 times the insulating value of conventional concrete
Foam board or rigid foam	Polystyrene Polyisocyanurate Polyurethane	Unfinished walls, including foundation walls Floors and ceilings Unvented low-slope roofs	Interior applications: must be covered with 1/2-inch gypsum board or other building-code approved material for fire safety Exterior applications: must be covered with weatherproof facing	High insulating value for relatively little thickness Can block thermal short-circuits when installed continuously over frames or joists
Insulating concrete forms (ICFs)	Foam boards or foam blocks	Unfinished walls, including foundation walls for new construction	Installed as part of the building structure	Insulation is literally built into the home's walls, creating high thermal resistance
Loose-fill and blown-in	Cellulose Fiberglass Mineral (rock or slag) wool	Enclosed existing wall or open new wall cavities Unfinished attic floors Other hard-to-reach places	Blown into place using special equipment, sometimes poured in	Good for adding insulation to existing finished areas, irregularly shaped areas, and around obstructions
Reflective system	Foil-faced kraft paper, plastic film, polyethylene bubbles, or cardboard	Unfinished walls, ceilings, and floors	Foils, films, or papers fitted between wood-frame studs, joists, rafters, and beams	Suitable for framing at standard spacing Bubble-form suitable if framing is irregular or if obstructions are present Most effective at preventing downward heat flow, effectiveness depends on spacing

Table 14.2 (cont.)

Type	Material	Where applicable	Installation methods	Advantages
Rigid fibrous or fiber insulation	Fiberglass Mineral (rock or slag) wool	Ducts in unconditioned spaces Other places requiring insulation that can withstand high temperatures	HVAC contractors fabricate the insulation into ducts either at their shops or at the job sites	Can withstand high temperatures
Sprayed foam and foamed-in-place	Cementitious Phenolic Polyisocyanurate Polyurethane	Enclosed existing wall Open new wall cavities Unfinished attic floors	Applied using small spray containers or in larger quantities as a pressure sprayed (foamed-in-place) product	Good for adding insulation to existing finished areas, irregularly shaped areas, and around obstructions
Structural insulated panels (SIPs)	Foam board or liquid foam insulation core Straw core insulation	Unfinished walls, ceilings, floors, and roofs for new construction	Construction workers fit SIPs together to form walls and roof of a house	SIP-built houses provide superior and uniform insulation compared to more traditional construction methods; they also take less time to build

Cooling Load Calculations

Conduction

$$\dot{Q}_{cond} = UA\,\Delta T \tag{14.1}$$

where U is the overall heat transfer coefficient, A is the heat transfer surface area, and ΔT is the temperature difference between the outdoor and indoor environment.

Radiation through Fenestrations

$$\dot{Q}_{glass} = (SHGF)(A)(SC) \tag{14.2}$$

SHGF: solar heat gain factor (W/m^2, Btu/ft^2)
A: surface area of the fenestration (m^2, ft^2)
SC: shading coefficient (–)

Ventilation/Infiltration

$$\dot{Q}_{vent} = \dot{Q}_{vent,sens} + \dot{Q}_{vent,lat} \tag{14.3}$$

$$\dot{Q}_{vent,sens} = \dot{m}_{vent}c_p(T_i - T_o) \tag{14.4}$$

$$\dot{Q}_{vent,lat} = \dot{m}_{vent}(W_i - W_o)h_{fg} \tag{14.5}$$

where h_{fg} is the enthalpy of evaporation.

$$\dot{m}_{vent} = \rho_{air}\dot{Q}_{vent} = \frac{\dot{Q}_{vent}}{v_{air}} \tag{14.6}$$

Amount of infiltrated air can be calculated using the hourly air change method:

$$\dot{m}_{inf} = \rho_{air}\dot{Q}_{inf} = \frac{\dot{Q}_{inf}}{v_{air}} \tag{14.7}$$

$$\dot{Q}_{inf} = \frac{(ACH)V}{C_T} \tag{14.8}$$

where ACH is the amount of air change per unit volume of the space in an hour and is typically $0 \leq ACH \leq 2.0$; V is the volume of the space, and C_T is a constant.

Internal (Occupants, Lights, Equipment)
- *Occupants:* Heat gains are determined depending on the degree of activity. Heat gain from people has both sensible and latent portions.
- *Lights:* Heat gains are determined either theoretically by accounting for the ballast factor of the light, or by a rule of thumb which assumes a heat gain per unit surface area (W/m^2 or $Btu/h\ ft^2$).
- *Equipment:* similar to lights, either a theoretical equation is used, or a rule of thumb to determine the heat gain from the equipment in the space that is to be conditioned.

Heating Load Calculations

Transmission Losses

Transmission losses can be applied to the whole building envelope, including the walls, ceiling/roof, windows, doors, and floor. As for heat losses through floors, if this is an entry-level floor or a basement, depending on the ground depth, the U-factor can be retrieved from relevant tables:

$$\dot{Q} = UA(T_i - T_o) \tag{14.9}$$

Infiltration Losses

Losses due to infiltrating air during the heating season can be calculated in a similar manner as for the cooling season:

$$\dot{Q}_{inf,sens} = \dot{m}_{inf}\,c_p(T_i - T_o) \tag{14.10}$$

$$\dot{Q}_{inf,lat} = \dot{m}_{inf}\,(W_i - W_o)\,h_{fg} \tag{14.11}$$

14.2.4 Lighting and Appliances

In addition to HVAC systems, lighting and appliances are two other energy consumption items in buildings, as well as in some industry and transportation sectors. Besides the internal heat loads these items add to the spaces and hence increase air-conditioning energy consumption in cooling applications, they also directly impact the energy consumption of the buildings due to their operating energy demands.

Table 14.3

Comparison between traditional incandescents, halogen incandescents, CFLs, and LEDs [6]						
	60 W traditional incandescent	43 W halogen incandescent	**15 W CFL**		**12 W LED**	
			60 W traditional	43 W halogen	60 W traditional	43 W halogen
Energy saved (%)	—	~25	~75	~65	~75–80	~72
Annual energy cost[*] ($)	4.80	3.50	1.20		1.00	
Bulb life (hours)	1000	1000–3000	10,000		25,000	

[*] Based on 2 hours/day of usage, and an electricity rate of 0.11 $/kWh

Lighting

To improve the energy efficiency in lighting of buildings, there are alternative lighting systems to traditional incandescent light bulbs that can used. These alternatives could be energy-saving incandescent light bulbs, compact fluorescent light (CFL) bulbs, or light emitting diodes (LED). Comparison between traditional light bulbs and the alternatives can be seen in **Table 14.3** [6].

Lighting (or luminous) efficiency is determined by the ratio of equivalent light power to the power input. The amount of light needed is called the *luminous flux* and is measured in terms of *lumens*; it is also referred to as *brightness*. *Luminous efficacy* is the ratio of lumen output to power input. This ratio can also be used as an efficiency term called the *lamp efficiency*. This, however, is an efficiency value with units of lumen per watt (lm/W). To find luminous efficiency as a percentage, on the other hand, the equivalent light power needs to be determined:

$$\text{Equivalent light power} = \frac{\text{Lumen output}}{\text{Efficacy}} \tag{14.12}$$

Luminous efficiency of the bulb is, then:

$$\eta_{\text{luminous}} = \frac{\text{Equivalent light power (W)}}{\text{Power input (W)}} \tag{14.13}$$

→ FURTHER LEARNING

ASHRAE Student Design Competitions
www.ashrae.org/communities/
student-zone/competitions

Example 14.2 Impact of Building Insulation on Energy Efficiency

A single-family house in Seoul, South Korea, has a base area of 10×12 m^2. Walls are 3 m high. The front door is 1.5×2.2 m^2. There are five windows in total in all four walls, with each window having dimensions 1.5×1.5 m^2. The R-value of the existing wall insulation is 2.64 m^2 K/W (R-15). Outdoor temperature is 19 °C, and the indoor temperature is set to 25 °C. Determine:

a. the rate of heat loss (W) through the walls for the existing conditions
b. the rate of heat loss (W) through the walls if the house is renovated and the wall insulation is replaced with a new insulation material having same thickness and an R-value of 3.7 m^2 K/W (R-21)

Solution

Net wall area is calculated by subtracting all door and window areas from the total area:

$$A_{\text{wall}} = A_{\text{total}} - (A_{\text{door}} + \Sigma A_{\text{windows}})$$
$$= 3[2(10 + 12)] - [(1.5 \times 2.2) + 5(1.5 \times 1.5)] = 117.45 \text{ m}^2$$

a. Rate of heat loss from the walls is calculated using **Equation 14.9**:

$$\dot{Q}_{\text{wall}} = U_{\text{wall}} A_{\text{wall}} (T_i - T_o)$$

and

$$U_{\text{wall}} = \frac{1}{R_{\text{wall}}} = \frac{1}{2.64 \, \frac{\text{m}^2\text{K}}{\text{W}}} = 0.38 \, \frac{\text{W}}{\text{m}^2 \text{ K}}$$

Then:

$$\dot{Q}_{\text{wall}} = \left(0.38 \, \frac{\text{W}}{\text{m}^2\text{K}}\right)(117.45 \text{ m}^2)(25 - 19)°\text{C} = 267.8 \text{ W}$$

b. With the improvement of the wall insulation, the new U-value and corresponding rate of heat loss will be:

$$U_{\text{wall}} = \frac{1}{R_{\text{wall}}} = \frac{1}{3.7 \frac{\text{m}^2\text{K}}{\text{W}}} = 0.27 \, \frac{\text{W}}{\text{m}^2\text{K}}$$

Then:

$$\dot{Q}_{\text{wall}} = \left(0.27 \frac{\text{W}}{\text{m}^2\text{K}}\right)(117.45 \text{ m}^2)(25 - 19)°\text{C} = 190.3 \text{ W}$$

Improvement of house insulation can have a significant impact on the energy performance of a building. As can be seen in this example, the heating load from the walls was reduced by 29% with improved insulation. Source: DonNichols/E+ via Getty Images.

Example 14.3 Lighting Energy Savings

Facilities management at a university decided to replace 15 energy-saving halogen light bulbs with compact fluorescent light (CFL) bulbs in a classroom in the Arts and Sciences Building. Halogen light bulbs have a rating of 43 W with a luminous flux of 620 lm. The

CFL blubs to replace the existing halogen light bulbs are rated 13 W and their luminous flux is 850 lm.

a. How many CFL light bulbs will be sufficient to maintain the same amount of brightness in the classroom?
b. How much energy would be saved annually if these lights are on 12 hours a day for 350 days of the year?

Solution

a. Total lumens provided by 15 energy-saving halogen light bulbs is:

$$15 \text{ bulbs} \times 620 \text{ lm/bulb} = 9300 \text{ lm}$$

One CFL bulb provides 850 lm luminous flux. Then the minimum number of CFL bulbs required to provide the same amount of brightness in the classroom will be:

$$\frac{9300 \text{ lm}}{850 \text{ lm/bulb}} = 10.94$$

Hence 11 CFL bulbs will be able to provide the same amount of lumens in the classroom.

b. First, let's determine the total number of hours each light bulb is on during the year:

$$\left(12 \; \frac{\text{h}}{\text{day}}\right)\left(350 \; \frac{\text{day}}{\text{yr}}\right) = 4200 \; \frac{\text{h}}{\text{yr}}$$

Total annual energy consumption with halogen bulbs:

$$15 \text{ bulbs} \times 4200 \text{ h/yr} \times 43 \text{ W/bulb} = 2709 \text{ kWh}$$

Total annual energy consumption with CFL bulbs:

$$11 \text{ bulbs} \times 4200 \text{ h/yr} \times 13 \text{ W/bulb} = 600.6 \text{ kWh}$$

Then, the annual savings will be:

$$2709 - 600.6 = 2108.4 \text{ kWh}$$

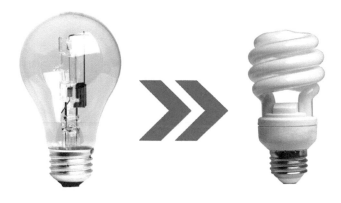

Appliances

Appliances are the major source of energy consumption in residential buildings and can also have a significant share of energy use in commercial buildings. Some standard appliances are refrigerator, dishwasher, washing machine, and dryer. Any energy efficiency improvements in these devices are directly reflected in utility bills in terms of reduced energy consumption. There are online tools to estimate the energy use of appliances, such as the U.S. Department of Energy's *Appliance Energy Calculator*. Energy savings through energy efficiency improvements can also be calculated using these tools with known values of new efficiency or reduced power input to the appliances.

FIGURE 14.11 Energy use of standard household appliances. Source: U.S. EPA.

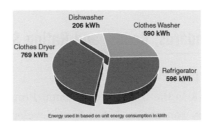

Example 14.4 Energy Efficiency of Domestic Refrigerators

Consider a domestic refrigerator with a cooling capacity of 420 W and a compressor power rating of 150 W. If the cooling capacity is determined by the heat gain of the refrigerator, under steady-state conditions determine:

a. the coefficient of performance (COP) for the refrigerator
b. the annual cost of electricity consumption by the refrigerator if the compressor runs for 14 hours a day and the cost of electricity is 7.6 cents per kWh
c. the annual cost of electricity, as in part **(b)**, if the overall heat transfer coefficient, U, is reduced by 10% with improvements in the refrigerator cabinet (i.e., the cabinet is insulated with *vacuum insulated panels*)

Solution

a. The COP of the refrigerator is determined by using the given cooling capacity and compressor power:

$$\text{COP} = \frac{\dot{Q}_{\text{cooling}}}{\dot{W}_{\text{compressor}}} = \frac{420 \text{ W}}{150 \text{ W}} = 2.8$$

b. Annual cost of electricity is determined by:

$$\text{Cost} = \left(\dot{W}_{\text{compressor}}\right)(\text{Annual operating hours})(\text{Unit price})$$
$$= (150 \text{ W})\left(14 \frac{\text{h}}{\text{day}}\right)\left(365 \frac{\text{days}}{\text{yr}}\right)\left(0.076 \frac{\$}{\text{kWh}}\right)\left|\frac{1 \text{ kWh}}{10^3 \text{ Wh}}\right| = 58.25 \frac{\$}{\text{yr}}$$

c. Cooling capacity is equal to heat gain rate of the refrigerator cabinet under steady-state conditions:

$$\dot{Q}_{\text{cooling}} = \dot{Q}_{\text{gain}} = UA(T_{\text{ambient}} - T_{\text{cabinet}})$$

In this case, the U-value is reduced by 10%. With other parameters remaining the same, this means the heat gain rate will be reduced by 10%, which also means a 10% reduction in required cooling capacity. With the same compressor COP, this also translates to 10% less energy consumption by the compressor. Then,

$$\text{Cost} = (0.9)(150\text{ W})\left(14\ \frac{\text{h}}{\text{day}}\right)\left(365\ \frac{\text{days}}{\text{yr}}\right)\left(0.076\ \frac{\$}{\text{kWh}}\right)\left|\frac{1\text{ kWh}}{10^3\text{ Wh}}\right| = 52.4\ \frac{\$}{\text{yr}}$$

Note: *Besides insulation improvements in the refrigerator cabinet, efficiency enhancement of the compressor can also reduce energy consumption of the refrigerator.*

14.2.5 Standards, Codes, and Rating Systems

ENERGY STAR

ENERGY STAR is the U.S. Environmental Protection Agency's (EPA) program most well-known by consumers. It has been developed to provide credible and unbiased information on energy efficiency of products that consumers can rely on. The EPA ensures that any product that has earned the ENERGY STAR label is certified to fulfill the energy efficiency expectation of the product.

As for buildings, the EPA has launched programs towards both residential and commercial sectors to enable and accelerate the integration of these programs to achieve energy saving for all buildings. **Figure 14.12** depicts a decision tree for new constructions to determine for which program they would qualify to earn the label.

LEED

Leadership in Energy and Environmental Design (LEED) is the green building rating system of the U.S. Green Building Council (USGBC) and Canada Green Building Council (CaGBC). Most of the certified green buildings in both of these countries were evaluated through this system. It can be used for all building types and all building phases including new construction, interior fit-outs, operations and maintenance, and shell and core fit-outs. Project accomplishment levels are rewarded based on the points collected. These levels, from lower to higher scores, are certified silver, gold, and

FIGURE 14.12 Multifamily New Construction (MFNC) Program decision tree. Source: U.S. EPA.

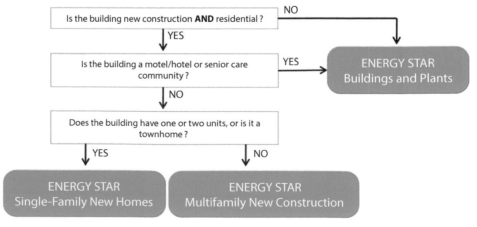

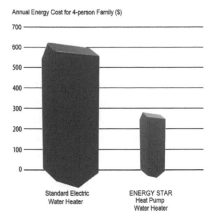

Annual Energy Cost for 4-person Family ($)

FIGURE 14.13 Water heater annual energy consumption comparison. Source: Data from Energy Star.

FURTHER LEARNING

Energy Efficiency Rating Systems
www.energystar.gov
www.usgbc.org/leed
www.breeam.com
www.greenglobes.com
www.ibec.or.jp/CASBEE/english
https://new.gbca.org.au

platinum. While LEED Platinum and LEED Gold buildings fit the definition of a "green building" or a "high-performance" building, there are LEED Silver buildings which still perform very well.

BREEAM (UK)

The Building Research Establishment Environmental Assessment Method (BREEAM) was launched in 1990, eight years before its competitor, LEED, was initiated. It is a scheme that was established to provide unbiased certification of the assessment of the sustainability performance of individual buildings, communities, and infrastructure projects. It is an international evaluation tool which has been widely adopted globally. Assessment and certification processes can be conducted at various stages, from design and construction to operation and renovation.

Green Globes (United States and Canada)

Green Globes is another green building rating and certification program that is mainly instrumented in Canada and the United States. The modules available in this program can be applied to new constructions, major renovations for existing buildings, and sustainable interiors including commercial or tenant interior projects or fit-ups. These modules can be applied to a diverse group of buildings in the residential, commercial, and public sectors, including multi-residential buildings, hotels, warehouses, hospitals, schools, and academic and industrial facilities.

CASBEE (Japan)

The Comprehensive Assessment System for Built Environment Efficiency (CASBEE) is a method developed by a committee through the collaboration of academia, industry, and national and local governments, which established the Japan Sustainable Building Consortium (JSBC). This tool is used for evaluating and rating not only the environmental performance of buildings, but also the performance of the built environment. As such, the program defines a dimensionless variable, Built Environment Efficiency (BEE), which is defined as:

$$BEE = \frac{\text{Environmental quality of the building}}{\text{Environmental load of the building}} \qquad (14.14)$$

where the denominator term evaluates the negative aspects of the environmental impact which extends to the neighborhood of the building.

Green Star (Australia)

Green Star is another international sustainability assessment tool for buildings, fit-outs, and communities. For building assessment, it encompasses the design, construction, and operation stages. The program was launched by the Green Building Council of Australia (GBCA) in 2003. Similar to other equivalent programs, performance evaluation is done based on the credits collected in different categories depending on the sustainability impact of the structure.

14.3 Industry

The share of energy consumption by the industry sector made up about 22.6% of overall energy usage in the OECD countries by 2018 [7]. Therefore, reducing energy consumption through energy efficiency improvements is one of the priorities of the energy stakeholders in these countries. Intergovernmental organizations such as the

United Nations through its regional bodies in the world or governmental entities such as the U.S. Department of Energy focus on industries with the purpose of addressing energy efficiency in various industrial sectors.

The United Nations Environment Programme's (UNEP) report *Energy Efficiency Guide for Industry in Asia* was developed to guide the industry sector in Asia, which is the main contributor to a globally growing energy demand [8]. The guide is an outcome of the Greenhouse Gas Emission Reduction from Industry in Asia and the Pacific (GERIAP) project and includes a methodology for industries in the Asian region to enhance energy efficiency. On the European side, the United Nations Economic Commission for Europe's (UNECE) *Industrial Energy Efficiency Action Plan* includes three key approaches in achieving goals by member states in Europe. These approaches are [9]:

1. Identifying specific partner resources which can be combined through collaboration efforts to improve communication on explaining the business case for industrial energy efficiency;
2. Working with energy-efficiency experts and supporting organizations to develop a platform that functions as a facilitator between businesses, policymakers, and energy efficiency initiatives;
3. Engaging with businesses through existing initiatives to support promoting awareness and usance of existing solutions and support.

The U.S. Department of Energy's *Office of Energy Efficiency and Renewable Energy* (EERE) aims to address the transition to a clean energy economy covering on all sectors including buildings, industry, and transportation. The Advanced Manufacturing Office (AMO), in support of the EERE, focuses on the industrial sector aiming to improve energy and material efficiency. The AMO works towards its goals through three subprograms, which are *Advanced Manufacturing R&D Projects*, *Advanced Manufacturing R&D Consortia*, and *Technical Partnerships*. The main goals of these programs are to support R&D projects focusing on high-impact and energy-efficient manufacturing technologies, bringing together manufacturers, suppliers, companies, academia, national laboratories, and state and local governments to form public–private consortia, and supporting adoption of energy-efficient technologies and practices [10].

When we look at the motherland continent, Africa (ancient name *Alkebu-lan*, meaning mother of humankind), we see many projects encouraging industrial energy efficiency improvements. In 2020, an energy-efficiency initiative in South Africa (*Industrial Energy Efficiency Improvement in South Africa through Mainstreaming the Introduction of Energy Management Systems and Energy Systems Optimization*) received the International Energy Project of the Year recognition awarded by the global Association of Energy Engineers (AEE). The project was led by the South Africa National Cleaner Production Centre (NCPC-SA) and the United Nations Industrial Development Organization (UNIDO) [11].

All these global efforts and initiatives aim to reduce the industrial energy consumption of various sectors including energy, water, construction, metal products, chemicals, paper, textiles, food, and many others.

FURTHER LEARNING

Industrial Decarbonization Efforts
www.energy.gov/eere/amo

14.3.1 Combined Heat and Power

Combined heat and power (CHP) or *cogeneration* plants utilize steam for both process heat and electricity generation. The term cogeneration defines production of more than

FIGURE 14.14 Combustion turbine or reciprocating engine CHP system. Source: U.S. EPA.

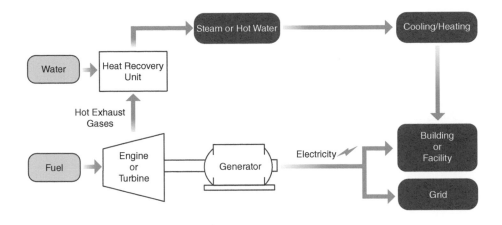

FIGURE 14.15 Steam turbine CHP system. Source: U.S. EPA.

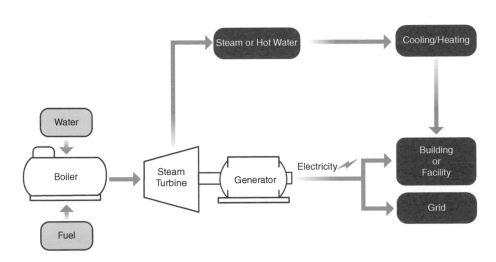

one useful form of energy (i.e., process heat and electricity) from the same energy source, which can be a fossil fuel such as natural gas, or a renewable source such as biomass. Use of CHP systems in industry introduces several benefits. The major advantage of these systems is enhanced overall efficiency, which can go above 80%. This is significantly higher than the average thermal efficiency of fossil fuel power plants in the United States, which is about 33%. It should be noted that the difference between the two power-plant configurations is mainly that in CHP systems less energy is released from the condenser as the process heater does not require heat rejection. Another benefit of CHP systems in industry is eliminating the transmission and distribution losses through power lines as the facility is producing its own electricity.

There are two common configurations of CHP systems: combustion turbine with heat recovery unit, and steam boiler with steam turbine. Schematics of both configurations are illustrated in **Figures 14.14** and **14.15**, respectively.

Efficiency of cogeneration plants is determined by the *utilization factor*, which is the fraction of input energy used for process heating and/or power generation. Utilization factor is given by:

$$\in_U = \frac{\dot{W}_{net} + \dot{Q}_{process}}{\dot{Q}_{in}} = \frac{\left(\dot{W}_{turb} - \dot{W}_{pumps}\right) + \dot{Q}_{process}}{\dot{Q}_{furnace}} \tag{14.15}$$

14.3.2 Industrial Recycling

According to the EPA, *recycling* is defined as collection and processing of materials that would otherwise be discarded and utilizing these materials for useful purposes. Recycling contributes to achieving reduced energy consumption, lower CO_2 emissions, and less waste to be disposed of. There are many industries that focus on energy efficiency. Some of these industries are:

- Aerospace
- Aluminum
- Asphalt
- Cement
- Dairy
- Fertilizer
- Glass
- Paper
- Petrochemical
- Pharmaceutical
- Plastics
- Steel

Among these sectors, the *aluminum*, *glass*, *paper*, *plastics*, and *steel* industries make up the majority and account for 53% of total waste products in the municipal solid waste (MSW) stream [12]. The amount of energy savings in these sectors can be significant with appropriate recycling measures taken.

Example 14.5 Industrial Energy Efficiency Improvement through CHP

An industrial facility uses conventional power and steam generation separately to generate 30 units of electricity and 45 units of steam. The efficiencies of the power plant and the boiler are 30% and 80%, respectively. The facility decides to invest in improving energy efficiency and switches to a CHP system that houses a natural gas combustion turbine and a heat recovery boiler. The new system being considered can produce the required amount of electricity and steam with an input of 100 units of fuel. Calculate the overall efficiencies of both the conventional and CHP systems.

Solution

Let's look at each system separately:

a. *Conventional system*
   i. Power generation: Efficiency for power generation is 33%. Then,

$$\text{Input} = \frac{\text{Output}}{\text{Efficiency}} = \frac{30 \text{ units}}{0.33} = 91 \text{ units}$$

   ii. Steam generation: Efficiency for steam generation is 80%. Then,

$$\text{Input} = \frac{\text{Output}}{\text{Efficiency}} = \frac{45 \text{ units}}{0.80} = 56.3 \text{ units}$$

Hence, total required input is $91 + 56.3 = 147.3$ units of fuel to achieve the targeted values of electricity and steam generation.

Overall efficiency for the plant in this case is:

$$\eta_{\text{conventional}} = \frac{\text{Output}}{\text{Input}} = \frac{(30 + 45)\ \text{units}}{147.3\ \text{units}} = 0.51\ \text{ or }\ 51\%$$

b. *CHP system*

$$\eta_{\text{CHP}} = \frac{\text{Output}}{\text{Input}} = \frac{(30 + 45)\ \text{units}}{100\ \text{units}} = 0.75\ \text{ or }\ 75\%$$

Note: *The overall efficiency increases from 51% to 75%, which results in a reduction in the cost of energy. The cost can be further reduced with appropriate implementation of additional measures such as industrial recycling and use of MSW as fuel.*

14.4 Transportation

Transportation energy consumption made up approximately 34.6% of the total energy consumption of the OECD countries in 2018. It was the single largest energy-consuming sector, with oil being the dominant source with a share of 92% feeding the transport sector [7]. This significant increase of share for transportation, mainly in the developed countries, has led governments to take actions towards energy-efficient transportation strategies.

14.4.1 Regulations and Programs

The *Office of Energy Efficiency and Renewable Energy* (EERE) of the U.S. Department of Energy aims to address energy-efficiency related topics in different fields such manufacturing, buildings, and transportation. The main topics of interest under sustainable transportation strategies include biofuels, hydrogen and fuel cells, and EVs. The EERE projects and investments have helped reduce fuel cell and EV battery costs over the past decade. The *Clean Cities Program* has resulted in a reduction in petroleum use by about 3.8 billion liters (1 billion gallons), decreasing greenhouse gase (GHG) emissions by 7.5 million tons [13].

In Europe, under the Energy Efficiency Directive (2021/27EU), each EU country needed to prepare a *National Energy Efficiency Action Plan* (NEEAP) outlining their estimated energy consumption, planned energy-efficiency measures, long-term renovation strategies, and the enhancements that each individual member state expects to achieve to attain the future energy goals of the Union [14].

Other countries have also developed similar reports. An example case is the National Energy Efficiency Action Plan (NEEAP) of Turkey, which has been effective since 2017. The report includes 55 action plans in the areas of buildings and services, energy, transport, industry and technology, agriculture, and cross-cutting (horizontal) sectors [15]. Nine action plans addressing transportation are as follows:

- Promoting energy-efficient vehicles
- Developing benchmarking on alternative fuels and new technologies
- Developing and improving bicycle and pedestrian transport

- Reducing traffic density in cities: Discouraging use of automobiles
- Promoting public transport
- Developing and implementing institutional restructuring for urban transport
- Strengthening maritime transport
- Strengthening rail transport
- Compiling transport data

Different states have similar measures along with some specific variations pertaining to the geography and the practices that are being performed.

14.4.2 Fuel Economy

Fuel economy is a tangible parameter in evaluating energy efficiency in transportation. It can be defined as the relationship between the distance traveled and the amount of fuel consumed. There are two commonly used units for fuel economy. In most countries around the world, liters per 100 km (L/100 km) is used. In the United States and the United Kingdom, miles per gallon (mpg) is used. Canada uses both unit systems. Conversion between mpg and L/100 km units can be expressed in two different forms depending on the volume being measured: US gallon (3.785 L) or imperial gallon (4.546 L):

$$1 \text{ mpg (US gallon)} = \frac{235}{\text{L/100 km}}$$

$$1 \text{ mpg (imperial gallon)} = \frac{282}{\text{L/100 km}}$$

The history of real-world fuel economy in mpg (US gallon) and CO_2 emissions in grams per mile traveled can be seen in **Figure 14.16**. The graphs demonstrate

FIGURE 14.16 History of real-world fuel economy and CO_2 emissions. Source: U.S. EPA.

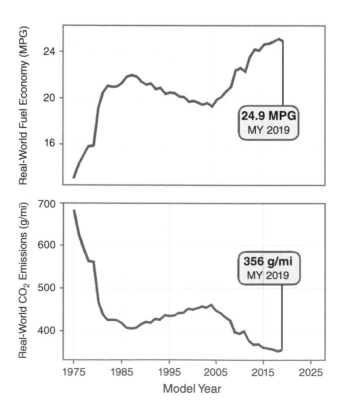

FIGURE 14.17 A sample fuel economy
label. Source: www.fueleconomy.gov

how technology has progressed in terms of fuel economy and the environmental impact of vehicles.

The U.S. Environmental Protection Agency and U.S. Department of Energy have developed a fuel economy guide to inform consumers about the fuel-efficiencies of vehicles. The objective of this report is to provide a dependable ground for comparing vehicles. Most of the vehicles included in the report are evaluated in three different categories for fuel economy estimates [16]. These are:

- *City mpg estimate* represents urban driving in which the vehicle is driven in stop-and-go traffic
- *Highway mpg estimate* represents a mixture of rural and interstate highway driving in which the vehicle is driven for longer distances in free-flowing traffic
- *Combined mpg estimate* represents a combination of city driving (55%) and highway driving (45%)

The EPA has also developed a fuel economy label format which can be found at *fueleconomy.gov*. A sample label is illustrated in **Figure 14.17**. The sections on the fuel economy label are:

1. Vehicle technology and fuel
2. Fuel economy
3. Comparison of fuel economy to other vehicles
4. Amount saved (or spent more) over 5 years compared to average vehicle
5. Fuel consumption rate
6. Estimated annual fuel cost
7. Fuel economy and greenhouse gas rating
8. CO_2 emissions
9. Smog rating
10. Details about conditions and assumptions

FIGURE 14.18 An example illustration from Alternative Fueling Station Locator. Source: Alternative Fuels Data Center, EERE, U.S. Department of Energy.

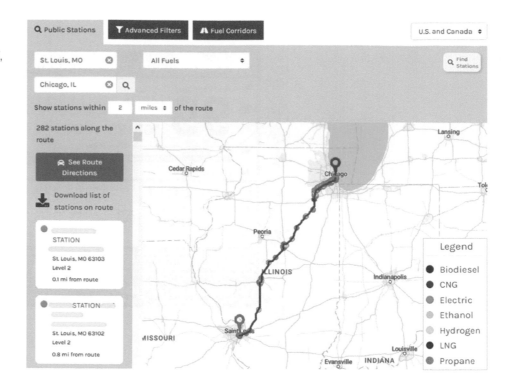

11. QR (quick response) code
12. Source of data
13. Driving range
14. Charge time

In recent years there has been a shift towards vehicles using alternative fuels. This transition in the earlier stages came with the challenge of limited fueling stations and information on their locations. Not only is there a growing network of these stations worldwide, but cartography of these stations has also evolved remarkably. The Alternative Fuels Data Center of EERE has an application that can locate fueling stations for alternative fuel vehicles for a given route in the United States and Canada. The application is called *Alternative Fueling Station Locator*. A sample illustration for a selected route from St. Louis to Chicago can be seen in **Figure 14.18**. The locator provides stations on the route for alternative fuels such as biodiesel, compressed natural gas (CNG), electric, ethanol, hydrogen, liquefied natural gas (LNG), and propane. There are other applications that can locate various types of alternative fueling stations around the world. These applications have a database of stations that can be located on a map.

FURTHER LEARNING

Alternative Fueling Station Locator
https://afdc.energy.gov/stations/

Example 14.6 Fuel Economy Comparison of Different Vehicles

Fuel consumption values for some vehicles are listed below based on data from the Fuel Economy Guide 2021 on *fueleconomy.gov*. The analysis assumes 15,000 miles of annual travel, with 55% city driving and 45% highway driving. Unit price for gasoline is assumed to be $3.66. For each model:

a. Determine the combined fuel economy (mpg)
b. Calculate the annual fuel cost ($/yr)

Manufacturer, model, configuration	mpg (city)	mpg (highway)
BMW Z4 sDrive30i / 2.0 L / 4 cyl	25	32
Chevrolet Corvette / 6.2 L / 8 cyl	15	27
Porsche 911 Carrera 4S / 3.0 L / 6 cyl	17	24
Nissan 370Z / 3.7 L / 6 cyl	17	26
Mini Cooper Convertible / 1.5 L / 3 cyl	26	37

Solution

Combined mpg can be calculated by:

$$\text{mpg}_{\text{combined}} = \frac{1}{\left(\dfrac{0.55}{\text{mpg}_{\text{city}}}\right) + \left(\dfrac{0.45}{\text{mpg}_{\text{highway}}}\right)}$$

Annual fuel cost is calculated as:

$$\text{Annual fuel cost}\left(\frac{\$}{\text{yr}}\right) = \left(\frac{\text{Annual mileage}}{\text{mpg}_{\text{combined}}}\right)(\text{Cost of fuel per gallon})$$

Generating a table on a spreadsheet file with input values and columns for the output values being sought with corresponding formulas yields the results as listed in the table below:

Manufacturer, model, configuration	mpg (combined)[*]	Annual fuel cost ($/yr)
BMW Z4 sDrive30i / 2.0 L / 4 cyl	28	1961
Chevrolet Corvette / 6.2 L / 8 cyl	19	2890
Porsche 911 Carrera 4S / 3.0 L / 6 cyl	20	2745
Nissan 370Z / 3.7 L / 6 cyl	20	2745
Mini Cooper Convertible / 1.5 L / 3 cyl	30	1830

[*] The calculated mpg values were rounded to the nearest integer.

Note: *In this example, US customary units have been used as the data from the Fuel Economy Guide is provided in US units.*

CLASSROOM DEBATE

Assume your institution received a grant to enhance sustainability. If you could only spend the grant towards either enhancing energy efficiency or adding renewable energy technology, which one would you be in favor of? Form groups and debate.

14.5 Summary

To achieve a sustainable environment, measures need to be taken focusing on both energy consumption and energy production. Energy production technologies, including fossil fuels and renewables, were discussed in earlier chapters. As for energy consumption, there are two terms that need to be highlighted: *energy conservation* and *energy efficiency*. Energy conservation is a means of energy saving by giving up one or more standards such as health, comfort, productivity, or quality. It does not need a high-tech approach. This is an easy way of saving energy. Energy efficiency, on the

FIGURE 14.19 *Pacific golden plovers* migrate from Alaska to Hawaii with a non-stop flight every year. The journey of over 4000 km takes 3–4 days. Considering the amount of body fat they have as their fuel for the flight and the required amount to complete the journey, scientists believe that they manage this flight with their v-shaped formation and strong navigation skills that enable a more direct flight route. Flying in a v-shape reduces the drag acting on each bird, which makes the flight more energy efficient. Being good navigators helps them avoid taking a longer route to the destination, which also saves energy. Source: Winfried Wisniewski/Photodisc via Getty Images.

other hand, requires research, technology enhancement, and product development. It requires engineering. In fact, as the efficiency of a system increases and gets closer to its saturation, improvement requires more rigorous engineering.

To achieve sustainability, energy efficiency should be improved in all areas of modern life including mainly buildings, industry, and transportation. Some other sectors are agriculture, health, entertainment, and power generation.

Buildings play a significant role in our lives, whether they are residential, business, educational, industrial, public, or health. They possess a considerable share of total energy consumption. HVAC systems make up a significant portion of total building energy consumption. Use of energy-efficient air-conditioners, refrigerators, HVAC units, furnaces, boilers, fans, blowers, and pumps will lead to less heating, cooling, and ventilation energy consumption. Improvement of building insulation is of great importance as it decreases the heating and cooling loads of a building remarkably. Research and development on sustainable insulation materials with high R-values is an on-going challenge. The insulation industry is a wide sector. There are many insulation materials and applications that address this challenge. Green roofs are one example for sustainable building insulation. This *live* insulation method benefits a building in multiple ways, including energy savings, stormwater retention, carbon sequestration, acoustic insulation, and reduced urban heat island effect which is a common problem of highly populated cities dominated by concrete and asphalt.

Industries including manufacturing, cement, steel, and petrochemicals consume significant amounts of energy in both developed and developing countries. Some of the major subsectors are food and beverages, textiles, construction, machinery, chemicals, paper, wood products, metals, electronics, and transport equipment. Improvements in industrial energy consumption can be achieved by taking measures such as employing energy-efficient machinery, regulating energy demand, implementation of combined heat and power (CHP) systems, and recycling.

Transportation energy consumption has increased significantly over the past two decades. Addressing energy efficiency in transportation is another necessity in reducing global energy consumption and CO_2 emissions. Electric vehicles, alternative fuels, modern public transportation, improved maritime and rail transportation, and expanded bicycle paths in cities are some of the steps being undertaken by governments and municipalities worldwide.

Global transitions to smarter buildings, smarter cities, and energy-efficient industry and transportation are of great necessity in achieving sustainable development goals. Measures mentioned in this chapter are some examples of courses of action embarked on in different parts of the world.

REFERENCES

[1] International Renewable Energy Agency, "Synergies Between Renewable Energy and Energy Efficiency: A Working Paper Based on REmap," IRENA, Abu Dhabi, 2017. Accessed: Dec. 24, 2020 [Online]. Available: www.irena.org/remap.

[2] U.S. Energy Information Administration, "Use of Energy Explained," EIA, 2016. www.eia.gov/energyexplained/use-of-energy/ (accessed Dec. 24, 2020).

[3] United Nations Economic Commission for Europe, "Work Plan of the Group of Experts on Energy Efficiency for 2020–2021, ECE/Energy/2019/8," UN, Geneva, 2019. Accessed: Dec. 25, 2020 [Online]. Available: https://unece.org/fileadmin/DAM/energy/se/pdfs/CSE/comm28.2019/ECE_ENERGY_2019_8_Final.pdf.

[4] World Green Building Council, "Our Green Building Councils," WGBC, 2017. www.worldgbc.org/ (accessed Dec. 25, 2020).

[5] U.S. Department of Energy, "Types of Insulation," DOE, 2017. www.energy.gov/energysaver/weatherize/insulation/types-insulation (accessed Jan. 27, 2021).

[6] U.S. Department of Energy, "How Energy-Efficient Light Bulbs Compare with Traditional Incandescents," DOE, 2012. www.energy.gov/energysaver/save-electricity-and-fuel/lighting-choices-save-you-money/how-energy-efficient-light (accessed Feb. 5, 2021).

[7] International Energy Agency, "World Energy Balances: Overview," IEA, Paris, Jun. 2020. Accessed: Feb. 3, 2021 [Online]. Available: www.iea.org/reports/world-energy-balances-overview.

[8] United Nations Environment Programme, Division of Technology, Industry, and Economics, Sweden, Styrelsen För Internationellt Utvecklingssamarbete, and I. Asia, *Energy Efficiency Guide for Industry in Asia*. Nairobi, Kenya: United Nations Environment Programme, Division of Technology, Industry, And Economics, 2006.

[9] United Nations Economic Commission for Europe, "Industrial Energy Efficiency Action Plan, ECE/Energy/GE.6/2020/3," UNECE, Geneva, 2020. Accessed: Feb. 3, 2021 [Online]. Available: https://unece.org/sites/default/files/2020-12/ECE_ENERGY_GE.6_2020_3e.pdf.

[10] U.S. Department of Energy, Office of Energy Efficiency and Renewable Energy, "Office Structure." www.energy.gov/eere/amo/office-structure (accessed Feb. 3, 2021).

[11] United Nations Industrial Development Organization, "South Africa: Industrial Energy Efficiency Project Wins International Award," Oct. 15, 2020. www.unido.org/news/south-africa-industrial-energy-efficiency-project-wins-international-award (accessed Feb. 3, 2021).

[12] U.S. Department of Energy, "Barriers to Industrial Energy Efficiency," DOE, Washington, DC, Jun. 2015. Accessed: Feb. 3, 2021 [Online]. Available: www.energy.gov/sites/prod/files/2015/06/f23/EXEC-2014-005846_6%20Report_signed_v2.pdf.

[13] U.S. Department of Energy, Office of Energy Efficiency and Renewable Energy, "EERE at a Glance," DOE, Apr. 2016. Accessed: Feb. 3, 2021 [Online]. Available: www.energy.gov/sites/prod/files/2016/07/f33/EERE%20At_A_GLANCE_072016.pdf.

[14] European Parliament and Council of European Union, "Directive (EU) 2018/2002 of the European Parliament and of the Council of 11 December 2018 Amending Directive 2012/27/EU on Energy Efficiency," Dec. 2018. Accessed: Feb. 3, 2021 [Online]. Available: https://eur-lex.europa.eu/legal-content/EN/TXT/?uri=celex%3A32018L2002.

[15] Asia Pacific Energy Portal, "National Energy Efficiency Action Plan (NEEAP) of Turkey, 2017–2023," Republic of Turkey Ministry of Energy and Natural Resources, Ankara, Mar. 2018. Accessed: Feb. 4, 2021 [Online]. Available: https://policy.asiapacificenergy.org/sites/default/files/National%20Energy%20Efficiency%20Action%20Plan%20%28NEEAP%29%202017-2023%20%28EN%29.pdf.

[16] U.S. Department of Energy, "Fuel Economy Guide," EERE, EPA, May 2021. Accessed: May 28, 2021 [Online]. Available: www.fueleconomy.gov/feg/pdfs/guides/FEG2021.pdf.

14.1 Define the phrases *energy conservation* and *energy efficiency*. Describe their differences. Give a real-world example in the fields listed below to highlight the differences:

a. Buildings

b. Industry

c. Transportation

14.2 Madrid in Spain is one of the leading cities in Europe in transitioning to energy-efficient lights for street lighting towards becoming a smart city. If 84,000 of the streetlights in the city were converted from 180 W low-pressure sodium (LPS) lamps to 75 W LED lights, calculate the annual energy savings with this transition. Assume the lights are on for an average of 10 hours per day, and the cost of electricity is $0.30 per kWh.

14.3 A family is considering switching to a more energy-efficient clothes washer. Annual energy and water use of the current washer they have and the new washer they are planning to buy are given below. If the cost of electricity is $0.12 per kWh and the cost of water is $0.003 per gallon, calculate the annual energy and water savings with the new washer.

Washer	Annual energy use (kWh/yr)	Annual water use (gallons/yr)
Current model	311	7640
Newer model	60	4059

14.4 An office facing south in a commercial building in Accra, Ghana, has only one wall exposed to the outside. The wall is fully glass, which is clear single glazing with $2.7 \times 9 \text{ m}^2$ dimensions. The building is considering a renovation which also includes replacing the glass with double-glazed tinted glass to save energy on cooling. Calculate the difference in cooling load through the glass by means of both conduction and solar radiation. The properties of the existing and proposed glass types are given in the table below. Indoor set temperature for the office is 20 °C and the outdoor temperature is 32 °C. Solar heat gain factor (SHGF) for the façade facing south is given as 550 W/m^2.

Glass type	U-value (W/m^2 K)	Shading coefficient (–)
Single glazing – clear	6.17	0.95
Double glazing – tinted	2.79	0.71

14.5 Iași is a university city in northeastern Romania and is known as the cultural capital of the country. A single-story campus building in Iași has a base area of $26 \times 30 \text{ m}^2$ with a wall height of 4.30 m. The building has two doors, at the front and back, each with dimensions of $2.5 \times 2.7 \text{ m}^2$. In total there are 18 windows all around the building, with each window having dimensions $1.8 \times 1.5 \text{ m}^2$. The R-value of the existing wall insulation is 2.3 m^2 K/W. Outdoor temperature is 4 °C, and the indoor temperature is set to 22 °C. Determine:

    a. the rate of heat loss (W) through the walls for the existing conditions

    b. the rate of heat loss (W) through the walls if the building is renovated and the wall insulation is replaced with a new insulation material having the same thickness and an R-value of $3.5\ m^2\ K/W$.

14.6 The owner of an apartment building in Chicago decides to replace its existing EPDM (ethylene propylene diene monomer) membrane roof with a green roof for improving building energy efficiency and stormwater retention. The roof area is $120\ m^2$. The U-value for the EPDM roof is constant throughout the year. For the green roof, the U-value differs in winter and summer due to the plants (which act as a live insulation layer) being in a dormant or active state. If the apartment at the top floor has a set indoor temperature of $20\ ^\circ C$ throughout the year, and outdoor design temperatures and U-values for winter and summer are as listed in the table, determine the amount of heating and cooling energy reduction through the roof achieved by switching to a green roof system.

Season	EPDM U-value (W/m^2 K)	Green roof U-value (W/m^2 K)	Outdoor temperature[*] (°C)
Winter	0.49	0.42	−15.4
Summer	0.49	0.35	32

[*] Outdoor temperatures are based on ASHRAE values for 99% design condition for heating and 1% design condition for cooling for Chicago.

A building with a green roof in Downtown Chicago. Source: stevegeer/ E+ via Getty Images.

14.7 A steel manufacturing plant is considering investing in becoming more energy efficient with conversion from traditional power and heat to a CHP system utilizing a 200 kW microturbine that operates with natural gas. Total fuel use for generating power and heat is 28,193 MMBtu with the existing system. Net electric efficiency is 32% with an electricity generation of 1700 MWh$_e$. Boiler efficiency is 80%. Determine:

    a. the process heat output in the existing system (MWh$_t$)

    b. the power-to-heat ratio in the existing system

    c. the overall efficiency of the existing system (%)

    d. the total fuel use for the same amount of electricity and heat generation if the overall efficiency of the new CHP system is 74%.

14.8 Annual waste material recycling and reuse potential from municipal waste for a country is provided in the table below. Energy and CO_2 savings potential per ton of waste recycled are also listed for different waste materials. Calculate the total energy savings in PJ and CO_2 savings in 1000 t of CO_2e for this country for the selected year.

Waste material	Material recycling and reuse potential (1000 tons)	Energy savings potential (GJ/ton waste)	CO$_2$ savings potential (ton CO$_2$e/ton waste)
Wood	1430	44.7	109.1
Glass	190	8.8	138.7
Plastic	173	42.7	258.2
Paper and cardboard	614	31.5	303.9
Metal	878	14.6	1908.7
Food	1612	2.1	3.5

14.9 The U.S. Department of Energy's *Transportation Energy Future* (TEF) project highlights the potential reduction in fossil fuel use and GHG emissions by 2050 by a combination of multiple strategies in transportation. The table below shows the projected effects of transportation efficiency enhancements, as well as increases in use on net energy consumption by 2050. According to the table below, (a) research the annual fossil fuel consumption in each transportation category and calculate the total change in fossil fuel use by 2050 based on the predictions in the TEF report, and (b) comment on possible reasons for significant increases in vehicle use in the aviation and ocean marine sectors.

	Light-duty vehicles (LDVs)	Trucks	Aviation	Inland marine	Ocean marine	Rail	Pipeline	Off-road
Energy efficiency improvements	61%	50%	65%	30%	75%	35%	20%	18%
Vehicle use increases	75%	87%	217%	32%	450%	47%	16%	20%
Net changes in total energy consumption	−32%	−17%	+11%	−8%	+38%	−4%	+1%	−6%

14.10 A family is considering renewing their car to switch to either a PHEV or a BEV for energy efficiency and environmental-friendliness purposes. They drive an average of 22,000 km/yr. Their current vehicle is an ICEV with a fuel consumption of 8.1 L/100 km. Cost of gasoline is $0.92/L. Calculate the payback for both the PHEV and BEV options considered. Cost of electricity is $0.14/kWh.

Vehicle type	Price ($)	Fuel economy
PHEV	64,000	1.65 L/100 km + 28 kWh/100 km
BEV	82,000	22 kWh/100 km

CHAPTER 15
Future Prospects

15.1 Introduction

Future prospects attributed to energy topics are covered in three sections in this chapter. First, energy consumption projections are reviewed to have some idea about what can be expected on the energy demand side. Then, prospects that can address the energy supply and demand concerns for the coming decades are examined from two perspectives. One of these points of view is centered around technology-based prospects that focus mainly on energy generation and storage solutions. The second perspective is sector-based prospects where possible future solutions on energy efficiency and reduced energy demand for the building, transportation, and industry sectors are communicated.

15.2 Energy Consumption Projections

Researching, interpreting, and brainstorming on new and future energy topics including technologies and sectors initially rely on a foundation of knowing what the history, current situation, and projections on energy demand look like. In **Figure 15.1**, global primary energy consumption over the past decade and the projected values for the next three decades by fuel types are illustrated [1]. **Figure 15.2** depicts US values on delivered energy across end-use sectors. The graph also shows the projections for the reference (2.1% GDP growth rate), high economic growth (2.6% GDP growth rate), and low economic growth (1.6% GDP growth rate) cases. Economic growth is another parameter besides population growth and changing end-user habits that affects energy consumption. History and projections of US energy consumption on a sector basis can be seen in **Figure 15.3** [2]. These projections show an increasing anticipated trend for the electric power and industry sectors. While the trends for transportation and buildings (residential and commercial) do not seem to portray increasing behavior, the characteristics of the trends for these sectors are expected to be quite different for developing countries. This can be attributed to an increase in the percentage of the population owning vehicles and buildings utilizing a more diverse fleet of appliances.

Africa is expected to become one of the key regions for global energy markets. It is among the fastest growing regions with the youngest population in the world. According to the IEA's Africa Energy Outlook, Africa will have a higher oil demand than China and the Middle East for a projected timeframe between 2018 and 2040. This projection can be seen in **Figure 15.4** for selected countries and regions [3].

FIGURE 15.1 Global primary energy consumption history and projections by fuel type. Source: U.S. Energy Information Administration, International Energy Outlook 2021.

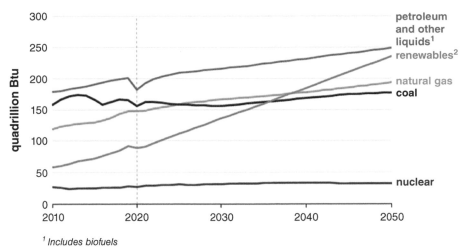

¹ Includes biofuels

² Electricity generation from renewable sources is converted to Btu at a rate of 8,124 Btu/kWh

FIGURE 15.2 US delivered energy across end-use sectors, AEO2021 economic growth cases. Source: U.S. Energy Information Administration, Annual Energy Outlook 2021.

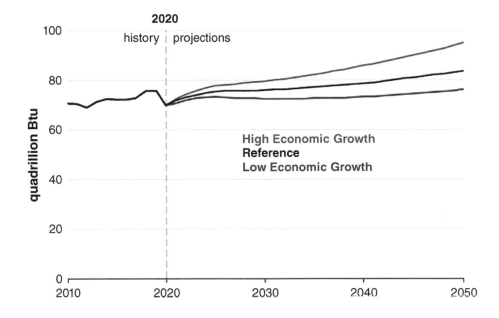

FIGURE 15.3 US energy consumption history and projections by sector, AEO2021 Reference case. Source: U.S. Energy Information Administration, Annual Energy Outlook 2021.

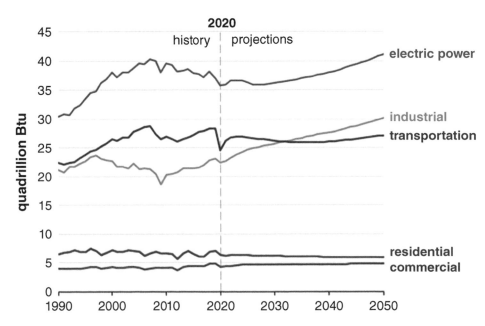

FIGURE 15.4 Oil demand in selected countries and regions, 2018–2040. Source: International Energy Agency (IEA), Africa Energy Outlook 2019.

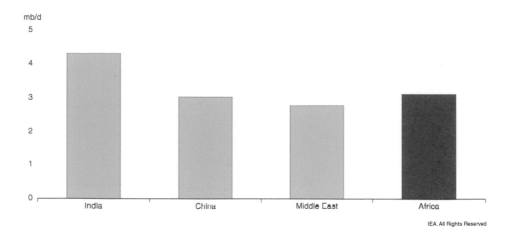

15.3 Technology-Based Prospects

Ongoing research and anticipated developments in renewable energy, nuclear energy, and energy storage technologies are discussed in this section. As for the R&D studies, while some of the work is limited to fundamental or conceptual levels, some is promising, with tangible outcomes in either materials or application fields.

Biomass

Global research and market trends indicate that the major focus on the future of biomass energy is on biofuels, mainly on jet fuel. One of the challenges with jet fuel is the cost of feedstocks, just as is the case for other types of biofuels. There is also an environmental concern with conventional fuel due to higher emissions. Research is being conducted on achieving **sustainable aviation fuels** (SAF) to address both environmental and economic concerns. These are fuels that can be extracted from renewable resources that are both cost-effective and have a low carbon footprint. Some of the promising research studies pertaining to biofuels are:

- Jet fuel or green diesel derived from **algae**
- Hydroprocessed esters and fatty acids (HEFA)
- Sourcing SAF feedstocks such as iso-alkanes and cycloalkanes from inexpensive resources
- Utilizing **kelp** as biofuel resource

Besides biofuels, **municipal solid waste** as a source of biomass energy is also expected to gain more momentum. Even though the growth of global waste production is at a decreasing rate, the increase in waste and the total amount is significant. With improvements in combustion technologies, incineration applications are also expected to increase worldwide even though other alternative treatment methods such as composting, recycling, and landfill will be gaining popularity.

Geothermal Energy

Geothermal energy technologies are comparatively at a more mature level than other energy saving and generation methods. Although not at a component or system level, developments should be expected at the application method level such as designing

FIGURE 15.5 *Turquoise salt flat lagoons* at the Atacama Desert in Chile. Lithium reservoirs lie under these salts. While the potential for lithium is very high, extraction of lithium from these underground reservoirs has also caused controversies due to the potential impacts of brine extraction on the ecosystem. Source: Germán Vogel/Moment via Getty Images.

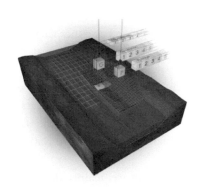

FIGURE 15.6 Standard modular hydropower (SMH) technology studied at Oak Ridge National Laboratory (ORNL) focuses on three important aspects of hydropower: generation, passage, and foundation. The design enables a modular and robust foundation, high-efficiency power generation, and passage for fish, water, sediment, and recreational craft, hence minimizing the impact on the natural flow of the stream. Source: Oak Ridge National Laboratory, Water Power Program.

hybrid geothermal energy systems. These hybrid systems are combined versions of a geothermal and another fossil or non-fossil plant. Some examples are hybrid geothermal–coal-fired power plants, hybrid geothermal–solar power plants, or hybrid geothermal–biomass power plants. These systems not only offer higher power output, but they also provide more flexibility. Technology improvements on any integrated system will elevate the overall plant efficiency.

Although not directly related to energy production, development of techniques for **lithium extraction** from geothermal brines (**Figure 15.5**) can benefit the battery industry. While the concentration of lithium can be very low in these brines, significant amounts of lithium can be extracted due to the large volumes of geothermal fluid being processed at geothermal power plants. Extraction methods differ based on the substances utilized, such as high-capacity selective sorbents, bioengineered rare earth-adsorbing bacteria, and solid-phase extraction with nanocomposite sorbents [4]. Hence, the significance of multidisciplinary research involving other disciplines such as materials science and chemistry should be highlighted, just as in many other engineering applications.

Hydropower

Energy generation using hydropower is another mature technology (**Figure 15.6**). Yet, developments are still anticipated mostly on turbine technologies, small-scale applications, and pumped storage. The last one is discussed under the energy storage topic in this section.

As for turbines, improvement is expected in efficiency, economics, and eco-friendliness. It is important to design efficient and economic turbines while ensuring that the designs are habitat-friendly. Next-generation **Archimedes hydrodynamic screw turbines** offer efficiency and fish-friendliness; however, the economic challenges remain to be addressed. The cost of manufacturing these turbines and challenges in transportation and installation steps push researchers towards alternative solutions. Use of composite materials is one of these solutions. Advanced manufacturing and materials science will be key players in satisfying the criteria for successful turbine designs.

On the applications side, **small-scale or very low head (VLH) systems** are expected to become more popular, mainly for mini-grid and rural electrification purposes. Reliability, cost, and the practicality of these applications have been some of the issues that are being addressed in the research and product development phases. Another concern that has been brought up is the environmental impact of these small systems. Although their impact is not as large as regular hydropower dams, they can still cause disfigurement of rivers, interfere with fish movements, and risk environmental aesthetics. There is ongoing research addressing all these concerns to come up with low-cost, efficient, and environmentally friendly systems.

Nuclear Energy

On the nuclear energy side, future research is to be expected mostly on alternative nuclear fuels, modular reactors, and nuclear fusion. While uranium reserves seem to be sufficient for almost two centuries with the current consumption rate, a search for alternative nuclear fuels for nuclear fission reactions is crucial. **Thorium (Th)** is one of the elements that is under research. While it is a fertile rather than a fissile element, it can be used along with a fissile fuel such as recycled plutonium, or it can be transmuted

FIGURE 15.7 A microreactor being transported. The reactor is attached to the semi-trailer by two shock-absorbent mounts. Center pieces from left to right are the turbine, heat exchanger, and the reactor. Source: U.S. Department of Energy, Microreactor Program. Idaho National Laboratory.

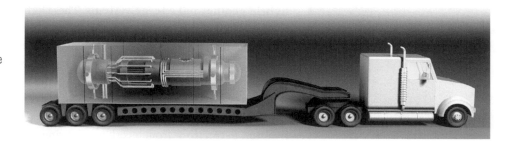

into a fissile fuel in nuclear reactors. It is more abundant in nature than uranium and could help with the nuclear fuel scarcity issue for the future.

Microreactor and **small modular reactor (SMR)** industries are anticipated to grow. There is global interest in small-sized reactors, some of which can also be modular. These systems offer a wider range of applications, lower capital costs, and the possibility for cogeneration or hybrid uses. Amongst the smaller sizes, a microreactor is a nuclear reactor that can provide up to 20 MW thermal energy for power generation and process heat. Most of these reactors are designed to be portable and can be transported by semi-trailer trucks, as seen in **Figure 15.7** [5]. Small modular reactors on the other hand can provide a range of 20–300 MW. They offer flexible power generation like the microreactors. The SMRs can be deployed as either single- or multi-module systems. This allows for synergetic hybrid power systems that can combine nuclear with alternative energy sources such as renewables.

Use of **3D-printing technology** is expected to grow in the nuclear industry as well. The Tennessee Valley Authority (TVA) has used 3D-printed fuel assembly brackets at its Browns Ferry nuclear power plant. The brackets were printed at the U.S. Department of Energy's Manufacturing Demonstration Facility at Oak Ridge National Laboratory [6].

Nuclear fusion does not seem to be realistic for the near future, mainly due to the need for enormously high temperatures (at least 100 million °C) and extreme pressures to fuse deuterium and tritium. Fusion reactions can be achieved in a laboratory environment; however, the net power output is nowhere close to making this technology feasible considering the amount of power consumption required to heat the plasma. Research on nuclear fusion is a challenging process. The ITER Tokamak is an encouraging initiative and a good example of international research collaboration addressing a global need.

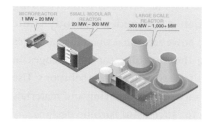

FIGURE 15.8 Comparison of reactor capacity ranges based on sizes. Source: U.S. Department of Energy, Office of Nuclear Energy.

Ocean Energy

A technology overview and the prospects for four different ocean energy technologies (OTEC, salinity gradient, tidal energy, and wave energy) are discussed in this section. Despite having the highest global potential amongst all ocean energy sources, OTEC technology is not expected to grow in the near future due to its low efficiency and costs. Salinity gradient energy is the least mature technology among the four ocean energy systems discussed. It is still at a conceptual phase and more research needs to be conducted for the application to become viable.

As for tidal energy, barrages are considered to be established technologies for harvesting energy through the height difference of low and high tides. Another approach is utilizing the energy from the flow of water caused by tidal currents. Horizontal-axis turbines in this category are also mature technologies. **Vertical-axis**

FIGURE 15.9 A kite moving in a figure-eight trajectory due to tidal currents generates electricity by its tidal energy converter. Source: M. Kaddoura, J. Tivander, and S. Molander, "Life cycle assessment of electricity generation from an array of subsea tidal kite prototypes," *Energies*, vol. 13, no. 2, p. 456, Jan. 2020.

FIGURE 15.10 Kite surfers also practice cruising their kites in a figure-eight trajectory. The kite pulls harder in some spots where the lift is greater. This is similar to tidal kites under water. Source: Merrill Images/Photodisc via Getty Images.

turbine, **venturi, oscillating hydrofoil, Archimedes screw spiral,** and **tidal kite** (Figure 15.9) are technologies that possess more room for development.

Wave energy applications have some technologies that are mature, while there are some for which there is more room for improvement. In 2021, the U.S. Department of Energy announced \$27 million in federal funding to support R&D projects on wave energy. This is an indicator that there will be investments in the wave energy field. **Oscillating water column (OWC), attenuator,** and **point absorber** technologies are comparatively more mature than some other technologies such as **overtopping devices, submerged pressure differential systems, oscillating water surge converters (OWSCs),** and **rotating mass units.** R&D and investments and are expected into these applications, which will complement wave energy technologies.

Solar Energy

In the solar energy industry, improvements are expected mainly in materials science. Enhancements with existing materials or new material technologies have already been the focus of R&D projects worldwide. **Multi-junction photovoltaic cells** built from III–V semiconductors are one of the promising directions. Solar cells using **gallium arsenide** (GaAs) grown on reusable **germanium** (Ge) are another alternative that is considered to be more cost-effective. Germanium is an element that exhibits properties resembling both metals and non-metals with characteristics between silicon and tin.

Another technology expected to grow is **thin-film photovoltaic cells.** The three most feasible materials to be utilized for this technology are **cadmium telluride** (CdTe), **copper indium gallium diselenide** (CIGS), and **amorphous silicon** (a-Si).

Perovskite solar cells are another type of next-generation photovoltaic technologies. They have lower price compared to some of the emerging materials in the PV industry, they require low-temperature processing, and they have higher absorptivity, which provides higher performance with lower light levels and diffuse light received on the solar cell.

Besides material technologies, improvements in solar technologies are expected to come from innovative and integrative applications as well. The technology of **bifacial solar panels** that enable energy harvesting from both sides of the panel has been around for a long time. However, this technology still has room for significant improvements, which makes it another candidate for future R&D investments. **Solar skin** and **solar fabric** technologies are two promising applications for buildings and humans. Smart buildings with **building-integrated solar systems** in the form of roofs, facades, and canopies will expand in numbers. Integration of solar cells with **wearable technologies** is another field to expect significant growth. Another applied field expected to spread for the solar industry is **green roofs** with photovoltaic panels. The synergy between the two technologies is fueled by the *symbiotic relationship* between them. Finally, **floating solar panels** need to be mentioned as a sector that has high potential to grow. Similar to the synergies between the green roofs and solar panels, floating systems also have a mutualistic relation with the water surface beneath them. The advantages they provide make them another strong candidate amongst solar energy applications for remarkable growth worldwide in the next decades.

Wind Energy

Wind energy technologies are mostly mature. There are still some aspects being studied to enhance the efficiency of these systems. Innovative drivetrain concepts are being investigated to contribute to this objective. **Smart rotors** have been studied for a while and research continues on this technology. The nonlinear characteristics of wind turbines can be addressed by **fuzzy logic control** (FLC). System optimization can be done with auxiliary methods such as **genetic algorithm**.

Computational fluid dynamics (CFD) for simulating air flow around the blades and understanding interactions of turbines with each other in wind farms is a method that has been in use for a while. Noise reduction and minimizing wake effects are two examples of scope for such analyses. CFD analysis is expected to grow to produce more accurate results with multiple variables included in these studies for product development.

On the production side, **innovative blade manufacturing methods** seem to attract major interest by both research institutions and the private sector. **Fiber-reinforced composite laminates** and **additive 3D printing** are two promising methods for both weight and cost reduction in the turbine blade industry.

As for the application side, **offshore wind** projects are expected become more popular. Major research and investment for these applications are anticipated to focus on reducing overall costs (construction, installation, and O&M), acting towards electrical infrastructure challenges, and addressing concerns about severe environmental conditions.

Energy Storage

Mechanical Considering that about 96% of the global energy storage in grid-scale applications is provided by pumped hydroelectric energy storage (PHES), it is reasonable to expect that R&D on **pump as turbine** (PAT) systems will continue. Variable frequency drives and PID (proportional–integral–derivative) controllers used with these systems are already mature technologies and a significant change in performance due to these components is not expected.

Thermal Existing thermal energy storage systems are mostly well established. Amongst sensible heat thermal energy storage (SHTES) systems, **thermophotovoltaic** applications should experience improvements with new TPV emitter materials. As for the latent heat thermal energy storage (LHTES) systems, the **phase change material** industry targeting the building and automotive sectors may grow with concurrent developments in these sectors.

Chemical The most remarkable changes in energy storage are expected to be seen with battery technologies. Lithium-ion batteries have been the most advanced and widespread electrochemical energy storage technology for a while. Although they offer a commendable performance index, there is ongoing research towards **post-lithium-ion batteries** which can offer better cost-effectiveness and less toxicity. It is projected that the battery industry may consider a shift to dual-ion batteries such as **graphene aluminum-ion** and **magnesium–sodium** batteries.

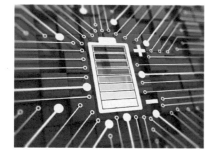

FIGURE 15.11 Improvements in battery technologies will determine the development rate of related industries, mainly electronics, residential use, and electric vehicles. Source: D3Damon/E+ via Getty Images.

15.4 Sector-Based Prospects

In this section, projected reflections of developments in energy efficiency, renewable energy, and energy storage technologies on different sectors are discussed. These sectors are buildings, transportation, and industry.

Buildings

The increase in building energy consumption due to globally rising urban density and changing habits will further motivate the building sector to transition to **smart**

FIGURE 15.12 Future home appliances: Energy-efficient smart domestic appliances will be saving energy while providing further convenience. Source: Donald Iain Smith via Getty Images.

buildings, where control over the building's environment and operations will either be given to the occupants or will be driven by **AI** and **machine learning**. This shift will involve a more holistic participation with governments, municipalities, investors, architects, engineers, constructors, facility managers, and building owners. From an energy-efficiency point of view, performance improvements for most of the HVAC (heating, ventilating, and air-conditioning), refrigeration, and lighting equipment may not be significant as these technologies are well established. Energy savings can be achieved by developments on the control side. Improved building insulation, **innovative window** and **shading systems**, and an increase in use of **heat pump units** will further contribute to achieving higher energy efficiency. As for renewable energy applications, the share of **solar-powered buildings** is expected to increase even more with the decrease in costs, increase in efficiency, and enhancement in battery technologies.

Transportation

Future energy strategies for the transportation sector can also be categorized into two fields: reducing energy consumption and use of alternative fuels. Energy consumption can be reduced by improvements in vehicle energy efficiencies. Strategies built towards building and promoting wider **public transportation** coverage also have a direct impact on energy savings. Considering that approximately 70% of the world's population will be living in urban areas by 2050, the importance of public transportation should be highlighted. The other field that addresses the future of transportation energy supply and demand is utilization of alternative fuels. The market share of **hydrogen vehicles** including both **HICEVs** and **FCEVs** is expected to increase. The rate and magnitude of this increase will depend on how the technological developments address some of the concerns about hydrogen. The lightest element, hydrogen, is also a candidate for heavy transport's future. Vehicles using **natural gas** are already in use, mainly in public transportation. The natural gas vehicle (NGV) market employing compressed natural gas (CNG) or liquefied natural gas (LNG) is estimated to grow as well. A major contribution to the transportation energy supply will be from **biofuels** and **electric vehicles**. Ethanol, biodiesel, green diesel, and biofuel from microalgae are anticipated to further grow in terms of research and production. Electric vehicle sales are expected to go up to 62 million vehicles per year by 2050. By that time, the global total number of EVs on the roads is predicted to be 700 million. BEVs will have a share of 48%, dominating ICE vehicles having a 44% share [7]. As for the United States, the EIA's delivered electricity and natural projections for the US transportation sector through 2050 are illustrated in **Figure 15.13** for different modes of transportation [2].

FIGURE 15.13 US delivered electricity and natural gas projections by mode for the transportation sector, AEO2021 Reference case. Source: U.S. Energy Information Administration, Annual Energy Outlook 2021.

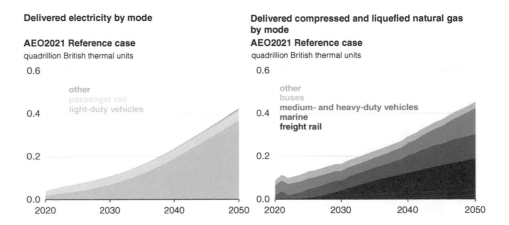

FIGURE 15.14 BEVs are expected to dominate ICEVs by 2050. With the addition of PHEVs, the difference will be higher in favor of EVs. Source: Viaframe/Stone via Getty Images.

FIGURE 15.15 An engineer inspecting the components of a machine in 3D in a virtual reality suite. Source: Monty Rakusen/Image Source via Getty Images.

Industry

Industry types can be categorized based on their energy consumption. In this case, the major categories can be energy-intensive manufacturing (basic chemicals, iron and steel, cement, food, paper, refinery, and non-metallic minerals), non-energy-intensive manufacturing (pharmaceuticals, machinery, electronics, transportation equipment, electrical components), and non-manufacturing groups (agriculture, construction, mining). While growth is expected in all subsectors towards 2050, non-energy-intensive manufacturing is anticipated to grow faster than the other subsectors. Overall, a global increase in energy demand for manufacturing industry comes with the necessity of addressing this potential issue. Manufacturing is one of the **exponential technologies** which will be experiencing transformation at an accelerated speed. **Advanced manufacturing** technologies aimed at performance enhancement, increased quality and flexibility, and reduced energy consumption are projected to provide solutions for the energy supply–demand concerns of industry. Use of **robots** and **automation** and employing **AI, IoT**, and **data analytics** will be promoting the manufacturing sector. Three-dimensional product and system simulation, and evaluation of these in **virtual reality** spaces (**Figure 15.15**) have already proven to save time and budget. More manufacturers will be shifting towards these methodologies. **Additive manufacturing** and **3D printing** technologies are also expected to grow and contribute to energy consumption and greenhouse gas emissions reduction.

15.5 Pathways to Sustainable Energy

In this book, different energy resources and technologies were covered from historic, theoretical, and applied perspectives. All information shared should be evaluated with the main goal of addressing the energy future of the globe. The three points that should be noted are (1) energy efficiency is the low-hanging fruit, (2) renewable energy is a must, and (3) improvements in non-renewable technologies are essential for a world that is free of energy shortage and any conflicts that may evolve out of such deficit. The principles and core doctrines for all these technologies have been scrutinized in an unbiased manner.

Pathways that will lead us to a sustainable energy future can include any of the future prospects discussed in the earlier sections of this chapter. The route that we will be taking globally will be determined by influences from several identifiers including governments, policymakers, consumers, R&D institutions, industry, environmental organizations, and the media. Depending on the geographic location, funds for research, political atmosphere and prioritization, investor interests, and public awareness, the weight of each identifier in affecting the decision-making process will vary.

The role of engineers and energy professionals in clearing the pathways for a sustainable energy future is critical. As introduced in Chapter 1, *Connecting the 3E: Energy, Environment, and Engineering* is an essential concept for every engineer to digest and apply in real life. This will not only help towards the Sustainable Development Goals (SDGs) that pertain to energy, but will also contribute to other SDGs, hence addressing all three pillars of sustainability, which are social, environmental, and economic. This highlights one of the skills that every engineer needs to have which is *being able to see the big picture*.

CLASSROOM DEBATE

Over the next 50 years, which of the four bodies below do you think will have a stronger influence on the direction of sustainable energy: governments, consumers, environmental organizations, or industry?

REFERENCES

[1] U.S. Energy Information Administration, "International Energy Outlook 2021," EIA, Washington, DC, Oct. 2021. Accessed: Oct. 10, 2021 [Online]. Available: www.eia.gov/outlooks/ieo.

[2] U.S. Energy Information Administration, "Annual Energy Outlook 2021," EIA, Washington, DC, Feb. 2021. Accessed: Sep. 15, 2021 [Online]. Available: www.eia .gov/outlooks/aeo/.

[3] International Energy Agency, "Africa Energy Outlook 2019," IEA, Paris, Nov. 2019. Accessed: Sep. 15, 2021 [Online]. Available: www.iea.org/reports/africa-energy-outlook-2019.

[4] National Renewable Energy Laboratory, "2021 U.S. Geothermal Power Production and District Heating Market Report," NREL, 2021. Accessed: Sep. 17, 2021 [Online]. Available: www.nrel.gov/docs/fy21osti/78291.pdf.

[5] Idaho National Laboratory, "Microreactors," INL, 2019. https://inl.gov/trending-topic/microreactors/ (accessed Sep. 17, 2021).

[6] U.S. Department of Energy, "TVA Installs 3D-Printed Fuel Assembly Brackets in a Commercial Reactor," DOE, Aug. 23, 2021. www.energy.gov/ne/articles/tva-installs-3d-printed-fuel-assembly-brackets-commercial-reactor (accessed Sep. 18, 2021).

[7] Wood Mackenzie, "700 Million Electric Vehicles Will Be on the Roads by 2050," Wood Mackenzie, Feb. 8, 2021. www.woodmac.com/press-releases/700-million-electric-vehicles-will-be-on-the-roads-by-2050 (accessed Sep. 27, 2021).

APPENDIX A
Energy Conversion Factors

	Btu	**Calorie**	**EJ**[1]	**MCF**[2]	**Quad**	**tce**[3]	**toe**[4]
Btu	1	252	1.06×10^{-15}	10^{-6}	10^{-15}	3.6×10^{-8}	2.48×10^{-8}
Calorie	3.97×10^{-3}	1	4.19×10^{-18}	3.97×10^{-9}	3.97×10^{-18}	1.43×10^{-10}	9.82×10^{-11}
EJ	9.48×10^{14}	2.39×10^{17}	1	9.48×10^{8}	0.948	3.41×10^{7}	2.35×10^{7}
MCF	10^{6}	2.52×10^{8}	1.06×10^{-9}	1	10^{-9}	0.036	0.0248
Quad	10^{15}	2.52×10^{17}	1.06	10^{9}	1	3.6×10^{7}	2.48×10^{7}
tce	2.78×10^{7}	7×10^{9}	2.93×10^{-8}	27.8	2.78×10^{-8}	1	0.688
toe	4.04×10^{7}	1.02×10^{10}	4.19×10^{-8}	40.4	4.04×10^{-8}	1.45	1

Note: To convert a value with a bold unit at the top row to another unit listed in the first column, multiply the quantity with the conversion factor in the intersecting cell.
[1] exajoule = 10^{18} J
[2] one thousand cubic feet of natural gas
[3] tonne of coal equivalent
[4] tonne of oil equivalent

Example unit conversions
1 kJ = 0.948 Btu
1 GJ = 947,817 Btu
1 GJ = 277.8 kWh
1 kWh = 3412 Btu
1 TWh = 0.086 Mtoe

1 W = 3.412 Btu/h
1 hp = 2545 Btu/h
1 hp = 746 W

1 therm = 100,000 Btu
1 toe (15 °C) = 41.85 GJ
1 toe (15 °C) = 39.66 MMBtu
1 quad = 1015 Btu
1 quad = 1.055 EJ

1 ton of refrigeration = 12,000 Btu/h
1 ton of refrigeration = 3.517 kW

APPENDIX B
Thermodynamic Tables (SI Units)

The tables in this appendix are adapted from Moran, Shapiro, Boettner, and Bailey (2014), *Fundamentals of Engineering Thermodynamics*, 8th edition, John Wiley & Sons, with permission.

Table B.1

Properties of saturated water: temperature table										
Pressure conversions: 1 bar = 0.1 MPa = 10^2 kPa		**Specific volume (m^3/kg)**		**Internal energy (kJ/kg)**		**Enthalpy (kJ/kg)**			**Entropy (kJ/kg K)**	
Temp. (°C)	**Press. (bar)**	**Sat. liquid $v_f \times 10^3$**	**Sat. vapor v_g**	**Sat. liquid u_f**	**Sat. vapor u_g**	**Sat. liquid h_f**	**Evap. h_{fg}**	**Sat. vapor h_g**	**Sat. liquid s_f**	**Sat. vapor s_g**
.01	0.00611	1.0002	206.136	0	2375.3	0.01	2501.3	2501.4	0	9.1562
4	0.00813	1.0001	157.232	16.77	2380.9	16.78	2491.9	2508.7	0.0610	9.0514
5	0.00872	1.0001	147.120	20.97	2382.3	20.98	2489.6	2510.6	0.0761	9.0257
6	0.00935	1.0001	137.734	25.19	2383.6	25.20	2487.2	2512.4	0.0912	9.0003
8	0.01072	1.0002	120.917	33.59	2386.4	33.60	2482.5	2516.1	0.1212	8.9501
10	0.01228	1.0004	106.379	42.00	2389.2	42.01	2477.7	2519.8	0.1510	8.9008
11	0.01312	1.0004	99.857	46.20	2390.5	46.20	2475.4	2521.6	0.1658	8.8765
12	0.01402	1.0005	93.784	50.41	2391.9	50.41	2473.0	2523.4	0.1806	8.8524
13	0.01497	1.0007	88.124	54.60	2393.3	54.60	2470.7	2525.3	0.1953	8.8285
14	0.01598	1.0008	82.848	58.79	2394.7	58.80	2468.3	2527.1	0.2099	8.8048
15	0.01705	1.0009	77.926	62.99	2396.1	62.99	2465.9	2528.9	0.2245	8.7814
16	0.01818	1.0011	73.333	67.18	2397.4	67.19	2463.6	2530.8	0.2390	8.7582
17	0.01938	1.0012	69.044	71.38	2398.8	71.38	2461.2	2532.6	0.2535	8.7351
18	0.02064	1.0014	65.038	75.57	2400.2	75.58	2458.8	2534.4	0.2679	8.7123
19	0.02198	1.0016	61.293	79.76	2401.6	79.77	2456.5	2536.2	0.2823	8.6897
20	0.02339	1.0018	57.791	83.95	2402.9	83.96	2454.1	2538.1	0.2966	8.6672
21	0.02487	1.0020	54.514	88.14	2404.3	88.14	2451.8	2539.9	0.3109	8.6450
22	0.02645	1.0022	51.447	92.32	2405.7	92.33	2449.4	2541.7	0.3251	8.6229
23	0.0281	1.0024	48.574	96.51	2407.0	96.52	2447.0	2543.5	0.3393	8.6011
24	0.02985	1.0027	45.883	100.70	2408.4	100.70	2444.7	2545.4	0.3534	8.5794
25	0.03169	1.0029	43.360	104.88	2409.8	104.89	2442.3	2547.2	0.3674	8.5580
26	0.03363	1.0032	40.994	109.06	2411.1	109.07	2439.9	2549.0	0.3814	8.5367
27	0.03567	1.0035	38.774	113.25	2412.5	113.25	2437.6	2550.8	0.3954	8.5156
28	0.03782	1.0037	36.690	117.42	2413.9	117.43	2435.2	2552.6	0.4093	8.4946
29	0.04008	1.0040	34.733	121.60	2415.2	121.61	2432.8	2554.5	0.4231	8.4739
30	0.04246	1.0043	32.894	125.78	2416.6	125.79	2430.5	2556.3	0.4369	8.4533
31	0.04496	1.0046	31.165	129.96	2418.0	129.97	2428.1	2558.1	0.4507	8.4329

Table B.1 (cont.)

Pressure conversions: 1 bar = 0.1 MPa = 10^2 kPa		Specific volume (m³/kg)		Internal energy (kJ/kg)		Enthalpy (kJ/kg)			Entropy (kJ/kg K)	
Temp. (°C)	Press. (bar)	Sat. liquid $v_f \times 10^3$	Sat. vapor v_g	Sat. liquid u_f	Sat. vapor u_g	Sat. liquid h_f	Evap. h_{fg}	Sat. vapor h_g	Sat. liquid s_f	Sat. vapor s_g
32	0.04759	1.0050	29.540	134.14	2419.3	134.15	2425.7	2559.9	0.4644	8.4127
33	0.05034	1.0053	28.011	138.32	2420.7	138.33	2423.4	2561.7	0.4781	8.3927
34	0.05324	1.0056	26.571	142.50	2422.0	142.50	2421.0	2563.5	0.4917	8.3728
35	0.05628	1.0060	25.216	146.67	2423.4	146.68	2418.6	2565.3	0.5053	8.3531
36	0.05947	1.0063	23.940	150.85	2424.7	150.86	2416.2	2567.1	0.5188	8.3336
38	0.06632	1.0071	21.602	159.20	2427.4	159.21	2411.5	2570.7	0.5458	8.2950
40	0.07384	1.0078	19.523	167.56	2430.1	167.57	2406.7	2574.3	0.5725	8.2570
45	0.09593	1.0099	15.258	188.44	2436.8	188.45	2394.8	2583.2	0.6387	8.1648
50	0.1235	1.0121	12.032	209.32	2443.5	209.33	2382.7	2592.1	0.7038	8.0763
55	0.1576	1.0146	9.568	230.21	2450.1	230.23	2370.7	2600.9	0.7679	7.9913
60	0.1994	1.0172	7.671	251.11	2456.6	251.13	2358.5	2609.6	0.8312	7.9096
65	0.2503	1.0199	6.197	272.02	2463.1	272.06	2346.2	2618.3	0.8935	7.8310
70	0.3119	1.0228	5.042	292.95	2469.6	292.98	2333.8	2626.8	0.9549	7.7553
75	0.3858	1.0259	4.131	313.9	2475.9	313.93	2321.4	2635.3	1.0155	7.6824
80	0.4739	1.0291	3.407	334.86	2482.2	334.91	2308.8	2643.7	1.0753	7.6122
85	0.5783	1.0325	2.828	355.84	2488.4	355.9	2296.0	2651.9	1.1343	7.5445
90	0.7014	1.0360	2.361	376.85	2494.5	376.92	2283.2	2660.1	1.1925	7.4791
95	0.8455	1.0397	1.982	397.88	2500.6	397.96	2270.2	2668.1	1.2500	7.4159
100	1.014	1.0435	1.673	418.94	2506.5	419.04	2257.0	2676.1	1.3069	7.3549
110	1.433	1.0516	1.210	461.14	2518.1	461.30	2230.2	2691.5	1.4185	7.2387
120	1.985	1.0603	0.8919	503.50	2529.3	503.71	2202.6	2706.3	1.5276	7.1296
130	2.701	1.0697	0.6685	516.02	2539.9	546.31	2174.2	2720.5	1.6344	7.0269
140	3.613	1.0797	0.5089	588.74	2550.0	589.13	2144.7	2733.9	1.7391	6.9299
150	4.758	1.0905	0.3928	631.68	2559.5	632.20	2114.3	2746.5	1.8418	6.8379
160	6.178	1.1020	0.3071	674.86	2568.4	675.55	2082.6	2758.1	1.9427	6.7502
170	7.917	1.1143	0.2428	718.33	2576.5	719.21	2049.5	2768.7	2.0419	6.6663
180	10.02	1.1274	0.1941	762.09	2583.7	763.22	2015.0	2778.2	2.1396	6.5857
190	12.54	1.1414	0.1565	806.19	2590.0	807.62	1978.8	2786.4	2.2359	6.5079

Table B.1 (cont.)

Pressure conversions: 1 bar = 0.1 MPa = 10^2 kPa		Specific volume (m³/kg)		Internal energy (kJ/kg)		Enthalpy (kJ/kg)			Entropy (kJ/kg K)	
Temp. (°C)	Press. (bar)	Sat. liquid $v_f \times 10^3$	Sat. vapor v_g	Sat. liquid u_f	Sat. vapor u_g	Sat. liquid h_f	Evap. h_{fg}	Sat. vapor h_g	Sat. liquid s_f	Sat. vapor s_g
200	15.54	1.1565	0.1274	850.65	2595.3	852.45	1940.7	2793.2	2.3309	6.4323
210	19.06	1.1726	0.1044	895.53	2599.5	897.76	1900.7	2798.5	2.4248	6.3585
220	23.18	1.1900	0.08619	940.87	2602.4	943.62	1858.5	2802.1	2.5178	6.2861
230	27.95	1.2088	0.07158	986.74	2603.9	990.12	1813.8	2804.0	2.6099	6.2146
240	33.44	1.2291	0.05976	1033.2	2604.0	1037.3	1766.5	2803.8	2.7015	6.1437
250	39.73	1.2512	0.05013	1080.4	2602.4	1085.4	1716.2	2801.5	2.7927	6.0730
260	46.88	1.2755	0.04221	1128.4	2599.0	1134.4	1662.5	2796.6	2.8838	6.0019
270	54.99	1.3023	0.03564	1177.4	2593.7	1184.5	1605.2	2789.7	2.9751	5.9301
280	64.12	1.3321	0.03017	1227.5	2586.1	1236.0	1543.6	2779.6	3.0668	5.8571
290	74.36	1.3656	0.02557	1278.9	2576.0	1289.1	1477.1	2766.2	3.1594	5.7821
300	85.81	1.4036	0.02167	1332	2563.0	1344.0	1404.9	2749.0	3.2534	5.7045
320	112.7	1.4988	0.01549	1444.6	2525.5	1461.5	1238.6	2700.1	3.4480	5.5362
340	145.9	1.6379	0.01080	1570.3	2464.6	1594.2	1027.9	2622.0	3.6594	5.3357
360	186.5	1.8925	0.00695	1725.2	2351.5	1760.5	720.5	2481.0	3.9147	5.0526
374.14	220.9	3.1550	0.00316	2029.6	2029.6	2099.3	0	2099.3	4.4298	4.4298

Table B.2

Properties of saturated water: pressure table										
Pressure conversions: 1 bar = 0.1 MPa = 10^2 kPa		Specific volume (m^3/kg)		Internal energy (kJ/kg)		Enthalpy (kJ/kg)			Entropy (kJ/kg K)	
Press. (bar)	Temp. (°C)	Sat. liquid $v_f \times 10^3$	Sat. vapor v_g	Sat. liquid u_f	Sat. vapor u_g	Sat. liquid h_f	Evap. h_{fg}	Sat. vapor h_g	Sat. liquid s_f	Sat. vapor s_g
0.04	28.96	1.0040	34.800	121.45	2415.2	121.46	2432.9	2554.4	0.4226	8.4746
0.06	36.16	1.0064	23.739	151.53	2425.0	151.53	2415.9	2567.4	0.5210	8.3304
0.08	41.51	1.0084	18.103	173.87	2432.2	173.88	2403.1	2577.0	0.5926	8.2287
0.10	45.81	1.0102	14.674	191.82	2437.9	191.83	2392.8	2584.7	0.6493	8.1502
0.20	60.06	1.0172	7.649	251.38	2456.7	251.40	2358.3	2609.7	0.8320	7.9085
0.30	69.1	1.0223	5.229	289.20	2468.4	289.23	2336.1	2625.3	0.9439	7.7686
0.40	75.87	1.0265	3.993	317.53	2477.0	317.58	2319.2	2636.8	1.0259	7.6700
0.50	81.33	1.0300	3.240	340.44	2483.9	340.49	2305.4	2645.9	1.0910	7.5939
0.60	85.94	1.0331	2.732	359.79	2489.6	359.86	2293.6	2653.5	1.1453	7.5320
0.70	89.95	1.0360	2.365	376.63	2494.5	376.70	2283.3	2660.0	1.1919	7.4797
0.80	93.5	1.0380	2.087	391.58	2498.8	391.66	2274.1	2665.8	1.2329	7.4346
0.90	96.71	1.0410	1.869	405.06	2502.6	405.15	2265.7	2670.9	1.2695	7.3949
1.00	99.63	1.0432	1.694	417.36	2506.1	417.46	2258	2675.5	1.3026	7.3594
1.50	111.4	1.0528	1.159	466.94	2519.7	467.11	2226.5	2693.6	1.4336	7.2233
2.00	120.2	1.0605	0.8857	504.49	2529.5	504.70	2201.9	2706.7	1.5301	7.1271
2.50	127.4	1.0672	0.7187	535.10	2537.2	535.37	2181.5	2716.9	1.6072	7.0527
3.00	133.6	1.0732	0.6058	561.15	2543.6	561.47	2163.8	2725.3	1.6718	6.9919
3.50	138.9	1.0786	0.5243	583.95	2546.9	584.33	2148.1	2732.4	1.7275	6.9405
4.00	143.6	1.0836	0.4625	604.31	2553.6	604.74	2133.8	2738.6	1.7766	6.8959
4.50	147.9	1.0882	0.4140	622.25	2557.6	623.25	2120.7	2743.9	1.8207	6.8565
5.00	151.9	1.0926	0.3749	639.68	2561.2	640.23	2108.5	2748.7	1.8607	6.8212
6.00	158.9	1.1006	0.3157	669.90	2567.4	670.56	2086.3	2756.8	1.9312	6.7600
7.00	165.0	1.1080	0.2729	696.44	2572.5	697.22	2066.3	2763.5	1.9922	6.7080
8.00	170.4	1.1148	0.2404	720.22	2576.8	721.11	2048	2769.1	2.0462	6.6628
9.00	175.4	1.1212	0.2150	741.83	2580.5	742.83	2031.1	2773.9	2.0946	6.6226
10.0	179.9	1.1273	0.1944	761.68	2583.6	762.81	2015.3	2778.1	2.1387	6.5863
15.0	198.3	1.1539	0.1318	843.16	2594.5	844.84	1947.3	2792.2	2.3150	6.4448

Table B.2 (cont.)

Pressure conversions: 1 bar = 0.1 MPa = 10^2 kPa		Specific volume (m³/kg)		Internal energy (kJ/kg)		Enthalpy (kJ/kg)			Entropy (kJ/kg K)	
Press. (bar)	Temp. (°C)	Sat. liquid $v_f \times 10^3$	Sat. vapor v_g	Sat. liquid u_f	Sat. vapor u_g	Sat. liquid h_f	Evap. h_{fg}	Sat. vapor h_g	Sat. liquid s_f	Sat. vapor s_g
20.0	212.4	1.1767	0.0996	906.44	2600.3	908.79	1890.7	2799.5	2.4474	6.3409
25.0	224.0	1.1973	0.0800	959.11	2603.1	962.11	1841	2803.1	2.5547	6.2575
30.0	233.9	1.2165	0.0667	1004.8	2604.1	1008.4	1795.7	2804.2	2.6457	6.1869
35.0	242.6	1.2347	0.0571	1045.4	2603.7	1049.8	1753.7	2803.4	2.7253	6.1253
40.0	250.4	1.2522	0.0498	1082.3	2602.3	1087.3	1714.1	2801.4	2.7964	6.0701
45.0	257.5	1.2692	0.0441	1116.2	2600.1	1121.9	1676.4	2798.3	2.8610	6.0199
50.0	264.0	1.2859	0.0394	1147.8	2597.1	1154.2	1640.1	2794.3	2.9202	5.9734
60.0	275.6	1.3187	0.0324	1205.4	2589.7	1213.4	1571.0	2784.3	3.0267	5.8892
70	285.9	1.3513	0.02737	1257.6	2580.5	1267.0	1505.1	2772.1	3.1211	5.8133
80	295.1	1.3842	0.02352	1305.6	2569.8	1316.6	1441.3	2758.0	3.2068	5.7432
90	303.4	1.4178	0.02048	1350.5	2557.8	1363.3	1378.9	2742.1	3.2858	5.6772
100	311.1	1.4524	0.01803	1393.0	2544.4	1407.6	1317.1	2724.7	3.3596	5.6141
110	318.2	1.4886	0.01599	1433.7	2529.8	1450.1	1255.5	2705.6	3.4295	5.5527
120	324.8	1.5267	0.01426	1473.0	2513.7	1491.3	1193.6	2684.9	3.4962	5.4924
130	330.9	1.5671	0.01278	1511.1	2496.1	1531.5	1130.7	2662.2	3.5606	5.4323
140	336.8	1.6107	0.01149	1548.6	2476.8	1571.1	1066.5	2637.6	3.6232	5.3717
150	342.2	1.6581	0.01034	1585.6	2455.5	1610.5	1000.0	2610.5	3.6848	5.3098
160	347.4	1.7107	0.009306	1622.7	2431.7	1650.1	930.6	2580.6	3.7461	5.2455
170	352.4	1.7702	0.008364	1660.2	2405.0	1690.3	856.9	2547.2	3.8079	5.1777
180	357.1	1.8397	0.007489	1698.9	2374.3	1732.0	777.1	2509.1	3.8715	5.1044
190	361.5	1.9243	0.006657	1739.9	2338.1	1776.5	688.0	2464.5	3.9388	5.0228
200	365.8	2.0360	0.005834	1785.6	2293.0	1826.3	583.4	2409.7	4.0139	4.9269
220.9	374.1	3.1550	0.003155	2029.6	2029.6	2099.3	0	2099.3	4.4298	4.4298

Table B.3

Properties of superheated water vapor								
T (°C)	v (m³/kg)	u (kJ/kg)	h (kJ/kg)	s (kJ/kg K)	v (m³/kg)	u (kJ/kg)	h (kJ/kg)	s (kJ/kg K)
	$p = 0.06$ bar $= 0.006$ MPa				$p = 0.35$ bar $= 0.035$ MPa			
	$T_{sat} = 36.16$ °C				$T_{sat} = 72.69$ °C			
Sat.	23.739	2425.0	2567.4	8.3304	4.526	2473.0	2631.4	7.7158
80	27.132	2487.3	2650.1	8.5804	4.625	2483.7	2645.6	7.7564
120	30.219	2544.7	2726.0	8.7840	5.163	2542.4	2723.1	7.9644
160	33.302	2602.7	2802.5	8.9693	5.696	2601.2	2800.6	8.1519
200	36.383	2661.4	2879.7	9.1398	6.228	2660.4	2878.4	8.3237
240	39.462	2721.0	2957.8	9.2982	6.758	2720.3	2956.8	8.4828
280	42.540	2781.5	3036.8	9.4464	7.287	2780.9	3036.0	8.6314
320	45.618	2843.0	3116.7	9.5859	7.815	2842.5	3116.1	8.7712
360	48.696	2905.5	3197.7	9.7180	8.344	2905.1	3197.1	8.9034
400	51.774	2969.0	3279.6	9.8435	8.872	2968.6	3279.2	9.0291
440	54.851	3033.5	3362.6	9.9633	9.400	3033.2	3362.2	9.1490
500	59.467	3132.3	3489.1	10.1336	10.192	3132.1	3488.8	9.3194
	$p = 0.70$ bar $= 0.07$ MPa				$p = 1.0$ bar $= 0.1$ MPa			
	$T_{sat} = 89.95$ °C				$T_{sat} = 99.63$ °C			
Sat.	2.365	2494.5	2660.0	7.4797	1.694	2506.1	2675.5	7.3594
100	2.434	2509.7	2680.0	7.5341	1.696	2506.7	2676.2	7.3614
120	2.571	2539.7	2719.6	7.6375	1.793	2537.3	2716.6	7.4668
160	2.841	2599.4	2798.2	7.8279	1.984	2597.8	2796.2	7.6597
200	3.108	2659.1	2876.7	8.0012	2.172	2658.1	2875.3	7.8343
240	3.374	2719.3	2955.5	8.1611	2.359	2718.5	2954.5	7.9949
280	3.640	2780.2	3035.0	8.3162	2.546	2779.6	3034.2	8.1445
320	3.905	2842.0	3115.3	8.4504	2.732	2841.5	3114.6	8.2849
360	4.170	2904.6	3196.5	8.5828	2.917	2904.2	3195.9	8.4175
400	4.434	2968.2	3278.6	8.7086	3.103	2967.9	3278.2	8.5435
440	4.698	3032.9	3361.8	8.8286	3.288	3032.6	3361.4	8.6636
500	5.095	3131.8	3488.5	8.9991	3.565	3131.6	3488.1	8.8342

Table B.3 (cont.)

T (°C)	v (m³/kg)	u (kJ/kg)	h (kJ/kg)	s (kJ/kg K)	v (m³/kg)	u (kJ/kg)	h (kJ/kg)	s (kJ/kg K)
	p = 1.5 bar = 0.15 MPa				p = 3.0 bar = 0.30 MPa			
	T_{sat} = 111.37 °C				T_{sat} = 133.55 °C			
Sat.	1.159	2519.7	2693.6	7.2233	0.606	2543.6	2725.3	6.9919
120	1.188	2533.3	2711.4	7.2693				
160	1.317	2595.2	2792.8	7.4665	0.651	2587.1	2782.3	7.1276
200	1.444	2656.2	2872.9	7.6433	0.716	2650.7	2865.5	7.3115
240	1.570	2717.2	2952.7	7.8052	0.781	2713.1	2947.3	7.4774
280	1.695	2778.6	3032.8	7.9555	0.844	2775.4	3028.6	7.6299
320	1.819	2840.6	3113.5	8.0964	0.907	2838.1	3110.1	7.7722
360	1.943	2903.5	3195.0	8.2293	0.969	2901.4	3192.2	7.9061
400	2.067	2967.3	3277.4	8.3555	1.032	2965.6	3275.0	8.0330
440	2.191	3032.1	3360.7	8.4757	1.094	3030.6	3358.7	8.1538
500	2.376	3131.2	3487.6	8.6466	1.187	3130.0	3486.0	8.3251
600	2.685	3301.7	3704.3	8.9101	1.341	3300.8	3703.2	8.5892
	p = 5.0 bar = 0.50 MPa				p = 7.0 bar = 0.70 MPa			
	T_{sat} = 151.86 °C				T_{sat} = 164.97 °C			
Sat.	0.3749	2561.2	2748.7	6.8213	0.2729	2572.5	2763.5	6.7080
180	0.4045	2609.7	2812.0	6.9656	0.2847	2599.8	2799.1	6.7880
200	0.4249	2642.9	2855.4	7.0592	0.2999	2634.8	2844.8	6.8865
240	0.4646	2707.6	2939.9	7.2307	0.3292	2701.8	2932.2	7.0641
280	0.5034	2771.2	3022.9	7.3865	0.3574	2766.9	3017.1	7.2233
320	0.5416	2834.7	3105.6	7.5308	0.3852	2831.3	3100.9	7.3697
360	0.5796	2898.7	3188.4	7.6660	0.4126	2895.8	3184.7	7.5063
400	0.6173	2963.2	3271.9	7.7938	0.4397	2960.9	3268.7	7.6350
440	0.6548	3028.6	3356.0	7.9152	0.4667	3026.6	3353.3	7.7571
500	0.7109	3128.4	3483.9	8.0873	0.5070	3126.8	3481.7	7.9299
600	0.8041	3299.6	3701.7	8.3522	0.5738	3298.5	3700.2	8.1956
700	0.8969	3477.5	3925.9	8.5952	0.6403	3476.6	3924.8	8.4391
	p = 10.0 bar = 1.0 MPa				p = 15.0 bar = 1.5 MPa			
	T_{sat} = 179.91 °C				T_{sat} = 198.32 °C			
Sat.	0.1944	2583.6	2778.1	6.5865	0.1318	2594.5	2792.2	6.4448
200	0.2060	2621.9	2827.9	6.6940	0.1325	2598.1	2796.8	6.4546

Table B.3 (cont.)

T (°C)	v (m³/kg)	u (kJ/kg)	h (kJ/kg)	s (kJ/kg K)	v (m³/kg)	u (kJ/kg)	h (kJ/kg)	s (kJ/kg K)
240	0.2275	2692.9	2920.4	6.8817	0.1483	2676.9	2899.3	6.6628
280	0.2480	2760.2	3008.2	7.0465	0.1627	2748.6	2992.7	6.8381
320	0.2678	2826.1	3093.9	7.1962	0.1765	2817.1	3081.9	6.9938
360	0.2873	2891.6	3178.9	7.3349	0.1899	2884.4	3169.2	7.1363
400	0.3066	2957.3	3263.9	7.4651	0.2030	2951.3	3255.8	7.2690
440	0.3257	3023.6	3349.3	7.5883	0.2160	3018.5	3342.5	7.3940
500	0.3541	3124.4	3478.5	7.7622	0.2352	3120.3	3473.1	7.5698
540	0.3729	3192.6	3565.6	7.8720	0.2478	3189.1	3560.9	7.6805
600	0.4011	3296.8	3697.9	8.0290	0.2668	3293.9	3694.0	7.8385
640	0.4198	3367.4	3787.2	8.1290	0.2793	3364.8	3783.8	7.9391
	p = 20.0 bar = 2.0 MPa				p = 30.0 bar = 3.0 MPa			
	T_{sat} = 212.42 °C				T_{sat} = 233.90 °C			
Sat.	0.0996	2600.3	2799.5	6.3409	0.0667	2604.1	2804.2	6.1869
240	0.1085	2659.6	2876.5	6.4952	0.0682	2619.7	2824.3	6.2265
280	0.1200	2736.4	2976.4	6.6828	0.0771	2709.9	2941.3	6.4462
320	0.1308	2807.9	3069.5	6.8452	0.0850	2788.4	3043.4	6.6245
360	0.1411	2877.0	3159.3	6.9917	0.0923	2861.7	3138.7	6.7801
400	0.1512	2945.2	3247.6	7.1271	0.0994	2932.8	3230.9	6.9212
440	0.1611	3013.4	3335.5	7.2540	0.1062	3002.9	3321.5	7.0520
500	0.1757	3116.2	3467.6	7.4317	0.1162	3108.0	3456.5	7.2338
540	0.1853	3185.6	3556.1	7.5434	0.1227	3178.4	3546.6	7.3474
600	0.1996	3290.9	3690.1	7.7024	0.1324	3285.0	3682.3	7.5085
640	0.2091	3362.2	3780.4	7.8035	0.1388	3357.0	3773.5	7.6106
700	0.2232	3470.9	3917.4	7.9487	0.1484	3466.5	3911.7	7.7571
	p = 40 bar = 4.0 MPa				p = 60 bar = 6.0 MPa			
	T_{sat} = 250.4 °C				T_{sat} = 275.64 °C			
Sat.	0.0498	2602.3	2801.4	6.0701	0.0324	2589.7	2784.3	5.8892
280	0.0555	2680.0	2901.8	6.2568	0.0332	2605.2	2804.2	5.9252
320	0.0620	2767.4	3015.4	6.4553	0.0388	2720.0	2952.6	6.1846
360	0.0679	2845.7	3117.2	6.6215	0.0433	2811.2	3071.1	6.3782
400	0.0734	2919.9	3213.6	6.7690	0.0474	2892.9	3177.2	6.5408
440	0.0787	2992.2	3307.1	6.9041	0.0512	2970.0	3277.3	6.6853

Table B.3 (cont.)

T (°C)	v (m³/kg)	u (kJ/kg)	h (kJ/kg)	s (kJ/kg K)	v (m³/kg)	u (kJ/kg)	h (kJ/kg)	s (kJ/kg K)
500	0.0864	3099.5	3445.3	7.0901	0.0567	3082.2	3422.2	6.8803
540	0.0915	3171.1	3536.9	7.2056	0.0602	3156.1	3517.0	6.9999
600	0.0989	3279.1	3674.4	7.3688	0.0653	3266.9	3658.4	7.1677
640	0.1037	3351.8	3766.6	7.4720	0.0686	3341.0	3752.6	7.2731
700	0.1110	3462.1	3905.9	7.6198	0.0735	3453.1	3894.1	7.4234
740	0.1157	3536.6	3999.6	7.7141	0.0768	3528.3	3989.2	7.5190
	p = 80.0 bar = 8.0 MPa				p = 100 bar = 10.0 MPa			
	T_{sat} = 295.06 °C				T_{sat} = 311.06 °C			
Sat.	0.0235	2569.8	2758.0	5.7432	0.01803	2544.4	2724.7	5.6141
320	0.0268	2662.7	2877.2	5.9489	0.01925	2588.8	2781.3	5.7103
360	0.0309	2772.7	3019.8	6.1819	0.02331	2729.1	2962.1	6.0060
400	0.0343	2863.8	3138.3	6.3634	0.02641	2832.4	3096.5	6.2120
440	0.0374	2946.7	3246.1	6.5190	0.02911	2922.1	3213.2	6.3805
480	0.0403	3025.7	3348.4	6.6586	0.0316	3005.4	3321.4	6.5282
520	0.0431	3102.7	3447.7	6.7871	0.03394	3085.6	3425.1	6.6622
560	0.0458	3178.7	3545.3	6.9072	0.03619	3164.1	3526.0	6.7864
600	0.0485	3254.4	3642.0	7.0206	0.03837	3241.7	3625.3	6.9029
640	0.0510	3330.1	3738.3	7.1283	0.04048	3318.9	3723.7	7.0131
700	0.0548	3443.9	3882.4	7.2812	0.04358	3434.7	3870.5	7.1687
740	0.0573	3520.4	3978.7	7.3782	0.0456	3512.1	3968.1	7.2670
	p = 120 bar = 12.0 MPa				p = 140 bar = 14.0 MPa			
	T_{sat} = 324.75 °C				T_{sat} = 336.75 °C			
Sat.	0.0143	2513.7	2684.9	5.4924	0.0115	2476.8	2637.6	5.3717
360	0.0181	2678.4	2895.7	5.8361	0.0142	2617.4	2816.5	5.6602
400	0.0211	2798.3	3051.3	6.0747	0.0172	2760.9	3001.9	5.9448
440	0.0236	2896.1	3178.7	6.2586	0.0195	2868.6	3142.2	6.1474
480	0.0258	2984.4	3293.5	6.4154	0.0216	2962.5	3264.5	6.3143
520	0.0278	3068.0	3401.8	6.5555	0.0234	3049.8	3377.8	6.4610
560	0.0298	3149.0	3506.2	6.6840	0.0252	3133.6	3486.0	6.5941
600	0.0316	3228.7	3608.3	6.8037	0.0268	3215.4	3591.1	6.7172
640	0.0335	3307.5	3709.0	6.9164	0.0284	3296.0	3694.1	6.8326

Table B.3 (cont.)

T (°C)	v (m³/kg)	u (kJ/kg)	h (kJ/kg)	s (kJ/kg K)	v (m³/kg)	u (kJ/kg)	h (kJ/kg)	s (kJ/kg K)
700	0.0361	3425.2	3858.4	7.0749	0.0308	3415.7	3846.2	6.9939
740	0.0378	3503.7	3957.4	7.1746	0.0323	3495.2	3946.7	7.0952
p = 160 bar = 16.0 MPa					*p* = 180 bar = 18.0 MPa			
T_{sat} = 347.44 °C					T_{sat} = 357.06 °C			
Sat.	0.0093	2431.7	2580.6	5.2455	0.0075	2374.3	2509.1	5.1044
360	0.0111	2539.0	2715.8	5.4614	0.0081	2418.9	2564.5	5.1922
400	0.0143	2719.4	2947.6	5.8175	0.0119	2672.8	2887.0	5.6887
440	0.0165	2839.4	3103.7	6.0429	0.0141	2808.2	3062.8	5.9428
480	0.0184	2939.7	3234.4	6.2215	0.0160	2915.9	3203.2	6.1345
520	0.0201	3031.1	3353.3	6.3752	0.0176	3011.8	3378.0	6.2960
560	0.0217	3117.8	3465.4	6.5132	0.0190	3101.7	3444.4	6.4392
600	0.0232	3201.8	3573.5	6.6399	0.0204	3188.0	3555.6	6.5696
640	0.0247	3284.2	3678.9	6.7580	0.0217	3272.3	3663.6	6.6905
700	0.0267	3406.0	3833.9	6.9224	0.0236	3396.3	3821.5	6.8580
740	0.0281	3486.7	3935.9	7.0251	0.0248	3478.0	3925.0	6.9623
p = 200 bar = 20.0 MPa					*p* = 240 bar = 24.0 MPa			
T_{sat} = 365.81 °C								
Sat.	0.0058	2293.0	2409.7	4.9269				
400	0.0099	2619.3	2818.1	5.5540	0.0067	2477.8	2639.4	5.2393
440	0.0122	2774.9	3019.4	5.8450	0.0093	2700.6	2923.4	5.6506
480	0.0140	2891.2	3170.8	6.0518	0.0110	2838.3	3102.3	5.8950
520	0.0155	2992.0	3302.2	6.2218	0.0124	2950.5	3248.5	6.0842
560	0.0169	3085.2	3423.0	6.3705	0.0137	3051.1	3379.0	6.2448
600	0.0182	3174.0	3537.6	6.5048	0.0148	3145.2	3500.7	6.3875
640	0.0194	3260.2	3648.1	6.6286	0.0159	3235.5	3616.7	6.5174
700	0.0211	3386.4	3809.0	6.7993	0.0174	3366.4	3783.8	6.6947
740	0.0222	3469.3	3914.1	6.9052	0.0184	3451.7	3892.1	6.8038
800	0.0239	3592.7	4069.7	7.0544	0.0197	3578.0	4051.6	6.9567
p = 280 bar = 28.0 MPa					*p* = 320 bar = 32.0 MPa			
400	0.0038	2223.5	2330.7	4.7494	0.0024	1980.4	2055.9	4.3239
440	0.0071	2613.2	2812.6	5.4494	0.0054	2509.0	2683.0	5.2327
480	0.0089	2780.8	3028.5	5.7446	0.0072	2718.1	2949.2	5.5968

Table B.3 (cont.)

T (°C)	v (m³/kg)	u (kJ/kg)	h (kJ/kg)	s (kJ/kg K)	v (m³/kg)	u (kJ/kg)	h (kJ/kg)	s (kJ/kg K)
520	0.0102	2906.8	3192.3	5.9566	0.0085	2860.7	3133.7	5.8357
560	0.0114	3015.7	3333.7	6.1307	0.0096	2979.0	3287.2	6.0246
600	0.0124	3115.6	3463.0	6.2823	0.0106	3085.3	3424.6	6.1858
640	0.0134	3210.3	3584.8	6.4187	0.0115	3184.5	3552.5	6.3290
700	0.0147	3346.1	3758.4	6.6029	0.0127	3325.4	3732.8	6.5203
740	0.0156	3433.9	3870.0	6.7153	0.0135	3415.9	3847.8	6.6361
800	0.0168	3563.1	4033.4	6.8720	0.0146	3548.0	4015.1	6.7966
900	0.0187	3774.3	4298.8	7.1084	0.0163	3762.7	4285.1	7.0372

APPENDIX C
Elements

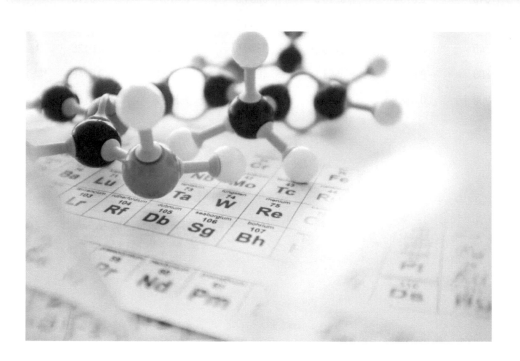

Element	Symbol	Atomic number (Z)
Actinium	Ac	89
Aluminum	Al	13
Americium	Am	95
Antimony	Sb	51
Argon	Ar	18
Arsenic	As	33
Astatine	At	85
Barium	Ba	56
Berkelium	Bk	97
Beryllium	Be	4
Bismuth	Bi	83
Bohrium	Bh	107
Boron	B	5
Bromine	Br	35
Cadmium	Cd	48
Calcium	Ca	20
Californium	Cf	98
Carbon	C	6
Cerium	Ce	58
Cesium	Cs	55
Chlorine	Cl	17
Chromium	Cr	24
Cobalt	Co	27
Copernicium	Cn	112
Copper	Cu	29
Curium	Cm	96
Darmstadtium	Ds	110
Dubnium	Db	105
Dysprosium	Dy	66
Einsteinium	Es	99
Erbium	Er	68
Europium	Eu	63
Fermium	Fm	100

Element	Symbol	Atomic number (Z)
Flerovium	Fl	114
Fluorine	F	9
Francium	Fr	87
Gadolinium	Gd	64
Gallium	Ga	31
Germanium	Ge	32
Gold	Au	79
Hafnium	Hf	72
Hassium	Hs	108
Helium	He	2
Holmium	Ho	67
Hydrogen	H	1
Indium	In	49
Iodine	I	53
Iridium	Ir	77
Iron	Fe	26
Krypton	Kr	36
Lanthanum	La	57
Lawrencium	Lr	103
Lead	Pb	82
Lithium	Li	3
Livermorium	Lv	116
Lutetium	Lu	71
Magnesium	Mg	12
Manganese	Mn	25
Meitnerium	Mt	109
Mendelevium	Md	101
Mercury	Hg	80
Molybdenum	Mo	42
Moscovium	Mc	115
Neodymium	Nd	60
Neon	Ne	10
Neptunium	Np	93

Element	Symbol	Atomic number (Z)
Nickel	Ni	28
Nihonium	Nh	113
Niobium	Nb	41
Nitrogen	N	7
Nobelium	No	102
Oganesson	Og	118
Osmium	Os	76
Oxygen	O	8
Palladium	Pd	46
Phosphorus	P	15
Platinum	Pt	78
Plutonium	Pu	94
Polonium	Po	84
Potassium	K	19
Praseodymium	Pr	59
Promethium	Pm	61
Protactinium	Pa	91
Radium	Ra	88
Radon	Rn	86
Rhenium	Re	75
Rhodium	Rh	45
Roentgenium	Rg	111
Rubidium	Rb	37
Ruthenium	Ru	44
Rutherfordium	Rf	104
Samarium	Sm	62
Scandium	Sc	21
Seaborgium	Sg	106
Selenium	Se	34
Silicon	Si	14
Silver	Ag	47
Sodium	Na	11
Strontium	Sr	38

Element	Symbol	Atomic number (Z)
Sulfur	S	16
Tantalum	Ta	73
Technetium	Tc	43
Tellurium	Te	52
Tennessine	Ts	117
Terbium	Tb	65
Thallium	Tl	81
Thorium	Th	90
Thulium	Tm	69
Tin	Sn	50
Titanium	Ti	22
Tungsten	W	74
Uranium	U	92
Vanadium	V	23
Xenon	Xe	54
Ytterbium	Yb	70
Yttrium	Y	39
Zinc	Zn	30
Zirconium	Zr	40

Abbreviations

AC	alternating current
AGR	advanced gas reactor
AI	artificial intelligence
ASHRAE	American Society of Heating, Refrigerating and Air-Conditioning Engineers
BEV	battery electric vehicle
BOS	balance of system
BTM	behind the meter
Btoe	billion tonnes of oil equivalent
Btu	British thermal unit
BWR	boiling water reactor
CAES	compressed air energy storage
CANDU	Canada deuterium uranium heavy-water reactor
CCGT	combined cycle gas turbine
CCS	carbon capture and storage
CFC	chlorofluorocarbon
CHP	combined heat and power
COP	coefficient of performance
CSP	concentrating solar power
DC	direct current
DOE	Department of Energy (USA)
DSM	demand-side management
EERE	Office of Energy Efficiency and Renewable Energy (U.S. DOE)
EF	ecological footprinting
EIA	Energy Information Administration (U.S. DOE)
EPA	Environmental Protection Agency
EU	European Union
FCEV	fuel cell electric vehicle
FTM	front of the meter
GCR	gas-cooled reactor
GDP	gross domestic product
GHG	greenhouse gas
GNP	gross national product
GNNP	green national net product
GWP	global warming potential
GW_e	gigawatt of electric power
GW_{th}	gigawatt of thermal power
HATT	horizontal-axis tidal turbine
HAWT	horizontal-axis wind turbine
HDI	Human Development Index
HHV	higher heating value
HICEV	hydrogen internal combustion electric vehicle
HTGR	high-temperature gas-cooled reactor
HVAC	heating, ventilating, and air-conditioning
HWR	heavy water reactor
IAEA	International Atomic Energy Agency
IAQ	indoor air quality
IEA	International Energy Agency
IoT	internet of things

IRENA	International Renewable Energy Agency
ITER	International Thermonuclear Experimental Reactor
LCOE	levelized cost of energy (or electricity)
LCOS	levelized cost of storage
LED	light emitting diodes
LHV	lower heating value
LNG	liquefied natural gas
LPG	liquefied petroleum gas
LWR	light water reactor
MBtu	thousand British thermal units
MMBtu	million British thermal units
MSW	municipal solid waste
Mtoe	million tonnes of oil equivalent
NOAA	National Oceanic and Atmospheric Administration
NREL	National Renewable Energy Laboratory
O&M	operation and maintenance
OECD	Organisation for Economic Co-operation and Development
OPEC	Organization of the Petroleum Exporting Countries
OTEC	ocean thermal energy conversion
OWC	oscillating water column
PCM	phase change material
PHES	pumped hydro energy storage
PV	photovoltaic
PWR	pressurized water reactor
RBMK	reaktor bolshoy moshchnosty kanalny (high-power channel reactor)
RED	reverse electrodialysis
SDG	sustainable development goal
SDS	Sustainable Development Scenario
SHDI	Sustainable Human Development Index
SMES	superconducting magnetic energy storage
TES	thermal energy storage
TWh	terawatt hour
UNDP	United Nations Development Programme
USGBC	U.S. Green Building Council
VATT	vertical-axis tidal turbine
VAWT	vertical-axis wind turbine
VOC	volatile organic compound

Index